USGS- Jurate Landwehr

STATISTICS: THEORY AND PRACTICE

Selected Papers by Sir Maurice Kendall

Other books of interest

See page xv in this book for a list of books by Sir Maurice Kendall

Statistical papers of George Udny Yule

George Udny Yule was, after Karl Pearson its founder, the greatest of the first exponents of modern statistical theory. Yule's centenary year was made the occasion for this republication of twelve of his most important papers – valuable background to modern method.

'Every statistician will benefit by reading this book.' – P. A. P. Moran, *Australian Journal of Statistics*

260 x 184 mm 455 pp 79 illustrations (ISBN 0 85264 201 6)

Experimental design: selected papers F. Yates

Very significant contributions to the theory of experimental design have been made by Dr. Yates. This selection of related papers, each with the author's later comments, provides a background for those concerned with experimental design.

'It was a sensible idea. . . to confine the set to a single field. The Past President of our Society is here at his best'. – *Computer Bulletin*

229 x 152 mm 308 pp 25 illustrations 112 tables (ISBN 0 85264 168 0)

Studies in the history of statistics and probability

In two volumes

Vol. 1. Ed. E. S. PEARSON and Sir MAURICE KENDALL

260 x 184mm 490pages Illustrations & tables 2nd Impr. (ISBN 0 85264 193 1)

Vol. 2. Ed. Sir MAURICE KENDALL & R. L. PLACKETT

260 x 184mm 437pages 2plates (ISBN 085264 232 6)

The selected papers of E. S. Pearson

This selection taken from his many papers was published to mark 30 years spent by the author as Editor of *Biometrika.*

273 x 189mm 334pp Bibliography 68 illustrations 84 tables (ISBN 0 85264 705 0)

Joint statistical papers of J. Neyman and E. S. Pearson

Many important advances in Statistics since the 1920s have been due to J. Neyman and E. S. Pearson working together or separately. This volume contains their joint work on the testing of statistical hypotheses.

273 x 189mm 307pp 47 illustrations 25 tables (ISBN 0 85264 706 9)

Ask for our statistical catalogues.

STATISTICS THEORY AND PRACTICE

Selected papers by Sir Maurice Kendall
(1907–1983)

Edited by

ALAN STUART, D.Sc. (Econ)

MACMILLAN PUBLISHING COMPANY
NEW YORK

Published in USA by
Macmillan Publishing Company
866 Third Avenue, New York, N.Y. 10022

Distributed in Canada by
Collier Macmillan Canada, Inc.

By arrangement with the originating publisher
CHARLES GRIFFIN & COMPANY LIMITED
London and High Wycombe

ISBN 0-02-853050-0

Printed in Great Britain by Pitman Press, Bath.

PREFACE

Maurice Kendall was a statistician whose published work covers an extraordinary range from intricate theory to detailed application. As well as his numerous papers, which are partially listed in the select bibliography that follows the obituary notice reproduced here, he published the many books also listed, which often incorporated and consolidated the work in his papers. While attempting to indicate the scope of his interests, this memorial selection has therefore concentrated on the less accessible of his papers.

The papers included here fall into four groups. First, there are four papers fundamentally concerned with applications, including his Inaugural Lecture as Professor, his Presidential Address to the Royal Statistical Society, the paper on the relation between votes cast and seats won in two-party elections and one on the analysis of economic time-series, with special application to stock market prices – a later paper, not reproduced here but listed in the select bibliography (1969, with Coen and Gomme) dealt with related matters.

The next group consists of six papers on the theory of symmetric functions of sample observations which proved and extended results intuitively arrived at by the great R. A. Fisher. In his obituary of Fisher (1963) Kendall told the extraordinary story of how Fisher apparently visualized the results on a train journey but subsequently was unable to remember how: they remained unproved for more than a decade until Kendall established them.

The third group includes a selection of Kendall's papers in the theory of ranking methods, including his 1938 discovery of the rank correlation coefficient since named after him and two of his important collaborations with Babington Smith, of which other fruits were the distribution theory of the other (Spearman's) rank correlation coefficient (1939) and the papers and tables of random sampling numbers (1938, 1939).

The last group of papers contains three of his early papers on the theory of time-series. The verse postlude displays his extraordinary talent in putting literary pastiche at the service of good statistical practice.

The papers are reproduced by permission of Mr. B. Babington Smith, OBE, M.A., and of the American Statistical Association, *The Annals of Human Genetics* (formerly *The Annals of Eugenics*), The Biometrics Society, The Biometrika Trustees, *The British Journal of Sociology, Economica*, The Institute of Mathematical Statistics and the Royal Statistical Society.

I hope the general statistical reader may profit from the applied papers here; the specialist should certainly find these and the theoretical papers valuable. This selection cannot hope to present a complete picture of Maurice Kendall's work, but it does, I believe, indicate the quality and the flavour of a very remarkable man.

London, January 1984 *A.S.*

CONTENTS

From *Journal of the Royal Statistical Society A*, Vol. **147** (1984)

SIR MAURICE KENDALL 1907–1983

An obituary notice in the
Journal of the Royal Statistical Society, Series A, Vol. **147** (1984)

Maurice George Kendall was born on September 6th, 1907 in Kettering, Northamptonshire, the only child of John Roughton Kendall, a local man, and of Georgina (*née* Brewer) of Standon in Hertfordshire. John Kendall's father George was the licensee of a Kettering inn, The Woolsack, and as a boy Maurice occasionally helped his grandfather there. Maurice survived cerebral meningitis, then frequently fatal.

During the First World War, John Kendall was working for Rolls Royce in Derby. Maurice's early education there was unremarkable, and he did not get a Grammar School place when he sat the then extremely selective examination; instead, he went to the Central School, intermediate between the Grammar and the predominant Senior (Secondary Modern). By the time he reached Matriculation age he was primarily interested in languages but then, apparently quite unexpectedly, began to show enough mathematical talent to persuade his headmaster to arrange his attendance at the Grammar School for Sixth Form work. His subsequent achievement of a scholarship to St John's College, Cambridge, justified these special arrangements, which were very unusual in the 1920s. John Kendall thought that his son would do better to leave school and train as an engineer, but Georgina, the more ambitious, prevailed and it was to her that Maurice was to dedicate his first book in 1943. Before going up to Cambridge, Maurice took up a short-term teaching appointment, and during this time he was in contact with the well-known mathematician Piaggio at Nottingham University College.

Life at St. John's was in striking contrast to that in Derby, and Maurice's naturally gregarious nature brought him several circles of friends apart from the group reading Mathematics at St John's: he played cricket for his college (and indeed was still a useful medium-paced bowler 25 years later); and he might have got a half-blue at chess had the company been less select –C.H.O'D. Alexander, later twice British champion, was one, and Jacob Bronowski another. Typically, Maurice enjoyed playing chess blindfold, and Bronowski and he played "negative chess", where the previous moves had to be deduced from a set postition. At St John's Maurice knew Major MacMahon, an elderly Fellow of St John's who had written the standard treatise on Symmetric Functions, and one wonders whether it was entirely fortuitous that one of Maurice's signal contributions to Statistics was to be the application of these to the theory of Fisher's k-statistics. He graduated a Mathematics Wrangler, and passed into the Administrative Class of the Civil Service in 1930. At the Ministry of Agriculture he was soon given responsibility for statistical work, and no doubt this led him to become a Fellow of the Society in 1934.

It is natural to assume, in the light of his revision of George Udny Yule's *Introduction to the Theory of Statistics* in 1937, that he had been tought by Yule, who was a Fellow of St John's, but in fact they never met until chance introduced them in 1935. Kendall decided to improve some of his vacation by reading in his old College library, but found on arrival that the Librarian was also on vacation. He made his way to the Fellow who was holding the Library keys, who asked him what the was going to do in the Library. "I thought," said Kendall, "that I would read some Statistics," and when the Fellow, who was Yule, decided soon after that his 25-year-old book needed revision, he remembered the young man from the Ministry and the Gods no doubt smiled stochastically. Kendall and Yule remained on close terms until Yule's death in 1951; Yule stood godfather to Kendall's second son.

The revision of Yule's book stimulated Kendall's interest in statistical problems and in the remaining pre-war years he attended advanced lectures at University College London, and published his first research papers, in the *Journal* and elsewhere. No advanced textbook was available then, and in 1938 it was proposed that one should be written by M. S. Bartlett, J. O. Irwin, Kendall, E. S. Pearson and J. Wishart. The project got no further than the preparation of a synopsis, for in 1939 the War dispersed the potential collaborators. Kendall, however, remained in London, and in 1940 left the Civil Service to become Statistician to the British Chamber of Shipping, where he had been preceded by another distinguished Fellow, Leon Isserlis. In wartime, this must have been a taxing job, and in 1945 he became Honorary Secretary of the Society; nonetheless he proceeded alone with *The Advanced Theory of Statistics*, which expanded into two volumes, published in 1943 and 1946. The Sc.D. was conferred on him by his old University.

The publication of this major work and a steady stream of research, mainly but not entirely in the theory of *k*-statistics, time-series and rank correlation methods, led to the offer of the new second Chair of Statistics at the London School of Economics, which Kendall held from 1949 to 1961. Here he founded a new Research Division which carried out a number of large sample survey projects in collaboration with Governmental and commercial agencies; most of the results were published in the Society's *Journal*. The organization of these and of more theoretical researches was supplemented by the preparation of the three-volume *Bibliography of Statistical Literature* from the earliest times up to 1958, using a grant from the Leverhulme Foundation, and of the *Dictionary of Statistical Terms*, with several foreign-language glossaries, for the International Statistical Institute.

These tasks alone make it clear that Kendall had a rare talent for the organization of research: he knew how, and to whom, to delegate, and he was distinguished among academics by his habit of delivering on time what he had promised. Though this was partly due to his long working day, it was primarily the expression of an orderly and organized mind. But although research was always at the centre of his interests, undergraduates and their teaching did not much appeal to him, and it was not altogether a suprise when he again changed the direction of his career, during his Presidency of the Society. He was successively Scientific Director, Managing Director and Chairman of the computer consultancy now called SCICON, and although he completed the rewriting of his major book into three volumes in 1966 he found himself moving away from Statistics. In 1972 Kendall reached retiring age, but characteristically undertook the most difficult assignment of his career, the Directorship of the World Fertility Survey, a huge, multinational sample survey project. Here his extraordinary organizational abilities were fully tested, although he was closer to statistical and demographic problems than previously, and his achievement in this post was recognized by the award of a United Nations Medal after illness compelled his retirement in 1980. Physical disability restricted his last years, but when he died on March 29th, 1983 he was planning to attend the I.S.I. conference in Madrid in September.

Maurice Kendall's most rare gift to the Statistics profession, and to the Society in particular, is his literary style, lucid, balanced, often ironical (see parts of his memoirs of R. A. Fisher in *Biometrika*, **50**, 1–15 and of Edgeworth in *Biometrika*, **55**, 269–275) and occasionally touched by deep feeling (as in his obituary of Yule in this *Journal*, **115**, 156–161). Essentially, however, he was a sceptic, especially of the pretentious, or of what he considered so. After attending a performance of Handel's "Rodelinda" in Stockholm in 1957 he translated a passage on the computation of moments in *The Advanced Theory of Statistics* into Italian, and had it set to a mock Handelian aria. The flavour may be obtained by singing to the tune of "O ruddier than the cherry" the words

Progressive iteration
Of frequency summation,
Of frequency summation,
Progressive iteration yields
The moment calculation.

Such enterprises were characteristic: he continually played with words. The unfinished story that gave Volume 2 of *The Advanced Theory of Statistics* its epigraph was written by K. A. C. Manderville, a latinate anagram of "Mavrice Kendall", and its heroine Lamia Gurdleneck was an anagram of "Maurice G. Kendall" as her aunt Sara Nuttal was of "Alan Stuart". A potential character was Emanuel D. Garlick, another anagram. His famous Longfellow pastiche "Hiawatha designs an experiment" was one of the happiest results of Maurice's word-play.

His scepticism sprang from a fundamental conservatism of character, which went hand-in-hand with his strong interest in technical innovation. He was always the most clubbable of men, as was exemplified by his long membership of the Statistical Dinner Club. For this, as for other traditions of the Society, he had high regard, and I suspect that he valued his Guy Medal in Gold as highly as the knighthood conferred on him for services to the theory of Statistics, or his Fellowship of the British Academy. It will be many years before our subject again has at its disposal his combination of talents.

Alan Stuart

A SELECT BIBLIOGRAPHY OF THE STATISTICAL WORKS OF SIR MAURICE KENDALL

BOOKS

1. (with G. U. Yule) *An Introduction to the Theory of Statistics* (11th edition) Charles Griffin, London, 1937.
2. *The Advanced Theory of Statistics* (2 Volumes) Charles Griffin, London, 1943, 1946.
3. *Rank Correlation Methods*, Charles Griffin, London, 1948.
4. (Editor) *The Sources and Nature of the Statistics of the United Kingdom* (2 Volumes), Oliver and Boyd, Edinburgh, 1952, 1957
5. *Exercises in Theoretical Statistics*, Charles Griffin, London, 1953.
6. (with W. R. Buckland) *A Dictionary of Statistical Terms*, Oliver and Boyd, Edinburgh, 1957.
7. *A Course in Multivariate Analysis*, Charles Griffin, London, 1957.
8. (with A. Stuart) *The Advanced Theory of Statistics* (3 Volumes) Charles Griffin, London, 1958, 1961, 1966.
9. *A Course in the Geometry of* n *Dimensions*, Charles Griffin, London, 1961.
10. (with A. G. Doig) *Bibliography of Statistical Literature* (3 Volumes) Oliver and Boyd, Edinburgh, 1962, 1965, 1968.
11. (with P. A. P. Moran) *Geometrical Probability*, Charles Griffin, London, 1962.
12. (with F. N. David and D. E. Barton) *Tables of Symmetric and Allied Functions*, Cambridge University Press, 1966.
13. (Editor with E. S. Pearson) *Studies in the History of Statistics and Probability*, Charles Griffin, London, 1970.
14. (Editor with A. Stuart) *The Statistical Papers of George Udny Yule*, Charles Griffin, London, 1971.
15. *Time-Series*, Charles Griffin, London, 1973.
16. *Multivariate Analysis*, Charles Griffin, London, 1975.
17. (Editor with R. L. Plackett) *Studies in the History of Statistics and Probability*, Volume II, Charles Griffin, London, 1977.

PAPERS

1938 Conditions under which Sheppard's corrections are valid. *J. R. Statist. Soc.*, **101**, 592–605.

A new measure of rank correlation. *Biometrika*, **30**, 81–93.

(with B. Babington Smith) Randomness and random sampling numbers. *J. R. Statist. Soc.*, **101**, 147–66.

1939 Geographical distribution of crop productivity in England. *J. R. Statist. Soc.*, **102**, 21–62.
(with B. Babington Smith)
Second paper on random sampling numbers. *J. R. Statist. Soc. Suppl.*, **6**, 51–61.
The problem of *m* rankings. *Ann. Math. Statist.*, **10**, 275–87.
Tables of random sampling numbers. *Tracts for Computers*, no. 24, x +60pp. (Cambridge University Press.)
(with S. F. H. Kendall and B. Babington Smith) The distribution of Spearman's coefficient of rank correlation in a universe in which all rankings occur an equal number of times. *Biometrika*, **30**, 251–73.

1940 Conditions for uniqueness in the problem of moments. *Ann. Math. Statist.*, **11**, 402–9 (correction: **12**, 464.)
Note on the distribution of quantiles for large samples. *J. R. Statist. Soc. Suppl.*, **7**, 83–5.
On the method of maximum likelihood. *J. R. Statist. Soc.* **103**, 388–99.
Some properties of *k*-statistics. *Ann. Eugen.*, **10**, 106–11.
Proof of Fisher's rules for ascertaining the sampling semi-invariants of *k*-statistics. *Ann. Eugen.*, **10**, 215–22.
The derivation of multivariate sampling formulae from univariate formulae by symbolic operation. *Ann. Eugen.*, **10**, 392–402.
(with B. Babington Smith) On the method of paired comparisions. *Biometrika*, **31**, 324–45.

1941 Effect of the elimination of trend on oscillations in time-series. *J. R. Statist. Soc.*, **104**, 43–52.
Relationship between correlation formulae and elliptic functions. *J. R. Statist. Soc.*, **104**, 281–3.
A theory of randomness. *Biometrika*, **32**, 1–15.
A recurrence relation for the semi-invariants of Pearson curves. *Biometrika*, **32**, 81–2.
Proof of relations connected with the tetrachoric series and its generalization. *Biometrika*, **32**, 196–8.

1942 On the future of statistics. *J. R. Statist. Soc.*, **105**, 69–80.
Note on the estimation of a ranking. *J. R. Statist. Soc.*, **105**, 119–21.
Partial rank correlation. *Biometrika*, **32**, 277–83.
On seminvariant statistics. *Ann. Eugen.*, **11**, 300–5.

1943 Oscillatory movements in English agriculture. *J. R. Statist. Soc.*, **106**, 91–124.

1944 On autoregressive time series. *Biometrika*, **33**, 105–22.
(with J. O. Irwin) Sampling moments of moments for a finite population. *Ann. Eugen.*, **12**, 138–42.

1945 On the analysis of oscillatory time-series. *J. R. Statist. Soc.*, **108**, 93–141.
Note on Yule's paper on studying time-series based on their internal correlations. *J. R. Statist. Soc.*, **108**, 266–30.
The treatment of ties in ranking problems. *Biometrika*, **33**, 239–51.

1946 Contributions to the study of oscillatory time-series. *Nat. Inst. Econ. Soc. Res. Occasional Papers IX*, 76pp. (Cambridge Univ. Press).

1947 Variance of τ when both rankings contain ties. *Biometrika*, **34**, 297–8.

(with H. E..Daniels) Significance of rank correlation where parental correlation exists. *Biometrika*, **34**, 197–208.

1948 Systematic sampling of continuous parameter populations. *Biometrika*, **35**, 291–6.

The estimation of parameters in linear autoregressive time series. *Econometrica Suppl.*, **17**, 44–57. (Also: *Proc. 25th Session Int. Statist. Inst. Conf. Washington, D. C.*, 1947 (Publ. 1951), **5**, 44–57.)

1949 On the reconciliation of theories of probability. *Biometrika*, **36**, 101–16.

Rank and product-moment correlation. *Biometrika*, **36**, 177–93.

Tables of autoregressive series. *Biometrika*, **36**, 267–89.

1949–1955 (with F. N. David) Tables of symmetric functions I-V. *Biometrika*, **36**, 431–49; **38**, 435–62; **40**, 427–46; **42**, 223–42 (correction, **45**, 292).

1950 The statistical approach. *Economica*, **17**, 127–45.

Factor analysis as a statistical technique. *J. R. Statist. Soc. B*, **12**, 60–73; 85–94.

(with A. Stuart) The law of the cubic proportion in election results. *Brit. J. Sociol.* **1**, 183–96.

1951 (with S. T. David & A. Stuart) Some questions of distribution in the theory of rank correlation. *Biometrika*, **38**, 131–40.

(with J. Durbin) The geometry of estimation. *Biometrika*, **38**, 150–8.

1951-2 Regression, structure and functional relationship. I, II. *Biometrika*, **38**, 11–25; **39**, 96–108.

1952 Moment-statistics in samples from a finite population. *Biometrika*, **39**, 14–16.

"George Udny Yule 1871–1951". *J. R. Statist. Soc. A*, **115**, 156–61.

1953 Analysis of economic time-series. I. Prices. *J. R. Statist. Soc. A*, **116**, 11–34.

(with R. M. Sundrum) Distribution-free methods and order properties. *Rev. Int. Statist. Inst.*, **21**, 124–34.

1954 Note on bias in the estimation of autocorrelation. *Biometrika*, **41**, 403–4.

Two problems in sets of measurements. *Biometrika*, **41**, 560–4.

1955 Further contributions to the theory of paired comparisons. *Biometrics*, **11**, 43–62.

1956 Studies in the history of probability and statistics. II. The beginnings of a probability calculus. *Biometrika*, **43**, 1–14.

(with D. N. Lawley) The principles of factor analysis. *J. R. Statist. Soc. A*, **119** 83–4.

1957 The moments of the Leipnik distribution. *Biometrika*, **44**, 270–2.

Studies in the history of probability and statistics. V. A note on playing cards. *Biometrika*, **44**, 260–2.

(with K. E. Gales) An inquiry concerning interviewer variability. *J. R. Statist. Soc. A*, **120**, 121–47.

1958 (with H. E. Daniels) Short proof of Miss Harley's theorem on the correlation coefficient. *Biometrika*, **45**, 571–2.

1959 Hiawatha designs an experiment, *Amer. Statistician,* **13**, (5) 23–4.

1961 Natural law in the social sciences, *J. R. Statist. Soc. A*, **124**, 1–18.

Studies in the history of probability and statistics.
XI. Daniel Bernoulli on Maximum Likelihood.
XII. The Book of Fate.
Biometrika, **48**, 1–18 and 220–2.

A theorem in trend analysis. *Biometrika,* **48**, 224–7.

1962 Ranks and measures. *Biometrika,* **49**, 133–7.

1963 "Ronald Aylmer Fisher 1890–1962". *Biometrika,* **50**, 1–15.

Studies in the history of probability and statistics. XIII. Isaac Todhunter's History of the Mathematical Theory of Probability. *Biometrika,* **50**, 204–5.

(with P. Wegner) An introduction to statistical programming. *Rev. Int. Statist. Inst.,* **31**, 307.

1964 (with D. Murchland) Statistical aspects of the legality of gambling, *J. R. Statist. Soc. A,* **127**, 359–91.

1966 Statistical inference in the light of the electronic computer. *Rev. Int. Statist. Inst.*, **34**, 1–12.

The possibility of determining Elizabethan authorship by statistical analysis. *Shakespeare Authorship Review,* **16**, 19.

1967 Discrimination and classification. In *Multivariate Analysis*, ed. Krishnaiah, Academic Press, New York.

(with E. M. L. Beale and D. W. Mann) Discarding redundant variables in multivariate analysis. *Biometrika,* **54**, 357–66.

1968 "Francis Ysidro Edgeworth (1845–1926)". *Biometrika,* **55**, 269–75.

1969 The early history of index numbers. *Rev. Int. Statist. Inst.*, **37**, 1–12.

(with P. J. Coen and E. D. Gomme) Lagged relationships in economic forecasting. *J. R. Statist. Soc. A*, **132**, 133–63.

From *Economica*, Vol. **17**, pp 127–145 (1950)

The Statistical Approach[1]

By M. G. Kendall

The statistician is popularly regarded as a man who knows nothing except facts; and facts, moreover, which are numerical in nature, that is to say, are of a particularly inhuman and specious kind. It is not easy for anyone reared among the humanities or in the practical world of industry and commerce to believe that a person who apparently spends his time merely in counting or measuring what he observes can, by that process alone, acquire any deep insight into what is really happening. That the statistician is a useful person on occasion is now accepted. That he can make great contributions to scientific methodology is becoming realised. But that he has a unified conception of the world at large, that he derives from his science a satisfaction of those desires for a grasp of the fundamentals of things which lie beyond pure reason—in short, that he has a philosophy—may come to you as a surprising claim.

I could think of no better purpose for this inaugural lecture from the chair to which I have had the honour of being elected than to elucidate this general aspect of my subject and to try to present it to you from the very broadest point of view. I therefore propose to speak, not about particular sets of statistics or particular statistical methods, but about the ideas and motives which lie behind them. I should like you, just for this one occasion perhaps, to see my subject as I see it myself, not as the pedestrian science of handling numerical data, not even as a comparatively new branch of scientific method, but as the matrix of quantitative knowledge of nearly every kind, as the principal instrument yet devised by man for bringing within his grasp the terrifying complexity of things and relations-between-things, and as a powerful illuminant of the process of rational thought itself. It is a great deal to attempt a sketch of the philosophy of the statistical approach in one lecture, but perhaps

[1] Inaugural Lecture given at the London School of Economics on January 17th, 1950.

an inaugural lecture is a suitable occasion and in any case there could be no better place to attempt it than the University where the greatest men of my subject, Karl Pearson, Udny Yule, Ronald Fisher and, may I say it with great pleasure, Sir Arthur Bowley, have done so much of their work. I speak with a profound sense of humility at the thought of following in the footsteps of such men. To maintain their standard of achievement is beyond my powers. But at least I can try to convey something of the impact of their work on our approach to the problems of the modern world.

Since the days when the word "statistics" had the limited connotation of data relating to statecraft my subject has found its most important applications in economic and social fields. It remains true, I think, that the work and welfare of mankind provide the most important domain for statistical inquiry. Certainly that is the common belief; for the layman the word "statistics" is practically synonymous with "economic statistics"—economic, that is, in a broad sense. Indeed, you might suppose that economic and social statistics are relatively much more important nowadays than they were. The advent of the planned economy is generally considered to have created a statistician's paradise, and however regrettable this creation may be it is accepted by most people as inevitable. The statistician, indeed, is now much more than a handmaiden to a planned society. Very often his work tells it what it must plan. I need not give more than one example; since 1939 the population of the world has probably increased by 6 to 7 per cent. and its food supply is not keeping pace, particularly as higher standards of nutrition are adopted in backward countries. The shadow of Malthus falls again across the world; and that great man, let us remember, was one of the first statisticians and a member of the small group which founded the Royal Statistical Society.

The emphasis laid on statistics by the economic and social sciences historically and practically has tended to obscure the fact that from the methodological viewpoint economic statistics forms only a small part of the whole field of the subject of statistics at the present time. To give an idea of the extraordinary range of my subject as it now exists I take some examples more or less at random from the work of the past twenty-five years. In agriculture the whole theory and practice of plant breeding and of field trials has been revolutionised. In industry the spread of statistical methods of quality control has been as rapid as it has been successful. In meteorology the statistician is at last getting to grips with the enigmatic behaviour of the weather. In nuclear physics the statistical approach is now a basic part of the subject. To take more particular instances, statistical methods are used in the study of epidemics, telephone traffic, industrial accidents, the standardisation of drugs, the measurement of human abilities, the migration of insects, the efficiency of examiners, the building-times of pre-fabricated houses, the distribution of blood-groups, factory production costs, flutter in aircraft structure, and in fact in almost every branch of science and industry. The statistician does not indeed stop there. He is extending his interests into domains which have not hitherto been considered as possible fields for the application of numerical methods. The chronology of ancient history, the dating of the books of the Rigveda, the credibility of witnesses, telepathic communication, have all been brought under the statistical microscope. Not long ago I heard of an enquiry into the distribution of intervals between successive notes in some of William Byrd's madrigals, and within the last week my attention was drawn to a thesis on the use of discriminant functions for handicapping horses. Some subjects, it is true, are not contaminated by the irritating exactitude of numerical expression. Poetry, Drama and Theology have a natural immunity to scientific enquiry. But one never knows what may happen.

Frankly, I enjoy making this kind of catalogue, but it is not the diversity of the applications of statistical methods to which I wish to direct attention. Their significant feature is the unity of the underlying method of approach. We are constantly finding that a technique evolved for one purpose can easily be adapted to another. Many years ago Karl Pearson was interested in the spread of mosquitoes under random migration from a malarial site. The problem of the random walk which he studied has today found applications to the spread of disease, the diffusion of gases, Brownian movement and most recently in three dimensions to investigations into the molecular structure of rubber. Some work by Cayley on a type of mathematical network known as a tree has been applied in sociology (to the study of the propagation of rumour) and in economics (to the study of systems of accounts). Three-quarters of the way through the nineteenth century Whitworth posed and solved a problem concerning the distribution of the size of pieces of a stick which is broken at random at a given number of points. His results have been applied to the design of traffic-light systems, bomb fragmentation, the detection of trade cycles, and the analysis of noise. Twenty years ago Yule wrote a famous study of sunspots. It is not surprising that his methods should have been applied to forecasting stock-market fluctuations and the study of variations in crop-yields ; but it is odd to find them turning up in studies of the standardisation of cotton thread, the preparation of polished metal surfaces, cosmic-ray showers and the design of gun turrets. Methods devised for comparing the performance of students in different subjects have been applied to studies of wool grading, social stratification, the efficiency of card-shuffling processes and the effect of bombing on civilian morale. From these examples you will not wonder that the statistician who contemplates his subject as a whole, who traces an idea or a method continuously from one end of science to the other, occasionally feels as if there lies beneath his hand something of the basic fabric of rational thought. And this is to say nothing of the problem of sampling, which brings him into contact with every type of inquiry where the data available are only a part of the data in existence. Wherever two or three are gathered together, experimentally speaking, there is scope for the application of the statistical method.

But I have not yet mentioned the distinguishing features of the statistical approach. What I have been saying, after all, amounts to no more than this, that the statistical method is widespread because numerical data arise in most branches of organised knowledge. Like mathematics it is a scientific method, and a method which is scientific is *ipso facto* capable of general application. The mere ubiquity of the statistical approach, then, striking as it may be, is not its most important feature. What marks it out for special attention is that it deals, not with individuals, but with aggregates. Statistics is the science of collectives and group properties. The statistician is interested in the individual only as the member of a group. It is from this basic fact that his strength and his weakness both arise. His strength because most, if not all, natural laws are group properties ; his weakness because he sometimes throws into a group individuals which are not homogeneous and misses something in his summarisations. His common sense usually saves him. That is why all good statisticians lay such emphasis on the importance of common sense. But we should in fairness admit that he is in danger of failing to see the trees for the wood. It is inevitable that he should be so, for one cannot watch the aggregate and at the same time keep every item of it in focus.

In general the laws obeyed by collectives are determined empirically and a large part of statistical technique is concerned with the setting up of a calculus of collective phenomena to handle observations in the aggregate. But there are also laws of a peculiar kind in mass action which permit of the prediction of the behaviour of an aggregate when it is quite impossible to frame laws concerning the individual. We are permitted, in a sense, to derive law from the actual absence of law. In fact, under certain conditions, the more numerous the perturbative influences at work on the individual, the more definite the law obeyed by the aggregate. We may go further still. There are certain aggregates, called random or stochastic, the members of which not only do not conform to law in the older sense of nineteenth-century determinism but are actually *conceived* of as not conforming to any such law. We say that they happen by chance. And yet we can derive quite definite laws which are obeyed by aggregates of such individuals and we can say without paradox that chance is subject to law.

The statistician, then, tends to think in terms of aggregates and the laws of group properties. This gives him a distinctive outlook on life which, without perhaps crystallizing into an explicit philosophy, is nevertheless formative of his attitude and his approach towards philosophical problems. Far from rendering him remote and inhuman it sometimes produces a sense of collective unity which is one of the features of the best religions. I give one or two illustrations.

There is a statistical law, known by the name of the French mathematician Poisson, expressing the fact that even relatively rare events happening at random have a definite pattern of behaviour. Practically every text-book written in the past forty years has exemplified this law by quoting some data collected by von Bortkiewicz concerning the number of officers in the Prussian cavalry corps who were killed by the kick of a horse. The example is so hackneyed that a bored reader once wrote to me imploring me to vary it. Now good statistical examples are hard to find, but on this occasion Mr. R. E. D. Clarke came to the rescue with the results of an examination of the fall of V 1 flying bombs in an area in South London. The object of the inquiry was to ascertain whether there was any ground for the belief that the bombs fell in clusters otherwise than by chance. The data showed that they did not and that the distribution was almost perfectly represented by the Poisson law. The statistician derives some pleasure from the reflection that while the science of war developed from the cavalry charge of 1865 to the flying bomb of 1945 the laws which interest him, the laws of homogeneous chaos, remained immutable.

I take a more serious illustration. At this moment there are alive and well in this country about a dozen people who will have been killed in road accidents by this time tomorrow evening. Who they are, where they are, or what they are it is impossible to foresee. But the statistician can foresee that they are somewhere. Doubtless the incidence of random deaths of this character could be prevented; but doubtless it will not be prevented, at least by tomorrow evening. The threat of the random event lies over us all. We flatter ourselves that we control our destiny, and in one sense perhaps we do; but most of us, I think, when we look back over the major incidents of our lives, are disturbed at the large element of chance which seems to control us.

It gives one a curious feeling of identity with the aggregate of humanity to reflect on this fact. One does not expect to find a quotation apposite to a statistical point in the sermons of a mystic, but I think John Donne was expressing this idea when he wrote: "Ask not, then, for whom the bell tolls. It is for *thee*." I mention this element of mysticism in the subject, not to develop the theme, for that would be out of place on this occasion, but to call attention to its existence; for there is, I hold, some such element at the root of all creative scientific endeavour, and statistics is no exception to the rule.

You must not misunderstand me. I am not attempting to draw a portrait of the statistician as a fatalist, resigned to inexorable random misfortune, or as a Pythagorean mystic looking for truth, beauty and virtue in the properties of numbers themselves. He is just as prosaic in ordinary life as anyone else, and more prosaic than some. Beneath the bludgeonings of chance his head is not picturesquely bloody or defiantly unbowed. It is much more likely to be working out the premiums on an insurance policy. At the same time, I must point out that his mind is not a mere collection of facts like a blue book. He has an attitude. Very definitely, he has an attitude.

The statistician even has a sort of fellow-feeling with Nature herself; not, I mean, in the flower-in-the-crannied-wall or the little-lamb-who-made-thee sense, but arising from the fact that Nature behaves in the statistical manner in some of her most important operations. She equips her creatures individually with some excellent weapons, but, recognising that she cannot hope to provide against every contingency, relies for the survival of the species on sheer weight of numbers. The herring and the oyster spawn in millions; the law of averages will see to it that some of their offspring survive. Nature, in fact, does not shoot at her targets with a weapon of high precision; she sprays them with a shot-gun. She has the statistical approach. It is of some interest to observe in passing that military authorities are catching up with her in this respect; on land the rifle is giving way to the tommy gun; at sea the submarine hunts in packs; in the air precision bombing makes way for a general plastering of the target area.

While we are on this high level, I might mention that the fate of the whole universe seems to be deciding itself on a statistical basis. So far as we can see the universe of things physical in its present form is rather improbable. Its matter and its energy are concentrated in heaps in a way which could hardly have happened if they had been scattered at random. But energy is being constantly radiated and matter appears to be transformed into energy. Consequently the universe is becoming more and more random, more and more unorganised. In technical terms, its entropy, the negative of the logarithm of the statistical probability of things, is increasing. There are theorems, known as ergodic, the effect of which is that under the laws of chance the ultimate state is more or less uniform independently of what the initial state might have been. The statistician, then, must look at the universe as a sort of Manichean struggle between the spread of chaos as energy is sprayed into space and the reverse process of organisation, which seems to be one of the characteristics of what we call life. It would be rash to draw conclusions from all this. The fact that the universe is unlikely does not prove that it must have been arranged from primordial chaos by an intelligence. The fact that it is dissipating its energy does not mean that it will always do so. Certainly it does not warrant on the statistician's part a pessimistic assumption that chaos is bound to win. Statisticians, in fact, are usually optimists; they have to be, to submit to the tedious discipline of their subject. But I need not elaborate the point. Statistical considerations will probably not affect our basic feelings on such matters which, I gather, are more dependent on the secretion of one's endocrine glands than on knowledge.

It is perhaps time that I came down to earth; but before I consider the statistical approach to a few of the more human current topics I should like to interpolate some comments on the relationship between statistics and mathematics, for there is an impression that one is a branch of the other which I feel it highly desirable to try to dispel. As is only natural for a science which depends so much on numbers, the theory of statistics is essentially mathematical in its modes of expression and even, to some extent, in its modes of thought. It draws freely on the language and theorems of mathematics. Indeed, it has stimulated a great deal of research into pure mathematics and there is hardly a branch of that subject which has not been overhauled and refurbished in recent years to meet some statistical requirement. In spite of this I must hold firm to the thesis that statistics is not a branch of mathematics. It is not a branch of any subject. It is a subject in itself. R. A. Fisher is often quoted as the authority for the statement that statistics is a branch of *applied* mathematics. But I think he meant—I hope he meant—that statistics is a subject in which extensive use is made of mathematics, as for example in astronomy or engineering. His dictum has not always been understood in that sense. It has become the custom to give the name "applied mathematics" to branches of pure mathematics dressed up in the guise of certain of the physical sciences. In that sense statistics is not a branch of applied mathematics. There are fundamental concepts in statistics which find no place in mathematics.

Indeed, in one respect the two are antithetical, for one is a theory of uncertain, the other of certain, inference.

Mathematics, in fact, like fire, is a good servant but a bad master. At its worst—or perhaps at its best, I make no ethical judgment—it is a form of escape from reality. The most recent developments in mathematical statistics give rise to some anxiety on the part of the theoretical statistician. I think we have to acknowledge quite frankly that some mathematicians enjoy being cryptic and that others are constrained to parade their learning in order to maintain a place in the struggle for scientific recognition. There is, of course, a need for rigorous mathematical exposition in statistical theory, and some of it can be put only in language which is intelligible only to the expert. Nevertheless I feel sometimes, in wrestling with papers on statistical matters with a full emblazonment of modern mathematics, sets of points, Lebesgue integration, asymptotic summability and so forth, that work which achieves a slight gain in generality at the expense of being incomprehensible is defeating its own purpose. The mathematician, I suggest, ought to have some small responsibility for making himself understood. One of the few doctrinal points on which all statisticians seem to agree is the necessity for keeping one foot firmly on the ground. If we look at the theory of time-series, for instance, we find that any theoretical results of importance have sprung directly from an attack on a practical problem. Statistics, like all progressive sciences, is experimental. You will forgive me, I hope, if I appear to labour this point but it is particularly important at the present time because we have reached a critical phase in the development of statistical method when pure mathematicians looking for something to research on are throwing up such a dust that the practical nature of the subject is being obscured and theoretical statistics is in danger of being discredited in the eyes of the more practical man.

I return to consider the statistical approach to more urgent human affairs ; I shall not attempt a comprehensive survey ; you are all aware of the importance of the statistical approach on such matters as national finance, the manpower problem, land utilisation and commercial organisation. In order to illustrate the statistical approach, as distinct from the collection and analysis of statistical data, I deliberately choose as illustrations three topics drawn from fields in which the statistical approach has not been regarded as particularly important, one from each of sociology, politics and social ethics. It must be made plain at the outset that the statistician—at least, this particular statistician—has no desire to poach on other people's preserves. He likes to be invited to inspect his neighbour's grounds but he does not trespass. The last thing I want to suggest is that he can teach their business to the experts in other fields. My point is rather that the statistician, in virtue of his subject, acquires an approach to certain problems of human organisation which might prove suggestive to his fellow workers. The statistician is not, and makes no claim to be, a sociologist, a politician or a moral philosopher ; but he may, perhaps, be allowed to suggest that the statistical approach, in the sense in which I am using the term in this lecture, should not be overlooked in the subjects with which they are concerned.

(*a*) First of all, an example from social economics or economic sociology, whichever you prefer. Since Pareto drew attention to the law of distribution of income the social trend has been towards greater equality in the distribution of the national product. Some people seem to want to go to the extreme. Others think that there are dangers that we have already gone too far in the direction of simple egalitarianism. But nobody seems to think it necessary to announce exactly what it is we are trying to do. Does anybody know, or has anybody tried to work out, what the distribution of incomes ought to be or could be made to be under different hypothesis ? It is not obvious, and it may not even be true, that the greatest good is assured to the greatest number by narrowing the dispersion of incomes. If we once admit that there should be inequalities, however restricted, between incomes the question immediately arises : how should incomes be distributed ? If a foreman is to receive more than a workman under his charge, how much more ? What determines the ratio of the salaries of a chairman and a vice-chairman of a company ? I am all too well aware that the answers to these questions are not to be sought in statistics alone ; but I think that the statistical approach may contribute something towards finding them if only by showing that certain solutions are impossible. It might, for example, be possible to show that the social pyramid must obey certain laws if it is to be stable and self-perpetuating. It might be found that a certain number of plums must be left in the national cake for private consumption if certain social policies are not to collapse, and if so, the statistician might be able to say how many there should be. My colleagues in sociology will, I hope, forgive me for putting forward these rather naïve ideas without showing just what the statistician can do to help. I cannot stay to develop them. My point is made if it is agreed that the hedonic calculus stands in need of a statistical overhaul.

(*b*) In the political field there are some interesting statistical conclusions to be drawn from the fact that we live under a system of government by committee. We all know what happens when one or two members of a decisional group are armed with a veto; perhaps we do not quite appreciate their power when they are not so armed. Professor Penrose recently has discussed what might be termed the Law of the Resolute Minority in organisations wherein members have equal voting rights. If the majority of a committee are indifferent the minority will nearly always get its way. Within the last few days interest has been aroused by the empirical law that the number of seats obtained in an election by a party varies as the cube of the number of votes cast for it. Again I cannot dwell on the theme, fascinating as it is. It leads to studies of such matters as the power of the floating vote, proportional representation, the laws of inertia of voting groups, and the optimum sizes of executive bodies—none of them statistical subjects, but all of them subjects on which the statistician has something to say in virtue of his expertise in the properties of aggregates.

(*c*) And now a rather extreme example. Strong feelings have been aroused in the last year or so by the proposal to abolish capital punishment. During the discussions both in Parliament and outside a suggestion that the death penalty be suspended for a fixed period was usually described as an experiment. No statistician, so far as I know, bothered to point out that it was nothing of the kind and that any data obtained during a period of temporary suspension would be valueless unless there were a control—if the murder rate went down the proponents of abolition would say that they told us so and the opponents would say that it would have gone down in any case; and if it goes up their positions would be interchanged. In either case the data are inconclusive. However, I am not referring to the difficulties of interpreting criminal statistics. Nor do I think the statistician has anything to say on the ethics of the question whether a community has a moral right to eliminate any of its members, for reasons of public safety as with traitors and murderers, for eugenic reasons as in the sterilisation of the unfit, or for economic reasons as in the case of proposals to put to death the hopeless idiot and the incurable lunatic. The point is that however one looks at such questions, as relating to economics, morals or ethics, the problem, if we can free ourselves from the emotional aura of the subject, is one of balancing the interests of the individual against those of his community. It arises directly from the inter-relationships of a group of individuals and as such it is statistical. Of course I do not mean that it can be solved on a calculating machine or by consulting tables of the normal integral. But *conceptually* it is statistical. If there is a rational solution it must be in statistical terms.

I turn now to a rather different field, the statistical approach in logic and scientific inference. I have already referred to certain laws obeyed by aggregates of a particular kind, such as the so-called Law of Large Numbers and the Principle of the Permanence of Statistical Ratios. Such laws are not categorically imperative in the sense that they operate inevitably or in the sense in which Eddington regarded certain non-dimensional physical constants as imposed by the intrinsic properties of our way of looking at space, time and matter. Strictly speaking there are no *a priori* laws in statistics any more than in any experimental science; and the modern form of the law of large numbers, to the purist, shares with the Holy Roman Empire, the English horn, the Jerusalem artichoke and several statistical concepts the interesting property of not being accurately described by any of the constituents of its own name, being neither a universal law nor confined in its application to large numbers. Nevertheless we do find that many aggregates conform in practice to the required pattern. There do exist aggregates of random occurrences and there do exist stochastic processes. The diversity of the applications of statistical methods which I mentioned at the beginning is sufficient testimony to the fact. In most sampling inquiries, of course, we arrange for them to appear by sampling at random.

In particular it appears that error itself behaves stochastically. This again is one of the main reasons why the statistical method has such widespread applications. Any system of physical measurement has its error function which is capable of statistical investigation. Any system of sampling has an inherent error, and the reason we try to arrange for it to incorporate a random element is that we then bring the error under statistical control. We can never predict the behaviour of the error on any given occasion, but we can predict the behaviour of aggregates of errors; and hence we can make exact statements, not about the behaviour of the individual but about its *probable* behaviour.

This brings me to another characteristic, perhaps the most characteristic, feature of the statistical method. It recognises frankly that outside the domain of pure mathematics, deductive logic itself and similar subjects which are concerned with logical relations between ideas or the syntactic construction of languages, few, if any, meaningful statements are certain. We all know this but it took mankind at least two thousand years to admit it officially; and it was very natural that there should be a reluctance to admit it because once we admit that everything is uncertain we appear to undermine any system of rational thought. Leibniz appears to have been the first to state explicitly the necessity for a probabilistic logic, but the belief that it was possible to derive by inward contemplation certain propositions about the natural world lived long after his time and still survives in some quarters. Even Descartes wrote in the deductive manner and Spinoza contemplated writing his ethics on the model of Euclid's geometry. And it is strange how dialectical materialism, which as I understand it is another name for the scientific method, lends itself to the assertion of the wildest dogmata.

In atomic physics the position has been reached wherein the measurement of position is *theoretically* impossible. A particle can be observed only by the emission of light-energy which itself can only occur if the particle has moved somewhere else. This principle of uncertainty has given the theory of probability a fundamental rôle in quantum mechanics. At the other end of the cosmic scale it is equally true that we can observe the outer stars only as they were millions of light-years ago when they emitted the signals now reaching us. What they are like now (if there is a cosmic Now) we cannot possibly determine. The statistician, as Pascal remarked of man in general, is poised between the infinitely small and the infinitely great. Normally his problems are not so essentially indeterminate as in quantum physics, but there is an element of uncertainty surrounding them. What used to be regarded as an exact measurement of a single magnitude is to him a distribution of observations with perhaps a mean but not a unique "true" value. In the nineteenth century the object of the scientist was to draw a picture of the universe in sharp deterministic lines. Nowadays those lines are blurred by error distributions and the determinism of the individual has given way to the determinism of the aggregate. We see things hazily not because our vision is defective but because it is so clear. Nevertheless we can still paint a picture of the universe which is accurate and æsthetically satisfying. The only point to remember is that, as with pictorial works of art generally, you must not stand too close to the canvas.

The theory of probability, as developed by the statistician, is in fact the theory of uncertain inference in fields of a special type. It admits that uncertainty is the general rule rather than the exception but does not despair on that account of setting up a system of rational inference. I must be a little careful here because whatever one says about probability is almost certain to be contradicted by someone; but I think I am on non-contentious ground in stating that the statistician is in a peculiarly advantageous position to utilise the theory of probability because of the explicit character of the sets of propositions which concern him. He can say nothing certain about the error on a particular occasion; but he can make a definite statement about its probability. He may be wrong on any one occasion, but he makes up for it on the other occasions when he is right. He can even *measure* the relative risks of being wrong on different hypotheses. And thus, accepting the tangled complexity of observed phenomena and the essential uncertainty of the future, he gauges his uncertainties, measures his degrees of confidence and does his best to provide an objective basis for rational judgment and rational action. It is not in mortals to command success but the statistician does more. He measures the risk of failure.

How he performs this feat, how he constructs his standard errors and his confidence belts, how he designs his inquiries so as to measure their own error, how he decides whether to attribute significance to an observed effect, I must refer to the text-books. But before I leave this point I have to correct one possible wrong impression. It is not claimed that all uncertain inference or all use of the theory of probability lies in the statistical field. There are important and uncertain propositions on which statistics can throw no light whatever. The statistician has, perhaps, been tempted to appropriate the whole domain of uncertain inference, if only because up to the present he has explored it most scientifically and has produced the most fruitful results from it. But I do not want to give the impression that he is trying to crowd out the logician and the philosopher from a region which is, or ought to be, peculiarly their own. He is lucky in that numerical expression gives him an advantage in precision, but he makes no claim to a monopoly of accurate inferential processes.

Apart from the theory of inference itself there is another field where the concept of randomness and stochastic happening may throw some light on the nature of reflective thought. I refer to the theory of machines. We have heard a good deal in the last year or so about the remarkable power of the electronic computer. It has been described, with no justification at all, as an electric brain. But up to the present any machine, electric or not, only does what its maker designs it to do. Remarkable as its achievements are, it is no more a brain in the customary sense than a magnifying glass is an eye. But it may be possible to construct machines which do not follow throughout a purposive routine laid down by a designer; they may incorporate a randomising element. In such circumstances the machine might do things which the designer did not foresee. It would still not be capable of independent thinking, for the results which it produced would only be judged by it as sensible against some criteria built into it by the designer. But it might achieve sensible results by means which the designer does not understand. I am inclined to think that this is the sort of thing which happens in our own minds when, as we say, we meditate on a question. The mind thrashes about in a more or less haphazard way until something clicks into place and a sensible result emerges. I make this suggestion very tentatively; but it seems to me a profitable line for the psychologist to follow to see whether we can explain the capacity of the reflective mind to jump to conclusions which afterwards turn out to be correct in terms of statistical rather than logically purposeful trains of thought.

If I have expressed myself intelligibly you will now understand what I meant in my opening sentences by saying that the statistician derives from his science an approach to many of the problems which are, for want of a better name, called philosophical. But on reaching this stage I am left with one rather uncomfortable feeling. The purpose of this lecture was to give you some account of my subject in relation to other branches of knowledge, but I am not sure whether what was intended as an objective statement of fact may not have sounded more like an éloge. If you notice, as you well may notice, a certain air of self-satisfaction about modern statisticians, I hope you will not be misled by it. It is one of the features of a youthful science that it tends to exaggerate its own importance, and the halo-effect among statisticians is to be regarded with indulgence as an occupational ailment like writer's scoliosis or computer's elbow, all of which afflict the statistician but all of which are amenable to treatment. There is, indeed, one thing which chastens him to the point of desperation. If he turns from a contemplation of the past achievements of his science to look at the future field awaiting his labours he needs no other discipline to bring home to him an urgent sense of inadequacy. In each of the main categories of theoretical statistics, for example, the theory of distribution, the theory of statistical relationship, the the theory of sampling and the theory of inference, it is possible to write down enough questions to keep a team of research workers going for the rest of their lives. But to describe in any detail what we do not know about statistics would need an encyclopædic ignorance. I shall merely refer to two particular fields which may give a sufficient idea of what lies in store for us.

In the past thirty years there have been astonishing developments in the mathematical theory of statistics, most of it being built on the basis of the theory of probability in random sampling. The mine is far from being worked out; but it is being found, particularly in those sciences where the human being enters as sampler or samplee, that the theory of random sampling is not enough. Substantial and peculiar forms of bias appear whenever the human judgment is given any latitude in selecting, measuring or recording a sample; the proneness of a human being not to tell the truth, the whole truth and nothing but the truth in response to a statistical inquiry is well known; and there also appear to be interaction effects between the human being as interviewer and the human being as interviewee. We know that these dangerous forms of bias exist; and in some cases we know a little about their form and magnitude. But on the whole we know regrettably little about them—certainly not as much as we should like to know or intend to know. I myself would expect that the next major development in our subject will be the accumulation of experimental work on the sources and nature of bias. Once the data are collected we may hope to be able to formulate the laws of bias, and mathematics can then take us a stage further in the theory of representative sampling. In practice, of course, the two things will develop concurrently; but if we must consider the subject as an alternation of practical and theoretical discovery, I think the next phase is a practical one. We must try to develop a theory of bias in sampling, but it will not be drawn from the air. I may remark in passing that this development, if it occurs, will accentuate difficulties in which we already find ourselves in connection with the teaching of statistics, the organisation of statistical research and the formation of research units. One of the problems which statisticians have not yet solved—or perhaps it would be better to say, have not been permitted by extraneous circumstances to solve—is how to distribute statistical work and statistical teaching. But that is a theme which I hope to develop on another occasion.

As a second example of future developments, this time in a theoretical field but with the greatest practical implications, I take the theory of experimentation. One very important branch of statistical theory is devoted to the design of experiments so as to economise in time and labour and to ensure that the statistician can answer the questions which are going to be put to him when the experiments are completed. Now the classical experiment was performed in a laboratory under very carefully controlled conditions. The investigator's ideal was to design an experimental set-up in which only the factors of immediate interest were permitted to vary, everything else being held constant. The statistical approach is quite different. It accepts, in the first place, the fact that a good many experiments cannot be performed under laboratory conditions, but it goes much further than that and makes a virtue of necessity by designing its enquiries so that all the errors likely to be met in practice can in fact occur and be measured as a single error term. In the laboratory errors are, so far as possible, excluded. In the field they are permitted full play and even, one might say, encouraged to show themselves.

This situation, as you will see, is likely to lead to a Great Schism in the scientific church unless it is carefully handled. The laboratory worker is apt to claim that the field worker is so beset by error that he cannot isolate his causative factors; the field worker would retort that the results of the laboratory worker cannot be applied outside the laboratory because the circumstances are not such as can be encountered in real life. This is much more than an academic argument. It came to a head in the war in connection with that undefined type of activity known as operational research when it was found, for example, that picking out aircraft pilots by their success in laboratory tests of ability to judge distance in depth did not have any appreciable effect in diminishing the number of landing crashes due to misjudgment. In one way the conflict looks more serious than it is. One easy way of resolving it is to do experiments both in the laboratory and in the field, which is what we should all do in practice if we had the time and money. Unfortunately we have to economise in both, and that is where the difficulties are likely to arise. What part the statistician has to play in a settlement of them I should hardly like to surmise; but it is clear that he cannot stand entirely aloof.

I have ranged today over a very wide field. You may well have thought, when I was speaking of the statistical approach to subjects such as inference in probability, the abolition of the death penalty or road accidents, that if that is what I mean by statistics we are all statisticians. And, indeed, if it is not an impertinence to say so, I think that is broadly true. It is almost impossible to be a responsible adult in the modern world without acquiring the statistical outlook to some extent, whether one spends one's time in speculation on the nature of things or in filling up football pool coupons. There is in essence nothing unusual or mysterious about the statistical approach. There is nothing *outré* or suspicious about the statistician. The only thing that distinguishes him from his fellows is that he specialises. And so, having imposed on you these general reflections, I return to my own specialised domain.

The London School of Economics.

From *Journal of the Royal Statistical Society A*, Vol. **124**, pp. 1–18 (1961)

Presidential Address

Natural Law in the Social Sciences

[The Inaugural Address of the President, Professor M. G. Kendall, delivered to the Royal Statistical Society on Wednesday, November 16th, 1960]

1. Introduction

I should like to begin this address by thanking my fellow members very sincerely for electing me to the highest office of the greatest statistical society in the world. It is an honour which can fairly be said to represent the summit of a British statistician's ambition and I enter my term of office with the liveliest pleasure; a pleasure tempered by a humility engendered by looking back over my long line of predecessors. However, half of us have to be below the median and I can only hope that I shall not fall so far below as to constitute an extreme value.

The theme of this address has its roots in the controversy which gave rise to the foundation of the Society. In 1832 a Statistical Section was added to the British Association for the Advancement of Science, itself only two years old at the time. The records are scanty, but it seems clear that from the outset there was some misgiving among natural scientists whether statistics should be admitted as a science at all. The task of defining the scope of this new and suspect section was entrusted to a committee of the Association, which met at Cambridge in 1833 and included Babbage, as chairman, Malthus, Hallam, Lubbock and Quetelet. The committee duly produced a definition restricting the section to "facts relating to communities of men which are capable of being expressed by numbers, *and which promise when sufficiently multiplied to indicate general laws*".

Maurice George Kendall (*b.* 1907) was educated at the Central School, Derby, and St. John's College, Cambridge, graduating as a Wrangler in 1929. He spent a fourth year at Cambridge reading mathematics and economics and entered the Civil Service in 1930. He worked in the Ministry of Agriculture and Fisheries (1930–41), and then as statistician, and in the later years as Joint Assistant General Manager, to the Chamber of Shipping (1941–49). Since 1949 he has been Professor of Statistics and Director of the Research Techniques Division at the London School of Economics.

Professor Kendall is a Fellow of the American Statistical Association, the Econometric Society, and the Institute of Mathematical Statistics, and a member of the International Institute of Statistics. He has been President of the Operational Research Society (1957–59) and is a Vice-President of the Market Research Society. He served on the Hinchliffe Committee on The Cost of Prescribing and the Hale Committee on The Superannuation of University Teachers. His activities within the Royal Statistical Society include Honorary Secretaryship (1945–48), the Chairmanship of the Research Section Committee (1949–51), and Joint Editorship of the *Journal*, Series B (1945–49). He was awarded the Guy Medal in Silver in 1940.

Professor Kendall's publications include the following books: (with G. Udny Yule) *An Introduction to the Theory of Statistics*; *The Advanced Theory of Statistics*; *Contributions to the Study of Oscillatory Time-Series*; *Rank Correlation Methods*; (as Editor) *The Sources and Nature of the Statistics of the United Kingdom*; (with W. R. Buckland) *A Dictionary of Statistical Terms*; *A Course in Multivariate Analysis*; and numerous papers on the theory of statistics and its applications in agriculture, economics, and psychology.

The italics are mine. The words must, I think, have been added by, or on behalf of, the natural scientists to indicate that the new section was interested, not just in numerical facts, but in discovering among them laws analogous to those which were bearing such fruit in the physical sciences. Innocent as they may appear, they were enough to cause a splinter movement. Quetelet, in particular, felt them to be so restrictive that he suggested to Babbage the foundation of a separate society. The result was the famous public meeting on the Ides of March, 1834, when the Statistical Society was formed "to procure, arrange and publish facts calculated to illustrate the condition and prospects of society". Our fathers were concerned with the production and dissemination of facts, to the exclusion of opinions or sentiment. There is no mention here of laws, and there never has been in subsequent amended versions of the Society's objects.

Undoubtedly our founders were right. None the less, even they permitted themselves to speculate about patterns of social behaviour; and ever since that time there has been a feeling that perhaps something analogous to the laws of the physical sciences might be found to operate in the behavioural sciences. The feeling grew stronger as it was realized that some physical laws were only manifestations of a statistical average, and as laws in the classical sense were brought to light in the biological sciences, which, in effect, occupy an intermediate position between the sciences of inert matter and the sciences of the social organism. It is a feeling which is deeply rooted in man's desire to perceive uniformity in his universe. On the other hand, it comes into sharp conflict with his vanity, which will not lightly surrender individual freewill to an overriding law of behaviour, and with his religion, which is reluctant to see the burden of personal guilt removed from his shoulders by a distribution function. It seemed to me that the time was ripe for a general airing of this subject. I propose, then, to discuss the extent to which we observe laws of behaviour in social phenomena, the nature of those laws and how we can make use of them. The time is not inappropriate, perhaps, for such a review. With the growing integration of human activities, problems of social organization and administration have become so complex that any systematic regularities which we can detect in the social sphere have an immediate application as instruments of government even if we do not fully understand the reasons for their existence.

2. The Character of Laws of Behaviour

By a "law" in the present context I mean a reproducible pattern of behaviour. I do not want to linger over subtle distinctions but to be clear I must spend a few minutes dismissing from the domain of discussion two kinds of statistical "law" which are not of concern here. The first arises from the operation of chance in giving "laws of averages" or "laws of large numbers". These belong to a class of propositions in pure mathematics such as the Central Limit Theorems. They are, of course, exceedingly useful, particularly in sampling enquiries. And it may be a law, in the behavioural sense, that the deductive theorems of probability are applicable to the real world. But I am not concerned directly here with the mathematical laws of the theory of probability.

The second class of "law" which I want to set aside from the discussion may be illustrated by the "laws" of collective decision-making. For example, Penrose (1946) drew attention a few years ago to what is sometimes referred to as the Law of the Resolute Minority: if some minority group on a committee is resolutely united and the other members are more or less indifferent, the minority will exercise an influence out of all proportion to its size. If, for instance, three members of a committee of 15 are implacably opposed to any resolution and a two-thirds majority is necessary to pass one; and if each of the other twelve has an even chance of voting either way, then a resolution will get through only about one time in fifty. Three resolute members on a committee of 23 can exercise as much influence as one on a committee of three, and a resolute bloc of a thousand can control the decisions of two million indifferent individuals. The rule can be given some precision: the same degrees of control are obtained in two populations by blocs proportional to the square root of the respective population numbers. Moreover, if we conduct elections hierarchically, the effect is enhanced. Suppose that each ward chose an elector, and that the electors of the set of wards chose the M.P. for a constituency, we might then find that a very small majority over the country as a whole would yield an almost solid House of Commons and instead of a cube-law, as at present, we might have a ninth-power law. The old maxim *divide et impera* has a new significance. A party with a small majority (or a resolute minority) has only to introduce enough democracy into an electoral system to extinguish all opposition. In a sense, I suppose, this could be described as a law of electoral systems. But it is not the sort of law I propose to discuss. By "law" I mean a pattern of a human aggregate which is observable, reproducible and, as a rule, quantifiable; perhaps only descriptive in character, perhaps explainable in terms of a model, but in any case related to observation.

Until quite recently very few laws in this sense were known in the social sciences. It is true that, from the gaming houses and the plague pits of the seventeenth century there arose the notion of the stability of certain kinds of statistical ratio. Some of them, such as death-rates, were measured with enough accuracy to allow of the foundation of an actuarial science. Eighteenth-century observers found a manifestation of supernatural agency in such phenomena. As early as 1710 Arbuthnot wrote a memoir entitled *An Argument for Divine Providence, taken from the constant Regularity observed in the Births of both sexes*, and in 1741 Süssmilch pursued the same theme among birth-rates, death-rates and sex-ratios in his *Göttliche Ordnung*. But it was recognized that these ratios fluctuated in time and in space, and the assertion of the existence of permanency among them did not pretend either to the precision or to the universality of a law of physics. They were the manifestations of Law rather than individual laws.

It was not until the latter part of the nineteenth century that laws of social phenomena began to be put forward, and even then they took second place to the regularities which Galton and Karl Pearson were revealing in biological variation. With the exception of some work by Edgeworth, whose name is not nearly so well known or honoured amongst us as it ought to be, the first notable law to be written down mathematically and checked against observation was, I think, Pareto's law of the distribution of incomes: the number of persons in receipt of income x is given by

$$y = \frac{A}{x^{1+\alpha}} \quad (x_0 \leqslant x \leqslant \infty), \tag{1}$$

where A and α are constants. Pareto showed that such a distribution held for practically every country which was developed enough to provide the statistics required to check it, with some (but not very much) variation in the parameter α. The law still holds today, at least as a good approximation, and considering what has been done by way of re-distributing income since Pareto enunciated the law in 1895, the pattern has proved to be remarkably stable.

And this, oddly enough, is both its strength and its weakness. Had it been a temporary phenomenon it would have passed into history as a curiosum. To impress itself on economic thought it had to be stable. But since it has proved so, and the dynamic nature of that stability was not understood, it furnished support for a belief that similar very general patterns might be detected in other economic or social phenomena. This, I think, was a mistake, although a natural one at the time. The success of physical scientists in arriving at unifying principles of great generality; of the applied mathematician in deriving all the equations of motion from Least Action; of the chemist in reducing the properties of matter to the inter-relationship of a few primary particles; of the physicist in bringing heat, light and sound together in a theory of wave motion—these and similar strides towards unification have led to the belief, or at least to the hope, that something of the same kind might be found in the social sciences.

One of the most comprehensive attempts to generalize in this field was made by the late G. K. Zipf (1949), whose book on *Human Behavior and the Principle of Least Effort* is quite one of the most fascinating quantitative studies ever made. It was Zipf who popularized the law which is known by his name, and fairly so, I think, though he had forerunners in Lotka (1926) and Pareto himself. Zipf's law is a kind of the-higher-the-fewer rule of which Pareto's law is a particular case. It says, in effect, that for certain kinds of activity with a measurable size x, the number y of individuals greater than or equal to x is given by

$$y = \frac{A}{x^\rho}, \qquad (2)$$

where ρ is a constant which is often quite close to unity. (An example is given later.) It is really remarkable how many measurable aspects of human activity seem to follow a law of this kind; apart from a number of linguistic and psychological phenomena, Zipf adduces a wide variety of economic and social distributions: salaries of executives, claims against insurance companies, sizes of towns, journeys on buses and railways, sizes of religious castes, companies by size of asset. The law does not stop with human beings. Cognate patterns have been observed in the distribution of biological genera by species, word frequencies in prose and the distribution of articles in scientific periodicals.

The body of evidence is impressive, but this work of Zipf, which I admire even where I disagree, seems to me to illustrate the danger of attaching too much importance to analogies with the physical sciences. For Zipf tried to deduce all this behaviour from a single principle, the minimization of effort. Now it has become the fashion among applied mathematicians—and their successes are undoubted—to seek for the ultimate determinants of physical law in some form of maximization or minimization. Matter moves along geodesics in a space-time continuum; direct currents in a network set themselves so as to minimize heat output; soap bubbles form so as to minimize surface area; and so on. The spectacle of a mathematician looking for something to optimize is, in fact, so familiar that it passes without comment. We have been doing the same thing in mathematical statistics and probability for some time, beginning with Daniel Bernoulli's formulation of maximum likelihood in 1778, continuing through the work of Legendre and Gauss on least squares and of Fisher on minimum variance and ending with various forms of optimization in the theory of decision functions. Personally, I think this has gone too far, even in mathematical statistics, and if I were ever to write a paper which has been much in my mind for the past ten years, entitled "Recent Advances in Retrogression", I should have much to say about procedures which carefully formulate a technique of optimization in situations where there are no known methods of setting up measures of the variables which are to be optimized. That, however, is by the way. I would not for a moment deny the value of optimization methods or of the notable theoretical advances to which they have led. The point, for present purposes, is that it may be (and probably is) chasing a will-o'-the-wisp to suppose that social behaviour is determined by principles of least something-or-other. Zipf made a gallant attempt, and in the same category I would place efforts by economists to explain business behaviour on the principle of maximizing profits or the tendency on the part of model-fitters to assume that Nature is obliging enough to minimize their errors. Some such principles may, indeed, exist. But if they are to be discovered, I think we must achieve them by synthesis at a later stage, and not be over-ambitious about the generality of the patterns which we are studying at the present time. The pathway of knowledge is littered with the wreckage of premature generalization.

3. Some Examples: Industrial Capacity

In the spirit of our founders, then, let us pass over matters of opinion and look at some of the facts. I proceed to give a few examples of what the patterns of behaviour are. In Table 1 I reproduce an interesting example due to Simon and Bonini (1958).

TABLE 1

Ingot capacities of ten leading steel producers in the U.S.A.

(*Millions of net tons per year, as at January* 1*st*, 1954)

(1) *Rank order*	(2) *Producer*	(3) *Actual capacity*	(4) *Theoretical capacity*	(5) *Column* (1) × *Column* (3)
1	U.S. Steel	38·7	34·3	38·7
2	Bethlehem	18·5	17·1	37·0
3	Republic	10·3	11·3	30·9
4	Jones and Laughlin	6·2	8·5	24·8
5	National	6·0	6·8	30·0
6	Youngstown	5·5	5·2	33·0
7	Armeo	4·9	4·8	34·3
8	Inland	4·7	4·2	37·6
9	Colorado Fuel and Iron	2·5	3·8	22·5
10	Wheeling	2·1	3·4	21·0
	Total	99·4	99·4	

As a rough approximation this can be regarded as a specially simple case of Zipf's law with the parameter of equation (2) equal to unity. It has the attractive property that if we write down the objects in rank order, the product of that order and the magnitude of the variable is constant. Column (5) shows this effect.

4. Distribution of Articles on Statistics

There are theoretical reasons, to which I refer later, for regarding the Zipf distribution (2) as an approximation to a more general distribution which Simon has christened the Yule distribution, with frequency function

$$f(x) = kB(x, \rho+1) \quad (x = 1, 2, \ldots), \tag{3}$$

or equivalently,

$$f(x) = \frac{k\Gamma(x)\,\Gamma(\rho+1)}{\Gamma(x+\rho+1)}$$

$$= \frac{k\rho!}{(x+\rho)(x+\rho-1)\ldots x}. \tag{4}$$

Table 2 may be of some special interest to statisticians. The first volume of the forthcoming bibliography of theoretical statistics and probability by Miss Alison Doig and myself contains about 9,000 references (in all Western languages) for the years 1950–58 inclusive. To save the trouble of analysing the whole set, I took the entries for the letters A–B and N–R inclusive, and show them in Table 2. Similar data for the letters N–R for the years 1940–49 are also shown.

TABLE 2

Analysis of articles on theoretical statistics and probability by journal, for certain years and initial letters of authors' names

	1940–49, *letters N–R*		1950–58, *letters N–R*		1950–58, *letters A–B*	
	No. of journals:		*No. of journals:*		*No. of journals:*	
No. of references	*Actual*	*Theoretical*	*Actual*	*Theoretical*	*Actual*	*Theoretical*
1	151	151·0	198	202·8	147	147·9
2	40	43·0	60	59·4	38	43·6
3	18	19·1	20	26·9	18	19·8
4	10	10·4	21	14·9	8	11·0
5	6	6·4	10	9·3	9	6·9
6	4	4·2	5	6·3	6	4·7
>6	22	16·9	32	26·5	28	20·1
Totals	251	251·0	346	346·1	254	254·0
	$\chi^2 = 1{\cdot}86$, $P = 0{\cdot}87$;		$\chi^2 = 5{\cdot}85$, $P = 0{\cdot}32$;		$\chi^2 = 5{\cdot}81$, $P = 0{\cdot}33$	

The pattern exhibited by the tables is typical. Something of the kind was noted by Bradford (1945), whose law of libraries I have shown elsewhere (Kendall, 1960) to be approximately equivalent to a Yule scheme.

The frequencies described as theoretical are those given by the Yule distribution of equation (3). The fits are obviously very good.

The same kind of distribution appears if we classify by author instead of by journal. At the end of Volume 2 of my *Advanced Theory* there are (excluding books and some occasional reports) 1,866 references covering 632 authors. This is a highly selected set and cannot be regarded as a random sample. There are all sorts of reasons why it should depart from the law quite substantially. Table 3, however, shows that the law still holds approximately.

Again the "theoretical" frequencies are given by fitting a Yule distribution. Considering the nature of the material the fit is surprisingly good.

TABLE 3

1,866 *references to statistics*, 1,700–1,943, *classified by numbers written by given authors*

x, *number of references*	*No. of authors mentioned x times: Actual*	*No. of authors mentioned x times: Theoretical*
1	364	380·4
2	95	108·3
3	50	48·0
4	36	26·1
5	15	16·0
6	11	10·7
7	12	7·5
8	8	5·5
9	5	4·2
>9	36	25·1
Totals	632	631·8

$\chi^2 = 14{\cdot}8$, $P = 0{\cdot}06$

5. CONSUMER PURCHASES

In an enquiry about the purchases of non-durable consumer goods, Ehrenberg (1959) obtained from some large samples of private consumers (individuals or households) data relating to a wide variety of products such as canned vegetables,

TABLE 4

Number of units r, of a commodity bought by number of consumers f_r
(26-*weekly data for a sample of* 2,000 *households*)

r	f_r *Observed*	f_r *Theoretical*	*r*	f_r *Observed*	f_r *Theoretical*
0	1,612	1,612	7	12	11
1	164	157	8	5	8
2	71	74	9	7	6
3	47	44	10	6	5
4	28	29	11	3	4
5	17	20	12	3	3
6	12	15	13 and over	13	12
			Totals	2,000	2,000

$\chi^2_9 = 3{\cdot}31$, $P = 0{\cdot}95$

cat food and coffee, soap, sausages and shampoos. All these products are sold in unit packages and it was possible to find how many consumers bought $0, 1, \ldots, r$ units in a certain space of time, usually 4 or 13 weeks. Ehrenberg has shown that these data are remarkably well fitted by a negative binomial distribution. Table 4 reproduces one of his examples.

It is of considerable importance here to note that the whole distribution can be generated from two parameters, which can be estimated from two quantities easily observed in a market survey: the size of the zero class and the mean number of purchases—in this case 1,612 and 0·636, respectively.

The negative binomial has, of course, been known to fit some kinds of data ever since Yule and Greenwood (1920) published their classical work on industrial accidents. Its appearance in economics, however, is something of a novelty and the closeness of Ehrenberg's fitting suggests that here is a law of some generality.

6. WARS

The late L. F. Richardson spent a good deal of his retirement in an attempt to detect regularities in the distribution of wars. He was, of course, asking for trouble. Of all human activities which mankind will steadfastly refuse to believe to be subject to statistical law, war is the foremost. For, at the worst, the conclusion might be that war is inevitable, and at the best it might be that the causes for which we go

TABLE 5

Numbers of outbreaks of war by year

No. of outbreaks in the year	*No. of years* Observed	Theoretical, Poisson
0	223	216
1	142	150
2	48	52
3	15	12
4	4	2
>4	0	0
Totals	432	432

$\chi_2^2 = 2{\cdot}75,\ P = 0{\cdot}26$

to war are only glosses composed to cover inherent pugnacity or a tendency for the strong to oppress the weak to the point of provoking violence. Some of Richardson's work has only just been published, and is in danger of being dismissed as a curiosity rather than a serious contribution to his subject. I am not concerned here with the rights or wrongs of this situation, but merely wish to call attention to the regularity exhibited by Richardson's law: the number of outbreaks of war in a given year follows the Poisson distribution. Table 5 shows Richardson's (1944) analysis of Quincy Wright's (1942) data for wars from A.D. 1500 to 1931. Anyone who can reflect on this picture without disquiet is, I think, inaccessible to statistical evidence.

7. STRIKES

It occurred to me that this work on fatal quarrels might have some application in the patterns followed by industrial disputes. I have therefore had an analysis made of major strikes in the U.K. for the years 1948–59. There are, of course, all sorts of difficulties in deciding the cover of the basic data in such cases. A "strike" for present purposes is defined as a major stoppage listed as such in the *Ministry of Labour Gazette*. (Sympathetic strikes, go-slow movements, working-to-rule and periodic stoppages such as the Friday-night strike, are omitted.) Table 6 shows the distributions according to number of outbreaks in particular weeks for industry as a whole, for the twelve years 1948–59.

TABLE 6

Distribution of major strikes in the U.K., 1948–59, *by number of outbreaks per week*

No. of outbreaks per week	*No. of weeks* Observed	Theoretical, Poisson
0	252	254
1	229	229
2	109	103
3	28	31
4 or more	8	9
Totals	626	626

$\chi_3^2 = 0{\cdot}77, P = 0{\cdot}86$

The agreement with the theoretical frequencies given by fitting a Poisson distribution with the same mean, 0·8994, is almost uncanny.† But there is an aggregation effect here. The Poisson law does not hold so closely for individual industries, as Table 7 shows.

TABLE 7

Number of outbreaks of strikes in four leading industries in the U.K., 1948–59

No. of outbreaks in 4-week period	*Coalmining* Observed	Poisson	*Vehicle manufacture* Observed	Poisson	*Shipbuilding* Observed	Poisson	*Transport* Observed	Poisson
0	46	58	110	104	117	113	114	114
1	76	57	33	42	29	36	35	36
2	24	29	9	9	9	6	7 (2 or more)	6 (2 or more)
3	9	9	4 (3 or more)	1 (3 or more)	1 (3 or more)	1 (3 or more)		
4 or more	1	3						
Totals	156	156	156	156	156	156	156	156

8. SWINGS AT GENERAL ELECTIONS

Finally, I give some data which are not of the J-shaped type, relating to the 1959 General Election. This has the additional interest of being one of the few bivariate examples available in the social field. All predictions of election results, both before the event and on the night of the election itself, are based on the assumption that the "swing" from one party to the other, as compared with the previous election, is uniform over the country. The extraordinary thing is that this assumption gives very good forecasts—in 1959 we were within a seat or two of the final figures after 15 results were to hand—but that it is, in fact, not borne out by the figures. Table 8

† There are many other interesting features of the strike distributions and Miss Joyce May, who supervised the collection and analysis of the data, will be publishing a more complete account of the results elsewhere.

TABLE 8

Swing at the 1959 General Election for 604 constituencies

Swing as percentage (central value of interval)

Ratio, Conservative to (Conservative plus Labour) votes (Central points of intervals)	−6·5	−5·5	−4·5	−3·5	−2·5	−1·5	−0·5	0·5	1·5	2·5	3·5	4·5	5·5	6·5	7·5	Totals
0·10								1								1
0·20			1		1	3	2	8	2			2		1	1	21
0·30				1	2	6	9	8	4	8	1	2	1	1		43
0·40	1	3	3	2	6	6	12	17	15	18	8	6			1	98
0·50			2	1	9	11	15	28	32	29	23	11	5	3		169
0·60		3		1	8	6	16	17	31	36	21	9	1	1	1	151
0·70				2		8	13	9	19	15	14	14	2	3	1	100
0·80		1				1	1	1	2	5	4	1	4			20
0·90										1						1
Totals	1	7	6	7	26	41	68	89	105	112	71	45	13	9	4	604

shows the distribution of swing for 604 constituencies. "Swing" means the ratio $C/(C+L)$ in 1959 less that in 1955, expressed as a percentage. (The 26 omitted constituencies comprise 18 for which the value $C/(C+L)$ did not exist in 1955 or 1959 or both; 2 seats won by Liberals; 6 seats for which the swing was outside the range given.)

It is to be noticed that swing is nearly independent of the ratio $C/(C+L)$ giving the political hue of particular constituencies, but is widely spread over them. There are, I believe, reasons why this distribution of swing can be summarized into an assumption of uniform swing, but I do not want to go into them on the present occasion. Again, I am mainly concerned to exhibit the definiteness of the pattern of behaviour which emerges from a set of acts of entirely free choice.

9. The Nature of Laws in the Social World

So much by way of illustration. It is not possible to doubt, I think, that patterns of behaviour exist in the social sciences in a quantifiable form. But it may legitimately be doubted whether they can be called laws in the sense used in the physical sciences, and this is the next point to examine.

Some laws of Nature, as usually stated, are tautologous: for example, Newton's first law of motion, that every body continues to move in a straight line until deflected by a force. Some of the laws of behavioural science are of similar type: for example, that savings equals investment in Keynesian economics. Oddly enough, they are by no means useless in spite of being empty; but their usefulness lies in shaping our thought rather than in describing the universe and I do not perceive in this respect any difference between the laws of natural and those of behavioural science.

Some laws, again, are meaningful but can be stated in non-quantified terms: that water tends to find its own level or that bad money drives out good. But these are usually particular cases of something more quantifiably expressed. What may be a law for the plumber or the Elizabethan economist may be only a manifestation of a much more general law for the hydraulic engineer or the modern financial expert. However, it may be going too far to require that every law should be quantifiable. The principal requirements of a law, I suppose, are that it shall have some generality extending beyond the observations on which it is founded and that its consequences shall be verifiable.

As to verifiability, there are certainly laws in the social sciences which can be checked experimentally; but we now begin to encounter differences between the natural and the behavioural domains. The laws of physics are almost entirely expressible in terms of the fundamental units of length, mass, time and electric charge. The laws of economics are often expressed in a manner which makes them difficult to reduce to units of such a type. For example, the apparently meaningful statement that an increase in demand will raise prices poses us the problem of measuring demand and of endeavouring to express price independently of fluctuations in the value of money. Moreover, although verifiability in the physical sciences is usually a matter only of finding a suitable laboratory technique, the difficulty of conducting controlled experiments in the social sciences makes a critical discrimination between alternative hypotheses a much more complicated operation.

One could spend a deal of time on these problems and their philosophical implications, but I prefer to leave them to the experts. Nowadays, in order to optimize my expenditure of effort, I divide intellectual difficulties into two classes: those which worry me and those which do not. Most of those concerning the scope and nature of law in science do not. But before I pass on, I must mention one which does: the problem of changes in a law through time.

Traditionally we have regarded the fundamental laws of Nature as permanent. In fact, this invariance is part of the immutability which seems to characterize the essence of law. Such modifications as may occur in our laws from time to time are rather to be regarded as discovered by man than invented by Nature. There have, admittedly, been suggestions that some of the entities which we believe to be absolute constants may alter; for example, the velocity of light *in vacuo* may be increasing. But nobody really doubts, I think, that Hooke's law and Boyle's law and Ohm's law and all the uniformities which we learnt at school were just as true a million years ago as they are now; whereas the laws of human kind are at any rate no older than humanity itself, are changing under our eyes, and to that extent may not be described as "true" laws. Zipf's law of urban population may have applied to Babylon; Bradford's law of libraries may have applied in Alexandria; Richardson's law of wars may have applied to Rome; Ehrenberg's law of consumer purchases may have applied in Venice. But we have no means of telling, and even if we knew that they did not, our faith in their current application would not be shaken. Nor should we discard them if we knew that they would cease to apply in a hundred years' time. Are they, then, really what we mean by laws?

The answer, I suggest, is affirmative. I see no reason why a law should not change over time. The mathematician, of course, would say that the matter is easily adjusted by putting a time-parameter into the model, but this is rather too slick a method of shoring up the crumbling wall of the Absolute. I would prefer to put it differently: man is a short-lived animal who is still, sociologically speaking, in a rapid phase of development. The laws which control his social actions and interactions may themselves be subject to rapid change. To such an organism the laws of Nature, with their vaster time-scale, appear unchanging, and may indeed be so. But the difference is possibly one of degree rather than of kind. It would be absurd to insist on the laws of social behaviour having the permanence of a physical law. I therefore see no reason for denying the use of the word "law" to describe the sort of social phenomena we have been considering. A precisian might prefer a different word, but let us not get involved in logomachy.

I come now to what appears to me to be the central difficulty. Suppose we observe a pattern of the type exemplified earlier. Does this mean that there are causative influences at work, or can it be due to chance? I have been insisting, possibly *ad nauseam*, on the reality of these patterns. But I must now insist on a point of interpretation: their existence does not entail a causal mechanism.

In fact, not only can choice mimic chance, but chance can mimic choice. Consider, for example, a number of persons with equal stakes playing at a fair zero-sum game. (A zero-sum game is one in which the losses of some players pass to others, so that no money is lost to the system.) Over a period of play, the distribution of their holdings will tend to an unequal pattern of the Pareto type; some people will be reduced to small or zero stakes and a few will accumulate large reserves. And if the game goes on long enough, ultimately all the money will be concentrated in a few hands. Thus chance, which might have been expected to level things out, will act in a very inegalitarian way and produce a distribution with a very purposive-looking outcome. It is not Providence but Chance which is on the side of the big battalions.

Again, if a number of people exert their knowledge and skill in a very purposeful activity, the outcome may be indistinguishable from a purely fortuitous mode of generation. Some years ago, Elderton (1945) pointed out that the scores of our greatest cricketers were distributed geometrically, that is to say, as if the chance of making a run were the same at any point of an innings. The batsman might equally well have used a set of dice (suitably loaded to match his ability) and not a bat, to determine how many runs he made on a particular occasion. This is obviously untrue, but the pattern does not tell us so. Another of our national sports was threatened when Mr. Hubert Phillips (1950) argued before the Royal Commission on Gambling that football pools were a lottery. Mr. Phillips's case was that the actual distribution of dividends was very much what it would have been on the basis of a purely random selection of "winners" by the competitors. The Commission's comments do not encourage complete assurance that they saw to the bottom of this argument, but they appear to have felt in their bones that there was a catch in it. In this they were, in my opinion, correct, although I am bound to say that, equally in my opinion, Mr. Phillips was right in his main thesis.

For certain purposes, of course, none of this matters. If a pattern is reproducible we can use it both for the control of the social organization and for prediction of the future, just as we can use Boyle's law as a macro-phenomenon notwithstanding that, in the kinetic theory of gases, the component molecules may differ in speed and direction just as much as individual cricketers or gamblers. For instance, in social survey work it is notoriously difficult to obtain information about the consumption of certain anti-prestige items such as beer and cigarettes, especially from heavy consumers, who are few in number but contribute a good deal to the total consumption. Personally I should be willing to use Ehrenberg's law to estimate the upper tails of a distribution in such a case. Sometimes our empirical rules are exact to the point of embarrassment. The prediction of election results over the past ten years has been so startlingly good that it has left the statistician very little to live for in psephology, a subject which, perhaps, it would be wise to abandon to the gilders of political lilies.

However, we clearly cannot be satisfied with the mere eliciting of pattern, even if we go about the process systematically and discover a good deal more regularity than has been observed in the past. For a number of very obvious reasons we must go behind the manifest phenomena to see whether some model of the generating process can be set up. Not until we have a satisfactory model can we be said to understand the process fully or to retain control over it in changing circumstances. And this brings us to the question of model building in the social sciences.

To construct anything approaching a realistic model of a social situation is a formidable undertaking. The first difficulty is one of aggregation. Sheer shortage of information and the intractability of multivariate complexes compel us to deal with summarized data. The result is that we are in some danger of generating by the averaging process patterns which are not really there, the classical example being periodic fluctuations in economic index numbers. Sometimes we may be reasonably certain that the aggregative process is not distorting constituent series; for example, the sum of a number of Pareto distributions has, as a rule, the shape of a Pareto distribution, so that aggregation may blur the edges but will not obscure the main outline. But in data like the Richardson distribution of wars there is bound to be doubt whether we are not scrambling together too many different things (a point which, I hasten to say, was quite clear to Richardson). Economic enquiries are particularly vulnerable to this difficulty and it is not always to be resolved by subdivision of the data—in one sense we have to aggregate to some extent to get a pattern at all. We must clearly take care not to aggregate too much. But how we know what is "too much" is a question I cannot answer.

A second difficulty arises from the degree of interlock between different parts of the social system. In considering a model for the strike data, for example, how much should we write in to allow for wage-rates, unemployment, cost of living, and all the other correlative movements which provide feed-back circuits in the economy? The natural appeal here is to a modernized version of Occam's dictum—we write in no more than is necessary to explain the data. But I have some serious doubts about Occam's razor in domains which are not open to simple experimental verification. The very concept of "simplicity" in economic or social hypotheses is extraordinarily difficult to analyse.

There is a very interesting theoretical problem here, as to what we mean when we say that one model is an approximation to another. Unless we can find some solution to this problem we might as well give up model-building altogether, for every model we write down is a simplification of real life. In the past there has been a tendency to be very slapdash about this question of specification, especially on the part of statisticians. The model-builder would take a few fairly simple ingredients, season with some error terms to allow for anything he happened to have left out and then assume that such terms behaved like random variables, so importing a new opprobrium into the word "epsilonology". This is not good enough. What is required here, I think, is a much profounder study of the sensitivity of models to deformation or excision.

A cognate difficulty in these fields is that some things are too easy to explain. The statistician is accustomed to distinguish two kinds of error in testing a hypothesis—rejecting it when it is true and accepting it when it is false. We forget, sometimes, that there are much more egregious errors open to us—asking the wrong question, failing to bring the right hypothesis to test, and specifying a model with so many degrees of freedom that it will fit almost anything. We find this latter effect in time-series of economic data, where the series are so short that the information in them, even when wrung out to the last drop, is insufficient to discriminate between different models or wide parametric ranges within the same model.

This is no occasion to depress our spirits by a long catalogue of all the difficulties we have to face. It is something, I hope, that we are beginning to realize their nature and extent. And we can derive much encouragement from the reflection that in recent years there have been several important advances in the direction of subsuming some types of behaviour under common laws.

The greatest step forward, I think, has been the realization that an apparently constant pattern may be only the steady state of a stochastic process. The point is well exemplified by recent work on the Pareto distribution. Early attempts ("early" on this time-scale means 20 years ago) to explain the law were naïve in the extreme. Snyder (1940) pointed out that it would hold if individuals (like Elderton's cricketers) had the same chance of making an incremental dollar however much they had made already. This must be one of the most implausible hypotheses ever propounded. The phenomenon looks quite different if we regard the distribution as moving through time, the individuals composing it moving about income-wise but in such a manner that the whole pattern stays more or less constant. The equilibrium then becomes dynamic, not static. Champernowne (1953) was, I think, the first to propose an explanation in stochastic terms, individuals moving from one income-group to another with certain transition probabilities. We can now write down models to explain the Pareto distribution which, to put it no higher, have considerable plausibility. Hart and Prais (1956) have applied similar ideas in studies of business concentration. Work of this type is in a phase of such rapid development that it is difficult to say where it will lead us; but the prospects are very encouraging.

Working along similar lines, Simon (1955) has shown that a good many observed patterns can be explained as the outcome of fairly simple stochastic schemes. As long ago as 1924, Udny Yule derived some early results in this field, but he was too soon and, oddly enough, he forgot about them himself in the studies of literary vocabulary of his later years where they have a natural application. Simon shows that distributions of word-frequencies, city sizes, incomes and biological species by genus are all expressible in terms of the Yule distribution of Section 4. In the same vein, Mr David Kendall (1948) showed that we obtain the logarithmic distribution (derived by Fisher as the limit of a negative binomial) from a stochastic birth-and-death process. I have already referred to the work of Hart and Prais (1956) on business concentration; and the lognormal distribution which they favour has been applied to a study of the size of trade unions by Hart and Phelps Brown (1957).

Thus a number of patterns of the types mentioned earlier can be generated by stochastic models. It may well be that such things as wars, strikes, epidemics and accidents can all be represented by a common model—a Statistical Theory of Disaster would make an attractive study. It is hard to resist the conclusion (and of course I should not try) that in unifying our observed patterns in this way we are approaching something akin to fundamental laws and moving towards that synthesis which is one of the main goals of scientific endeavour.

Nobody would claim that we have got very far. Undoubtedly more elaborate models are necessary and it may well be that they are mathematically intractable and have to be explored on an electronic computer. But there is enough achievement behind us to justify the belief that we should press on. In fields involving multivariate analysis and time-series difficulties are scarcely ever resolved at a single stroke; the problems have, as it were, to be teased out slowly with patience and insight, often by the joint action of a team rather than by an individual. If I look back over the tedious road from 1832 to the point we have reached today, I feel myself that we have surmounted the foothills and are at last beginning to discern the outlines of the mountainous problems which await us in the exploration of the laws of the social organism. We have ten times as many Fellows as in 1834, a hundred times as much knowledge, a million times as much facility in numerical analysis: what the multiplying factor is in terms of wisdom and enthusiasm I will not attempt to specify. If this address ever falls under the notice of some President of the next century who is still grappling with the problems of reducing social activity to scientific order I hope that he may understand that, difficult as it all is and little as we have been able to accomplish, we did not lose heart.

ACKNOWLEDGMENTS

I am indebted to Mr A. S. C. Ehrenberg for permission to reproduce Table 4; to Dr H. A. Simon and Mr C. F. Bonini for permission to reproduce Table 1; to Mr C. Robinson and members of the staff of the Central Computing Service, English Electric Company, for Table 8; and to Miss Joyce May and members of the computing unit of the Research Techniques Division, London School of Economics, for carrying out the analyses resulting in Tables 2, 3, 6 and 7.

References

BRADFORD, S. C. (1948), *Documentation.* London: Crosby Lockwood.

CHAMPERNOWNE, D. G. (1953), "A model of income distribution", *Econ. J.*, **63**, 318–351.

EHRENBERG, A. S. C. (1959), "The pattern of consumer purchases", *Appl. Statistics*, **8**, 26–41.

ELDERTON, Sir WILLIAM P. (1945), "Cricket scores and some skew correlation distributions (an arithmetical study)", *J. R. statist. Soc.*, **108**, 1–11.

HART, P. E., & PHELPS BROWN, E. H. (1957), "The sizes of Trade Unions: a study of the laws of aggregation", *Econ. J.*, **67**, 1–15.

—— & PRAIS, S. J. (1956), "The analysis of business concentration", *J. R. statist. Soc.* A, **119**, 150–180.

KENDALL, D. G. (1948), "On some modes of population growth leading to R. A. Fisher's logarithmic series distribution", *Biometrika*, **35**, 6–15.

KENDALL, M. G. (1960), "The bibliography of operational research", *Operat. Res. Quart.*, **11**, 31–36.

LOTKA, A. J. (1926), "The frequency distribution of scientific productivity", *J. Wash. Acad. Sci.*, **12**, 317–323.

PENROSE, L. S. (1946), "Elementary statistics of majority voting", *J. R. statist. Soc.*, **109**, 53–57.

PHILLIPS, H. (1950), Memorandum on Football Pools, *Evidence before Royal Commission on Betting, Lotteries and Gaming, 25th Day.* London: H.M. Stationery Office.

RICHARDSON, L. F. (1944), "The distribution of wars in time", *J. R. statist. Soc.*, **107**, 242–250.

SIMON, H. A. (1955), "On a class of skew distribution functions", *Biometrika*, **42**, 425–440.

—— & BONINI, C. P. (1958), *The Size Distribution of Business Firms.* Reprint 20 of the Graduate School of Business Administration, Carnegie Institute of Technology, Pittsburgh.

SNYDER, C. (1940), *Capitalism the Creator.* New York: Macmillan.

WRIGHT, Q. (1942), *A Study of War.* Chicago: University of Chicago Press.

YULE, G. U. (1924), "A mathematical theory of evolution, based on the conclusions of Dr J. C. Willis, F.R.S.", *Phil. Trans.* A, **213**, 21–87.

—— & GREENWOOD, M. (1920), "An inquiry into the nature of frequency distributions representative of multiple happenings", *J. R. statist. Soc.*, **83**, 255–279.

ZIPF, G. K. (1949), *Human Behavior and the Principle of Least Effort.* Addison: Wesley Press.

Proceedings of the Meeting

SIR HUGH BEAVER: The difficulty of one of your intercalary non-statistical past Presidents in following such a brilliant and worthwhile paper must be obvious to all of you. It is a privilege, but the difficulty is great. I feel perhaps like some dark sombre background against which you can see the real lightning more clearly or hear those "noises-off" they used to have as stage directions in Elizabethan drama. Even if I appear before you in that way, nevertheless, I am prepared that it be so, because this is a paper which deserves very considerable and enthusiastic praise.

I was interested to hear, not myself being aware before, of the origins of both the statistical section of the British Association and of our own Society. I noted the quotation that our Society was formed "to procure, arrange and publish facts calculated to illustrate the condition and prospects of society"; society as a whole. I do not know whether that was natural in the conditions of thinking of that time, or whether it was real farsightedness, but, certainly, however much it may have been appropriate then, now it is vastly more so. I suppose the population of the world now is at least double what it was in 1832 and 1834, and if the present trends go on unchecked—and I am not statistician enough to know whether that is likely or not, but should they go on unchecked—the population will again be doubled before the end of this century. So, however necessary and however desirable and however rewarding it could then be to study the behavioural procedure of society and endeavour to find and identify laws, it is now becoming vitally important; as we go forward in this increasingly scientific world we are becoming more and more aware that it is the science of managing men and managing society, and what society and the men that make society will do, that is the vital problem.

And it is full time that our Society, and a President so able to do the work, should have taken on this attempt—a most successful attempt—to open our eyes and to re-illumine and revitalize our thinking and our imagination in this respect.

The subject is one on which I would find myself not adequately qualified to indulge in criticism. Fortunately on these occasions the President's paper is not meant to be criticized, therefore I shall not attempt it. I was sorry, as I read through it and as I again listened to the President, that he did not follow up this fascinating question of the ability of the "resolute minority" to dominate a committee or situation. Having been in my life a member of many committees, and having chaired a number, I know how true that can be; and to those of you who may have the doubtful privilege of being in chairs where you find such a position, I can only say that eventually I have discovered that two can play at the game of dividing and ruling. The most resolute minorities I have ever met have in themselves the seeds of disaffection and disagreement. You can cultivate those seeds and this re-adjusts the equilibrium, and enables you to get something done and not merely one in fifty of the resolutions passed.

The question he poses is one which will, as the President has pointed out, ever elude us in its complete answer. That is satisfactory, I think, because then we shall always be striving. It may be that we have now got to the top of the first of the foothills, and can see something of the range of hills ahead. I remember years and years ago when I was in India, how approaching the Himalayas you came to one row of hills and saw more hills ahead of you; then you came to the next row—I suppose by that time small mountains—and saw another skyline ahead of you. You surmounted that, if you had the opportunity, and saw another skyline—and whether it ever ended I do not know because by then my leave was up and I came back. But it may be that in the next century a President will be addressing our successors and finding he has got some way further into the foothills still without ever seeing the final skyline.

I do think this has been a most illuminating and brilliant essay on the part of our President, and it is a great privilege to me, as well as a pleasure, to be able to propose a vote of thanks to him.

Professor E. S. PEARSON: One of my predecessors, Professor Bradford Hill, apologized for having on a third occasion to take part in welcoming a new President at the beginning of his period of office and I fear that I must repeat this apology. The fact is that Sir Harry Campion is in New Zealand and cannot therefore fulfil the traditional role of seconding the vote of thanks to-night.

At any rate, my threefold experience leads me to remind you that our last three Presidents have been drawn in turn from those three main sources of statistical supply—and of course of statistical demand—the Civil Service, Industry, and the Universities. But beyond this, Professor Kendall's status is possibly a unique one in that, while we regard him at present as an academic personage, his thirty years of active public life have been almost equally divided between the Civil Service, Industry, and a University.

Accepting him as a Professor, I was drawn by the title of his Address to examine whether there was any kind of natural law in the frequency with which our Society has chosen a Professor as its President. If there has been or is any law, however, we have here an example of a dynamic rather than a stationary stochastic process. Examine the record—15 of our first 17 Presidents were Peers of the Realm (2 of them each had three, separated, two-year periods of office!). After that there was indeed a break—we had Mr Gladstone—and his presidency was immediately followed in the 60's and 70's by those of three noteworthy Williams: William Newmarch, William Farr, and William Guy, all Fellows of the Royal Society. But in going on through the nineteenth and twentieth centuries I find no one listed as a Professor among our Presidents until we come to Professor Edgeworth, elected in 1912. Since then, the gaps between academic Presidents certainly appear to be shortening. Here are the intervals in presidential periods as unit: 6, 5, 2, 7, 1, 2, 3.

Our last year's President, in proposing the vote of thanks, has apologized with what I am sure was unnecessary humility for holding office last year. I think it would be found that the intervals between our Presidents drawn from Industry has also been shortening. In all this I have no doubt that we can find a sign of the changing interests and activities of our Society.

The Presidential Address has often provided our new Presidents with the opportunity of showing us in an interesting and reflective way the direction in which their minds were turning. Professor Kendall has seized this opportunity. After ranging over the whole field of mathematical statistics and making his mark on it among other ways, through the production of two, and shortly three, volumes of *The Advanced Theory of Statistics*, we have heard here today how he is turning his expertise in theory towards a study of the laws of behaviour in social phenomena. This interest was evident also in his recent Stamp Memorial Lecture.

In reading my advance copy of the Address, I found a wealth of fascinating and suggestive ideas thrown out in that characteristic way of his, particularly in the last three pages of the galley. I will mention just one line of thought of a somewhat philosophical nature which came to me.

I have been reading recently the outstanding book by Professor E. H. Gombrich, the Director of the Warburg Institute across the way, on *Art and Illusion*, and could not help being struck by a number of close analogies between the scientific model-maker and the artist, as Gombrich sees him. Both work through a process of "making and matching and remaking", to introduce a phrase which Gombrich illustrates again and again in his study of the history of representational art. Then, too, in both fields we have the fact, emphasized by Professor Kendall in his examples of Elderton's cricket scores and the Pareto distribution of incomes, that the patterns which we observe can have many interpretations, interpretations which may change and expand in the minds of successive generations grappling with the meaning of things. There are other parallels, too.

What I fancy emerges here, not yet fully grasped, is some innate law of unity in the processes of human thought and endeavour, though a law which is at present quite certainly unquantifiable.

Finally, Sir, on behalf of the Fellows, may I second this vote of thanks and welcome you very warmly to your Chair of office.

The vote of thanks was put to the meeting and carried unanimously.

As a result of the ballot taken during the meeting, the candidates named below were elected Fellows of the Society:

AHERN, Ronald Albert
AITCHISON, John, M.A.
ATIQUALLAH, M., Ph.D.
AUSTWICK, Kenneth, M.Sc., Dip.Ed.
BARNETT, Victor David, B.Sc.
BARROW, Gerald Ernest
BAYLEY, Judith Mary, B.Sc.(Stats.)
BEREPIKI, Clifford I.
BLOIS, Keith John, B.A.
BLUNDELL, Alan John
BOOTHROYD, Michael Roger, B.A.
BRETHERTON, Michael Hesketh, B.Sc.
BRIGHT, Harold Frederick, M.S., Ph.D.
BROWN, Susannah Ames, M.Sc.
BUCK, Stephan Frank, B.Sc.(Stats.)
BULL, Kenneth Walter John
BURROWS, Peter Michael, B.Sc.
CLARKE, Stanley Peter
COKER, Raymond, B.Sc.
COLLINS, Barry John Knox, B.Sc.
CREET, Armen John, B.Com.
DE, AMITABHA, B.A.
DOHERTY, William Eugene, M.Econ.Sc.
EDWARDS, Hilda V., B.Sc.(Maths.)
EKAMBARAM, S. K., B.A., B.Sc., M.A.
ELLIS, David Morgan, B.A.
FAULDER, Geoffrey Herbert Arthur, B.A.
FEARNLEY, William John

From *The British Journal of Sociology*, Vol. 1, pp. 183–96 (1950)

The Law of the Cubic Proportion in Election Results

M. G. KENDALL *and* A. STUART

[*Division of Research Techniques, London School of Economics*]

---

1. A GREAT deal of interest was aroused in connection with the 1950 General Election in the United Kingdom by an article in the *Economist* newspaper for 7 January 1950, calling attention to an empirical law which appears to operate in two-party contests in recent times. The law, briefly, states that the proportion of seats won by the victorious party varies as the cube of the proportion of votes cast for that party over the country as a whole. It has been found that in the British elections of 1935 and 1945 the law was remarkably closely followed, and there was equally striking agreement for the New Zealand election of 1949. In this paper we examine the theoretical and empirical bases of the law.

2. The law itself appears to have been suggested for the first time by the Rt. Hon. James Parker Smith in evidence before the Royal Commission on Systems of Elections of 1909 (Cmd. 5352). Smith, who had himself been a wrangler and a Smith's prizeman and had contested six elections (being successful in four of them, attributed the law to P. A. MacMahon. Smith emphasized the disparity between the strength of the representation of a party in the House of Commons and its share of the votes in the whole country. He was an advocate of proportional representation, although it appears from his evidence that his views on the subject were undergoing modification. This is what he actually said about MacMahon's law (Minutes of Evidence before the Royal Commission, 19 May 1909):

> Chairman: Although they polled more votes they had fewer members?—Yes, you cannot get exact uniformity or proportionality in these things. In 1906 the Liberals had the greatest majority of any. The Liberals only had 56·8 per cent of the votes, but they got 75·5 per cent of the representation, that is, more than three to one. That shows you the very constant excess proportion. I may say in parenthesis that that is not a thing that depends at all on the constituencies being of different sizes. It would not be affected by a redistribution that made every constituency of exactly the same size. It is just as likely to be true if some constituencies are bigger and some smaller as it is if they are all uniform; but that is not what it depends on. I think I may give an illustration that would make it obvious what I mean. Sup-

posing you had a great box—a great bin of marbles of two colours, red and blue—and supposing one sort is in a majority. Supposing there are 11,000 blue marbles and 9,000 red marbles all stirred up in the box together; it is quite obvious that if you put in your hand without looking and draw out one, the chance that you would get a blue marble would be 11 to 9. The marbles are the electorate, each marble representing a voter. But supposing you take out a whole shovelful of the marbles, then it is quite clear, if the marbles are effectively mixed all through the box, the chances are that the distribution of marbles in the shovel corresponds to the distribution of marbles in the whole box. Therefore the chance is much greater, if you take out a shovelful and then count the blue and red marbles, that the majority of marbles will be the same as the majority of the whole number than when you pick out a single marble. Supposing you take out the whole box and spread them out and count them, then we know the majority would be the blue majority. 11,000 against 9,000 *ex hypothesi*. If you divide it into two halves the chance that each half would have a majority of the blue would be enormous. If you took out a quarter, or just a big shovelful, the chance would still be considerably larger than the chance in regard to one single marble. You can call a shovelful a constituency. You can dig out shovelfuls of different sizes, call them constituencies, count the number of blue and red balls in each of them, and let each side score one for the number of shovelfuls in which it has a majority. I have been going into the mathematics of it, if you are willing to take it from me, and the chances of a single marble being drawn blue is 11 to 9; but the chance of a whole shovelful having a blue majority is something very near 2 to 1. I have had the help in this of my friend Major MacMahon, who is one of the leading mathematicians of the day, and he gives me this as the formula: that if the electors are in the ratio of A to B then the members will be at least in the ratio of A^3 to B^3. That is to say, in the present case, if the electors are as 11 to 9, then the members will be in the ratio of 11^3 to 9^3. That is not quite 2 to 1, but rather less. This formula applies when the two parties are nearly equally divided. If there is any great disparity, the exaggeration of the majority increases at a much more rapid rate.

3. The law as enunciated by Smith may then be put in this form: In a two-party contest between, say, Whites and Blacks, let p_0 be the proportion of votes cast for the winning party, say White, over the whole electoral area. Then if the number of seats won by White and Black are W and B respectively

$$\frac{W}{B} \geqslant \frac{p_0^3}{q_0^3} \quad . \quad . \quad . \quad . \quad . \quad (1)$$

where $q_0 = 1 - p_0$. The equality operates only in the neighbourhood of $p_0 = \frac{1}{2}$. We shall call this "the MacMahon law" and reserve the name of the "law of the cubic proportion" for the equality.

4. How MacMahon arrived at his result is a mystery. An examination of a bibliography of his work suggests that he himself did not publish anything on the subject. One should not, perhaps, attach too much weight to the sequence of ideas presented in oral evidence before a Royal Commission, but if Smith's evidence is any guide to MacMahon's train of thought the mystery deepens. If MacMahon regarded the law as a phenomenon of simple sampling of the kind mentioned by Smith then it could not have been derived mathematically; and in fact it is not true, because in virtue of a well-known theorem known as Bernoulli's, for any fixed value of p_0, however close to $\frac{1}{2}$, the ratio W/B tends to infinity as the size of the constituencies increases so

that the winning party is virtually certain of gaining all the seats. On the other horn of the dilemma, if MacMahon was appealing to observed distributions of the proportion p, say, among constituencies then he would not have found the law closely verified for the experience of his time [1] and would also have found that its operation is not restricted to the neighbourhood of $p_0 = \frac{1}{2}$. It is unthinkable that a man of MacMahon's calibre should have made an error in mathematics; but we do not know how much significance he himself attached to the result. Smith's evidence is obscure and the Royal Commission was very probably out of its depth and did not pursue the matter in further questions. All we can say, therefore, is that on the face of it the law seems to have been reached on very doubtful premisses. It might well have remained a forgotten mathematical curiosum had the *Economist* article to which we have referred not revived interest in it.

5. Before leaving the historical aspect, we might mention a paper by Edgeworth (1898) which was known to Smith in 1909. Edgeworth noted that the distribution of proportions p in English (he did not consider Scottish or Irish) constituencies was approximately of the type which the statistician calls "normal" with standard deviations varying from 0·0898 in 1886 to 0·0750 in 1895, and used the fact to calculate the proportion of Unionist seats from the proportion of Unionist votes for England as a whole. For instance, his method gave a percentage of 64·9 in 1886 as compared with an observed value of 66·0, and in 1895 a percentage of 66·7 as compared with an observed 69·2. Edgeworth made no mention of any such law as MacMahon's, and to apply his method it is necessary to know not only the global proportion of votes but the dispersion of proportions p among the various constituencies.

6. Since Edgeworth and MacMahon considered the relation of seats to votes, the rise of a third major political party has complicated the workings of simple majority voting in the U.K., but in the 1935, 1945 and 1950 general elections the two-party system effectively returned and, despite some hopes to the contrary, seems likely to remain in force in the foreseeable future. In the analysis below we adopt the following method of dealing with the embarrassments of third-party candidatures:

(*a*) Only those seats in which both major parties fought, and in which one of them was successful, are considered as falling within the scope of the law of the cubic proportion.

(*b*) The proportion calculated for each constituency is of the votes cast for the party winning a majority in the House of Commons to the combined votes of the major parties in that constituency.

(*c*) In constituencies returning two members (which were abolished in 1948) the candidates of each party were matched off and the constituency regarded as two.[2]

[1] Smith's written evidence applies the law to seven elections between 1885 and 1906. The agreement was good enough to establish his point about proportional representation, but not very good. The smallest error was six seats in 1892, the largest 48 seats in 1895.

[2] If one major party nominated only one candidate, this made the constituency single for our purposes.

By these arrangements, Northern Ireland was largely excluded from consideration. The twelve university seats (which were abolished in 1948), where voting was by proportional representation, were also excluded.

7. It is an empirical fact that in the elections of 1935 and 1945 (the first won by Conservatives and the second by Labour) the distributions of constituencies according to the proportion p of votes cast for the winners are very close to the normal form. Table I, (*a*) and (*b*), gives the actual distributions and in Fig. 1 we show the distribution function for 1945 together with the distribution function of a normal curve with the same mean and variance.[1] The agreement is evidently very good and that for 1935 was, if anything, better.

8. We may therefore take it that, at least for these elections, the distribution of p is approximately normal. This is not a sampling effect in the usual sense; that is to say, it does not arise from the fact that in samples from a homogeneous population of Conservative and Labour votes the distribution of p is binomial and therefore nearly normal. On the contrary, fluctuations of simple sampling are very small for constituencies of the size here discussed. The standard error in simple sampling in groups of, say, 60,000

TABLE I

(*a*) *Conservative Vote as a proportion of (Labour and Conservative) Vote,* 1935, *in the* 510 *constituencies contested by both parties and won by one of them.*

Proportion p	*No. of Constituencies*	*Cumulated Percentage of total no. of Constituencies*
·15–	3	·59
·20–	4	1·37
·25–	8	2·94
·30–	12	5·29
·35–	19	9·02
·40–	30	14·90
·45–	57	26·08
·50–	68	39·41
·55–	84	55·88
·60–	65	68·63
·65–	53	79·02
·70–	49	88·63
·75–	37	95·88
·80–	11	98·04
·85–	9	99·80
·90–	1	100
·95–	—	100
Total . .	510	

[1] The "distribution function" gives, for a specified p, the proportion of constituencies exhibiting that *or a lower value* of p. It is sometimes called the "cumulative frequency function". The actual figures are given in the last column of Table I. The means and variance were calculated from the grouped data of Table I and a Sheppard correction was applied to the latter.

(b) *Labour Vote as a proportion of (Labour and Conservative) Vote, 1945, in the 576 constituencies contested by both parties and won by one of them.*

Proportion p	*No. of Constituencies*	*Cumulated Percentage of total no. of Constituencies*
0–	1	·17
·05–	—	·17
·10–	—	·17
·15–	—	·17
·20–	3	·69
·25–	8	2·08
·30–	29	7·12
·35–	28	11·98
·40–	59	22·22
·45–	62	32·99
·50–	64	44·10
·55–	89	59·55
·60–	90	75·17
·65–	60	85·59
·70–	37	92·01
·75–	19	95·31
·80–	20	98·78
·85–	5	99·65
·90–	2	100
·95–	—	100
Total . .	576	

(c) *Labour Vote as a percentage of (Labour and Conservative) Vote, 1950, in the 607 constituencies contested by both parties and won by one of them.*

%	*No. of Constituencies*	*Cumulated Percentage of total no. of Constituencies*
15–	2	·33
20–	4	·99
25–	19	4·12
30–	53	12·85
35–	61	22·90
40–	70	34·43
45–	85	48·43
50–	84	62·27
55–	69	73·64
60–	61	83·69
65–	40	90·28
70–	26	94·56
75–	14	96·87
80–	13	99·01
85–	4	99·67
90–	2	100·00
Total . .	607	

(about the size of the average electoral area) is $\sqrt{(p_0 q_0/60{,}000)}$ and this cannot exceed $\frac{1}{2}/\sqrt{60{,}000}$, i.e. 0·002. The observed variation of p among constituencies is due to the fact that voters of similar political views tend to occur

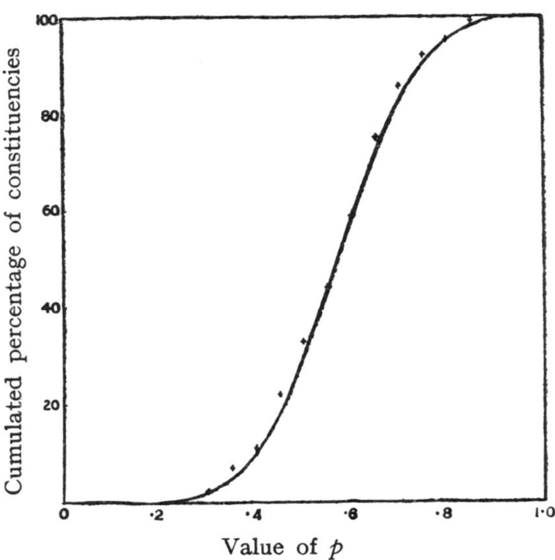

FIG. 1.—1945 Election. Crosses show actual data and continuous curve a normal distribution with the same means and variance.

in groups. Voters cannot be regarded as scattered at random over the various constituencies. We revert to this point in paragraph 20 below.

9. Consider now a distribution of values of p such as that illustrated in Fig. 2. Let the ordinate at the mean value of p, say p_0, be AB and the

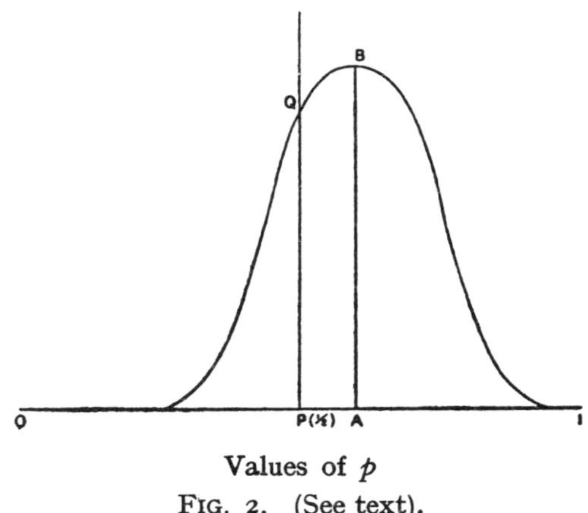

FIG. 2. (See text).

ordinate at the point $p = \frac{1}{2}$ be PQ. Then the number of winning seats is represented by the area to the right of PQ, say y, and the ratio W/B is equal to $y/(1 - y)$. Let the distance PA be x.

We now suppose that in a series of elections the proportion p_0 may vary but that the dispersion of the distribution of p remains constant; thus the curve in Fig. 2 may be regarded as displaced one way or the other but not altered in form as the overall proportion of votes cast for one party varies. This is equivalent to regarding the curve as held fixed and the line LM slid along its length. We then require the relationship between $y/(1-y)$ and the distance x. Now OA $= p$ and hence

$$\frac{p}{q} = \frac{\text{OA}}{\text{A1}} = \frac{\frac{1}{2}+x}{\frac{1}{2}-x}$$

Any law of the form

$$\frac{\text{W}}{\text{B}} = f\left(\frac{p}{q}\right) \qquad . \quad . \quad . \quad . \quad . \quad (2)$$

is therefore equivalent to

$$\frac{y}{1-y} = f\left(\frac{\frac{1}{2}+x}{\frac{1}{2}-x}\right). \qquad . \quad . \quad . \quad . \quad . \quad (3)$$

In order to determine a law of type (2) we then merely have to fit a curve of type (3) to the distribution curve (the cumulative frequency curve) of the observed distribution of p, y representing the proportionate cumulated frequency and x the excess of p over $\frac{1}{2}$.

10. Consider now the particular case represented by the law of the cubic proportion

$$\frac{y}{1-y} = \left(\frac{\frac{1}{2}+x}{\frac{1}{2}-x}\right)^3 \qquad -\tfrac{1}{2} \leqslant x \leqslant \tfrac{1}{2} \quad . \quad (4)$$

which may be written in the equivalent form

$$y = \frac{(1+2x)^3}{2(1+12x^2)} \qquad . \quad . \quad . \quad . \quad . \quad (5)$$

with a corresponding frequency distribution

$$f = \frac{dy}{dx} = \frac{3(1-4x^2)^2}{(1+12x^2)^2} \qquad -\tfrac{1}{2} \leqslant x \leqslant \tfrac{1}{2} \quad . \quad (6)$$

The mean of this distribution is zero and its variance is found, on integrating $x^2 f$ over the range $-\frac{1}{2}$ to $\frac{1}{2}$, to reduce to

$$\mu_2 = \frac{4\pi}{27\sqrt{3}} - \frac{1}{4} = 0{\cdot}018{,}711{,}0$$

In a similar manner we find

$$\mu_4 = \frac{11}{240} - \frac{2\pi}{81\sqrt{3}} = 0{\cdot}001{,}048{,}2$$

so that

$$\beta_2 = \frac{\mu_4}{\mu_2^{\,2}} = 2{\cdot}993{,}874$$

The kurtosis is thus almost exactly that of a normal distribution.

We had intended at this point to insert a diagram comparing the distribution function of the law as given by equation (5) with that of the normal distribution with the same mean (zero) and variance (0·018,711,0). The fit, however, is so good (not only near the mean but for a wide range of values of x corresponding to p, say, from 0·15 to 0·85) that the two curves are barely distinguishable on a diagram of practicable size. We therefore give some values of the ordinates and of the distribution function in Table II as a basis of comparison.

TABLE II

FREQUENCY FUNCTIONS AND DISTRIBUTION FUNCTIONS OF THE NORMAL CURVE AND THE CUBE-LAW COMPARED

x	$\frac{x}{\sigma} = \frac{x}{\cdot 136788}$	*Ordinates*		*Distribution functions*	
		Normal	*Cube*	*Cube*	*Normal*
− ·5	− 3·6553	·0037	0	0	·0000
− ·45	− 3·2897	·0132	·0091	·0001	·0001
− ·4	− 2·9242	·0404	·0456	·0014	·0017
− ·35	− 2·5587	·1105	·1279	·0055	·0053
− ·3	− 2·1932	·2633	·2839	·0154	·0141
− ·25	− 1·8276	·5490	·5510	·0357	·0338
− ·2	− 1·4621	1·0015	·9664	·0730	·0719
− ·15	− 1·0966	1·5986	1·5403	·1350	·1364
− ·1	− 0·7310	2·2326	2·2041	·2286	·2324
− ·05	− 0·3655	2·7280	2·7715	·3539	·3574
0	0	2·9165	3	·5	·5
·05	0·3655	2·7280	2·7715	·6461	·6426
·1	0·7310	2·2326	2·2041	·7714	·7676
·15	1·0966	1·5986	1·5403	·8650	·8636
·2	1·4621	1·0015	·9664	·9270	·9281
·25	1·8276	·5490	·5510	·9643	·9662
·3	2·1932	·2633	·2839	·9846	·9859
·35	2·5587	·1105	·1279	·9945	·9947
·4	2·9242	·0404	·0456	·9986	·9983
·45	3·2897	·0132	·0091	·9999	·9999
·5	3·6553	·0037	0	1·0000	1·0000

11. Since the observed distribution of p is very close to the normal form and the normal form itself is closely represented by the law of the cubic proportion, it follows that the observed distribution of p may be represented by the cube-law *provided that it has the correct variance.* For the elections of 1935, 1945 and 1950, surprisingly enough, this turned out to be so. The values for the means and standard deviations are

	Mean	*Standard deviation*
1935 election	0·577	0·133
1945 election	0·572	0·135
1950 election	0·513	0·138
Law of the cubic proportion		0·137

Since the distribution functions of the cube-law and the normal curve are so close to each other and the standard deviation for 1945 is so close to that of the cube-law, Fig. 1 effectively shows the fit of the cube-law to the 1945 data.

12. The validity of the law of the cubic proportion then depends on three things, (*a*) the empirical fact that the distribution of proportions p at an election is nearly normal, (*b*) the mathematical fact that the cubic-proportion law very closely approximates to a normal form with the same variance, and (*c*) the empirical fact that the variance of the cubic-proportion law is very closely approximated by the variance of the observed distributions. The law is thus not universal. It appears to be a feature of British elections (though not, as we note below, of U.S.A. elections) that the distributions of p are nearly normal and one might assume with some confidence for predictive purposes that they would continue to be so. Thus bases (*a*) and (*b*) are fulfilled. Whether (*c*) is likewise fulfilled depends on a variety of practical circumstances but it appears to be true that the variance of the distribution changes fairly slowly with time so that one would not expect a very marked change between one election and the next.[1]

13. We can evidently derive more general laws of the same character as the cube-law, but unless they have the same attractive simplicity there would not be much point in doing so. If we are attempting to predict the result of an election, the most straightforward approach would be to draw the distribution curves of the previous elections and use them to predict the curve for the coming election by a graphical synthesis ; the forecast can then be read off from the curve when a forecast of p_0 for the whole country has been obtained by surveys of public opinion or other means.

14. It is perhaps worth examining how far the cubic-proportion law is likely to be in error for small changes in the variance from its current value. If the standard deviation increases by 100δ per cent, the law will give

$$\frac{W}{B} = \left(\frac{1 + 2x}{1 - 2x}\right)^3$$

when it ought to give

$$\left(\frac{W}{B}\right)' = \left(\frac{1 + \dfrac{2x}{1 + \delta}}{1 - \dfrac{2x}{1 + \delta}}\right)^3$$

The difference reduces, to the first order in δ, to

$$\frac{W}{B} - \left(\frac{W}{B}\right)' = \frac{12x}{1 - 4x^2}\,\delta \quad . \quad . \quad . \quad . \quad . \quad (7)$$

Now in practical cases x usually lies between 0 and 0·1 and if the variance increases by 5 per cent δ will be about 0·025 and the corrective term given by (7) is then $0{\cdot}3x$ approximately. For a closely fought contest with x, say, 0·02 (a 52 : 48 split of votes) the term is 0·006 in a ratio of $(52/48)^3 = 1{\cdot}271$, and would amount only to less than one seat. For $x = 0{\cdot}1$ (a 60 : 40 split) the term is 0·03 in a ratio of 3·375 and again amounts to less than one seat.

[1] Between 1945 and 1950 the electoral areas in the United Kingdom were altered substantially and this had some effect on the normality of the 1950 distribution.

15. Consider generally the effect of variation in the standard deviation of the normal curve. If σ becomes smaller, then for any distance x (OA in Fig. 2) $\frac{x}{\sigma}$ will be larger, and the area of the curve lying to the right of PQ (proportion of seats won) will be larger than if the cubic-proportion law holds exactly. Conversely, if δ becomes larger $\frac{x}{\sigma}$ will be smaller, and the proportion of seats won smaller than if the cubic-proportion law holds exactly. At any election, therefore, in which the distribution of p is normal, variation in the standard deviation of p from 0·136788 will produce the effect that a higher or lower power than the third will be required to raise the votes ratio to the seats ratio. It does not, however, follow that a different power will give an equally good fit to the distribution curve over its whole course or, in consequence, that the appropriate law would therefore be a power law. But we may note in passing that if δ is less than 0·136788, as it was when MacMahon formed his law, a higher p_0 requires a higher power than 3 to produce an accurate prediction. This tallies with Smith's statement to the Royal Commission.

16. We have considered whether the cubic-proportion law operates in the U.S.A. elections for the House of Representatives. The distribution of proportions p (Democratic to total votes) is shown in Table III and it will be clear on inspection that there are important differences between the American

TABLE III

Proportion of major party vote going to Democratic Party in 1944 Congressional elections in the 371 seats contested by both major parties and won by one of them.

Proportion	*No. of Districts*	*Ten Southern States*[1]	*Adjusted no. of Districts*
·20–	2		2
·25–	4		4
·30–	34		34
·35–	39		39
·40–	52		52
·45–	51		51
·50–	42		42
·55–	41	4	37
·60–	27	1	26
·65–	23	9	14
·70–	13	5	8
·75–	8	1	7
·80–	7	3	4
·85–	4	2	2
·90–	10	10	—
·95–	14	14	—
Total . .	371	49	322

[1] The States excluded are Alabama, Arkansas, Florida, Georgia, Louisiana, Mississippi, N. and S. Carolina, Texas, Virginia.

and the British distributions. The American experience, in fact, is bimodal. This is due to the peculiar circumstances of the Southern States which return members to Congress with exceptionally large majorities. The Southern vote, in fact, is so solidly democratic that the elections are hardly contests. It is very remarkable that if we ignore the ten states in the South which returned only Democratic Congressmen (regarding these elections, perhaps, as virtually uncontested) the distribution returns to a form closely resembling the British type. Moreover, it has a standard deviation close to the value required by the law of the cubic proportion, being 0·132 for the 1944 experience.

We have not investigated foreign elections further because we feel that a satisfactory inquiry would need a more detailed knowledge of foreign electoral systems than we possess. Nevertheless it appears not improbable that some law such as the cube-law might be found to operate in the United States (and possibly elsewhere) when allowance is made for the special position of the South; we hope that some workers overseas may be interested enough to pursue the topic further.

17. In working on the British data for 1935 and 1945 we were compelled, as noted in paragraph 6, to exclude certain constituencies from the scope of the cubic-proportion law. To be logical we should also take as the proportion p_0 only the proportion of White to total (White + Black) votes in the constituencies to which the law is applied. The ratios calculated by the writer of the *Economist* article, however, are based on the overall vote for the country as a whole, which is much more easily accessible. The effect of the change to the exact basis would be slight and it has not seemed worth while to recalculate the 1935 and 1945 ratios. For 1950, however, we did so, with the following results:

Votes cast for Labour (including Irish Labour) in 607 seats = 13,169,551
Votes cast for Conservative and National Liberal in 607 seats = 12,327,106

The votes ratio is therefore 1·068341 and the cube of this is 1·219353. Applied to a total of 607 seats, this gives a division of 333 : 274 as against the actual result of 313 : 294, an error of 20 in the seats going to either party, and therefore of 40 in the majority.

18. This result looks worse than it is because the parties are so very close together, but it nevertheless provides a salutary reminder of the limitations of the cube-law in prediction. Despite the remarkable constancy of the standard deviation, the whole apparatus of prediction is endangered if the distribution of the proportion p in constituencies departs appreciably from the normal distribution. In 1950, the departure was sufficient to produce an error of 40 seats in the majority predicted. It is unlikely that, with a standard deviation so close to the prescribed value, the error would be greater than this unless there were some violent distortion, but it is as well to bear in mind, when making predictions based on opinion poll figures that, in addition to the sampling error and non-sampling biases automatically introduced into the prediction, which should be small with good polling methods, there is an additional range of uncertainty produced by non-normal distribution of the

proportion. Fig. 3 shows the frequency distribution of the election as given in Table I(c) fitted to a normal curve of the same standard deviation (0·138).

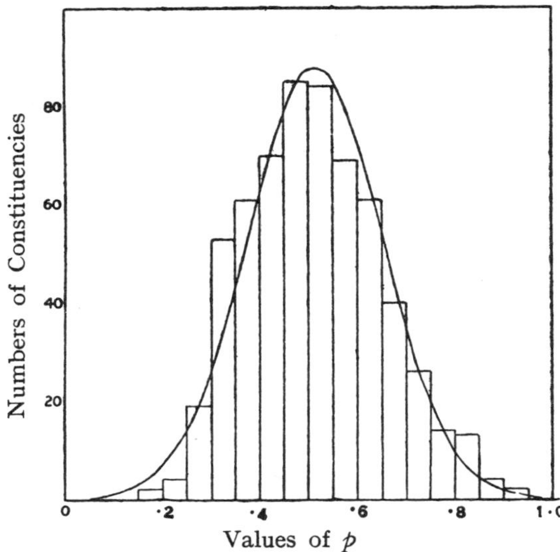

FIG. 3.—Election of 1950. Frequency-distribution of constituencies and normal curve of same mean and standard deviation.

19. In any case, of course, the use of the cubic-proportion law in prediction depends on obtaining an accurate forecast of the proportion p_0. We ourselves attempted a prediction of 1950 results on the basis of the Gallup poll figures published two or three days before the election. The results were amazingly close, being:

	Our forecast	*Actual*
Labour	316	315
Conservative and allies . . .	293	298
Liberals	9	9
Others	7	3
	625	625

This is almost too good to be true but we freely admit that our success was quite undeserved. The Gallup poll figures (45 per cent Labour, 43·5 per cent Conservative), though accurate within the limits of error which might have been expected, slightly underestimated the Labour vote; this error compensated almost exactly the deviation from the law of the cubic proportion; and we were lucky in our guesses as to what was going to happen in the score or so of seats not covered by the law. The calculations illustrate again the sensitivity of forecasts to small errors when the election is a very close one.

20. The fidelity with which the normal distribution is reproduced by the distribution of values of p naturally raises the question whether there is any underlying theoretical reason for its appearance. We have noted above that the distribution cannot be explained as a central limit effect in simple sampling. If the model mentioned by Smith were the correct one, that is, if we could

regard White and Black votes as scattered at random over the country, the distribution would be normal but its variance would not exceed 1/240,000, namely 0·000,004,167, for constituencies of 60,000. The variance observed at the present time is about 0·018, which is 4,320 times as great. If any stochastic scheme can account for the phenomenon at all, there must be a very high internal correlation between voters in the same constituency.

21. Now it is clear that the distribution of Conservative and Labour votes over Great Britain cannot be accounted for by any simple scheme. It depends on powerful geographical and historical as well as on political and demographic factors and no probabilistic scheme could hope to predict on prior grounds what the combined effect of these multitudinous interlocking influences might be. Nevertheless, it is possible to set up fairly simple schemes which reproduce the observed phenomena closely. We do not say that these are the correct schemes. They grossly oversimplify the actual circumstances. But it is instructive to consider them, if only because they diminish the natural surprise with which one encounters the normal distribution in electoral systems.

22. Consider first of all a system consisting of a large number of White and Black balls, and a sampling process such that there is a correlation between successive drawings; that is to say, if a White ball has been drawn on the previous trial it is more probable that a White ball will be drawn on the next trial than if a Black ball had been previously drawn. Let t be the probability that if a White ball has been drawn the next ball will be White, and t' the probability that if Black has been drawn the next will be White. We have by hypothesis $t - t'$ $(= \varepsilon$, say$) > 0$. This scheme is what the probabilist calls a Markoff chain, the chances t and t' being constants so that the result of any drawing depends on the result of the previous drawing but only on that result. The theorems concerning this process which we require are given by Uspensky (1937, pp. 223, 301). In fact

(*a*) If p_0 is the proportion of White balls in the parent population, assumed large, the mean value of the observed p's in samples of n is p_0;

(*b*) the distribution of p tends to normality as n tends to infinity, that is to say $x = p - p_0$ is distributed in the form

$$dF = \frac{1}{\sigma\sqrt{(2\pi)}} e^{-\frac{1}{2}x^2/\sigma^2} dx \qquad . \quad . \quad . \quad . \quad (8)$$

with a variance given by

$$\sigma^2 = \frac{p_0 q_0}{n} \frac{1 + \varepsilon}{1 - \varepsilon} \qquad . \quad . \quad . \quad . \quad . \quad (9)$$

Whereas in the case of independent drawings the variance is p_0q_0/n, which, as we saw above, is much too small, for dependent drawings in a Markoff chain it is given by (9). For our present case we have

$$\frac{1 + \varepsilon}{1 - \varepsilon} = 4{,}320$$

giving

$$\varepsilon = 0{\cdot}9995 \text{ approximately}$$

o

The observed variance could therefore be accounted for by a very high correlation between successive drawings expressed by the fact that $t - t'$ is 0·9995. This involves a higher degree of political cohesion than seems probable. We must remember, however, that the inhabitants of a constituency are not put down one by one as if they were drawn from a population under a Markoff scheme. There are cases where the correlation between members will be very high indeed (e.g. between husband and wife or between members of the same trade union) and cases where it is less marked.

23. As another possibility, suppose that we regard our sample as drawn under what the statistician calls a Lexian scheme. We now consider the samples as taken from several groups—not one bin of balls as suggested by Smith, but from several bins with different proportions of Black and White balls. If there are k groups and the probability of a member of the ith group voting for the winning party is p_i, the variance of proportions p obtained in samples of n (each sample chosen entirely from one group but all the groups contributing equally to the total number of samples) is given by

$$\frac{p_0 q_0}{n} + \frac{n-1}{n} \operatorname{var} p_i \qquad . \quad . \quad . \quad . \quad (10)$$

where var p_i is the variance of the proportions p_i among the k groups (Yule and Kendall, 1950, p. 401). The second factor in (10) will, in the circumstances we are here considering, be large compared with the first factor and can account for the large observed variance of p. This scheme also will account for the observed normality.

As to the plausibility of this scheme, suppose we were able to attach to each constituency an index of average social, economic or political class, and then to group all constituencies according to this index. It would then be possible to regard the resulting groups as the populations from which the constituencies are drawn as samples and perhaps this does not depart too far from reality. Constituencies are geographical areas and, for various reasons, a given social, economic or political class does not—and indeed cannot—regard all areas as equally eligible for residential purposes. In this sense, our Lexian sampling scheme has a certain plausibility, though we do not claim too much for it.

REFERENCES

Economist, (1950). 7 January.

Edgeworth, F. Y. (1898). "Applications of the Theory of Probability", *J. R. Statist. Soc.* **51**, p. 534.

Minutes of Evidence before the Royal Commission on Systems of Elections (1910). Cmd. 5352. B.P.P. XXVI.

Uspensky, J. V. (1937). *An Introduction to the Mathematical Theory of Probability,* McGraw-Hill.

Yule, G. U. and Kendall, M. G. (1950). *An Introduction to the Theory of Statistics,* 14th edition revised, Griffin & Co.

From *Journal of the Royal Statistical Society A*, Vol. **116**
pp 11–34 (1953)

The Analysis of Economic Time-Series—Part I: Prices

By M. G. Kendall

Division of Research Techniques, London School of Economics

(Read before the Royal Statistical Society, December 17th, 1952, Professor A. Bradford Hill, C.B.E., Ex-President, in the Chair)

Introduction and Summary

1. It has been customary to analyse an economic time-series by extracting from it a long-term movement, or trend, for separate study and then scrutinizing the residual portion for short-term oscillatory movements and random fluctuations. The assumption latent in this procedure is that the long-term and short-term movements are due to separate causal influences and therefore that the mathematical process of analysis corresponds more or less roughly to a real distinction of type in the generative system. When it began to appear that the classical method of analyzing stationary economic fluctuations into cycles and residual elements broke down, and that better accounts of such movements were given by autoregressive schemes in which the disturbance element played an integral part, it was an easy step to proceed one stage further, and to enquire whether the so-called trend in a series was in fact separable from the short-term movements, or whether it should be regarded as generated by a set of forces which also gave rise to the short-term movements.

2. The work described in the present paper sprang from this idea. It was necessary in the first place to find suitable material on which to test it and I give below particulars of 22 price-series, ranging from 486 terms at weekly intervals to 2,387 terms at weekly intervals, which were chosen for the purpose. I began to construct models and to fit them to these data and also to other series which are not mentioned in this paper; but before the work had gone very far it appeared that the pattern of events in the price series was much less systematic than is generally believed. This is perhaps the most significant part of the present paper. I shall not have space to deal with the corresponding phenomenon in production data, and shall mention only incidentally the problem of trend fitting.

3. Broadly speaking the results are these:

(*a*) In series of prices which are observed at fairly close intervals the random changes from one term to the next are so large as to swamp any systematic effect which may be present. The data behave almost like wandering series.

(*b*) It is therefore difficult to distinguish by statistical methods between a genuine wandering series and one wherein the systematic element is weak.

(*c*) Until some way has been found of circumventing this difficulty, trend fitting, and perhaps the fitting of any model, is a highly hazardous undertaking. It may be possible for an econometrician to test whether the data agree with a hypothesis suggested by prior analysis, but it may be impossible to discriminate between quite different hypotheses which all fit the data.

(*d*) There is experimental evidence and theoretical support for the belief that aggregative index numbers behave more systematically than their components. This *might* be due to the reduction of the random elements by averaging and the consequent emergence of systematic constituents; but it could equally well be due to chance. If it is, there will appear spurious time-correlations in aggregative series and the use of index-numbers in econometric work needs extensive reconsideration.

(*e*) An analysis of stock-exchange movements revealed little serial correlation within series and little lag correlation between series. Unless individual stocks behave differently from the average of similar stocks, there is no hope of being able to predict movements on the exchange for a week ahead without extraneous information.

The Data

4. For the purposes of this study I used 22 economic series as follows:

Actuaries' Index of Industrial Share Prices

	Period	*Interval*	*Number of Terms*
1. Banks and Discount Companies	1928–1938	Week	486
2. Insurance Companies	,,	,,	,,
3. Investment Trusts	,,	,,	,,
4. Building Materials	,,	,,	,,
5. Coal	,,	,,	,,
6. Cotton	,,	,,	,,
7. Electric Light and Power	,,	,,	,,
8. Gas	,,	,,	,,
9. Iron and Steel	,,	,,	,,
10. Oil	,,	,,	,,
11. Total Industrial Productive		,,	,,
12. Home Rails	,,	,,	,,
13. Shipping	,,	,,	,,
14. Stores and Catering	,,	,,	,,
15. Total Industrial Distributive	,,	,,	,,
16. Breweries and Distilleries	,,	,,	,,
17. Miscellaneous	,,	,,	,,
18. Total Industrial Miscellaneous	,,	,,	,,
19. Industrials (all classes combined)	,,	,,	,,
20. Basic cash wheat at Chicago in cents per bushel	Jan., 1883–Sept., 1934 (excluding 1915–1920 inclusive; one missing value interpolated for 10.3.1933)	,,	2,387
21. Monthly average of preceding series	Jan., 1883–Sept., 1934 (excluding 1915–1920)	Month	548
22. Spot cotton at New York in cents per pound	Aug., 1816–Jan., 1951 (excluding 1861–1866 and 1914–20; one missing value for Oct., 1857, inserted as average of the two neighbouring months)	,,	1,446

Sources

For the actuaries' indices I am indebted to the Institute and Faculty of Actuaries who allowed me to extract the figures from the Institute's records.

Series 20 and 21 are taken from *Wheat Studies of the Food Research Institute*, vol. 11, No. 3, November, 1934 (Stanford University, California).

Series 22 is taken from *Statistics on Cotton and Related Data*, U.S. Department of Agriculture, Bureau of Agricultural Economics, Statistical Bulletin, No. 99, Washington, D.C., June, 1951.

5. The golden rule in publishing work on time-series is to give the original data. To do so in this case would have required about thirty pages of journal. I shall, however, be glad to make available copies of the series or the cards on which they have been punched. A good many of the secondary statistics are presented below but here also the working sheets are available to any responsible person who is interested. So far as possible I have not "edited" any of the series even where there was some case for making the attempt; but in some instances I have omitted war periods and should have been compelled to do so by the lack of data. All such omissions are indicated above.

Analysis of the Price Data

The Chicago Wheat Series (20 *and* 21)

6. Although the wheat-price series was not the first I examined it is useful to begin an expository account by considering it because it is a long series of prices wherein little or no element of aggrega-

tion enters. The Stanford Wheat Studies from which the data were taken give graphs of the series. They are unfortunately too long to reproduce on a page of this size.*

From the primary data the series of 2,379 first differences were constructed and a frequency distribution formed which is given as the marginal column in Table 1. It was here that the first fact of significance emerged, for the resulting distribution had nearly perfect symmetry and an appearance of approximate normality. The bivariate distribution of one difference against the next succeeding difference is even more illuminating. Table 1 is a condensed version and presents on the face of it an excellent picture of independence; on a regression diagram the two lines would almost coincide with the variate axes.

7. The decision to use first differences instead of the original series was not an arbitrary one. When a price is fixed on a free market both parties know what was the price of the previous transactions and use that price as a starting point for negotiation. It is the change, not the absolute value, which constitutes the fundamental element in the price determination. One can think of exceptions, perhaps, but for the commodities I am discussing they are not relevant.

8. At first sight the implications of these results are disturbing. If the series is homogeneous, it seems that the change in price from one week to the next is practically independent of the change from that week to the week after. This alone is enough to show that it is impossible to predict the price from week to week from the series itself. And if the series really is wandering, any systematic movements such as trends or cycles which may be "observed" in such series are illusory. The series looks like a "wandering" one, almost as if once a week the Demon of Chance drew a random number from a symmetrical population of fixed dispersion and added it to the current price to determine the next week's price. And this, we may recall, is not the behaviour in some small backwater market. The data derive from the Chicago wheat market over a period of fifty years during which at least two attempts were made to corner wheat, and one might have expected the wildest irregularities in the figures. To the statistician there is some pleasure in the thought that the symmetrical distribution reared its graceful head undisturbed amid the uproar of the Chicago wheat-pit. The economist, I suspect, or at any rate the trade cyclist, will look for statistical snags before he is convinced of the absence of systematic movements. And he will be very right to do so.

9. One possible source of error is that, in treating as homogeneous a price-series extending from 1883 to 1934, we may be oversimplifying the hypotheses. The series was therefore divided into two sections covering the periods 1883–1914 and 1921–1934; and a bivariate distribution of first differences on the lines of Table 1 constructed for each. I omit the figures to save space, but the story told by each section was the same as for the series taken together.

There are seven widely outlying values in the whole series, and it did not seem to me to be sophisticating the data to omit them from the calculation of moments.† The following constants were computed from a more finely grouped distribution and are corrected by Sheppard's formulae for the grouping effect.

	Period 1883–1914	*Period* 1921–1934	*Both Periods Together*
μ_1'	0	0·0964	0·0289
μ_2	7·7576	22·7789	12·3294
μ_3	12·6787	34·5882	20·2709
μ_4	657·3073	2745·8873	1293·0607
β_1	0·344	0·101	0·219
β_2	10·856	5·292	8·506

The distributions are accordingly rather leptokurtic.

10. The general pattern of independence present in these data makes it very unlikely that there are large serial correlations present but to make sure they were worked out to the tenth order and are given in Table 2. For computing purposes it is easier to work out serial covariance

* The weekly series are, so far as possible, closing quotations on Fridays and are given to the nearest $\frac{1}{8}$ cent. The monthly figures are averages of four or five weekly figures.

† Three values are due to a jump from September 21st to September 28th, 1888, of 53 cents and a recession the following week of 38 cents; four values are due to a similar movement in May, 1898.

TABLE 1

Frequency Distribution of Differences of Series 20, *Difference between Weeks* t *and* t + 1 *against Difference between Weeks* t + 1 *and* t + 2.

(Condensed to save space from original table for which the interval was one unit. Values falling on the border line of intervals were allotted ½ to each.)

Difference, (t + 1)th *Week Less* (t + 2)th *Week; Cents per Bushel, Mid-point of Interval*

Difference, tth week less $(t + 1)$th week; cents per bushel.	−21	−19	−17	−15	−13	−11	−9	−7	−5	−3	−1	1	3	5	7	9	11	13	15	17	19	21	*Totals*
−17	—	—	—	—	—	—	—	—	—	—	½	—	—	—	—	—	—	—	—	—	—	—	½
−15	—	—	—	—	—	—	—	—	—	1	½	1	—	—	—	2¼	¼	—	—	—	—	—	5
−13	—	—	—	—	—	—	—	—	1	1	1	2	—	—	½	¾	¼	—	—	1	—	1	8½
−11	—	—	—	—	—	—	—	—	1½	—	½	½	2	1	—	—	1	1	—	—	—	—	7½
−9	—	—	—	1	—	—	—	2	2½	3	3½	4	2	3	2	—	—	—	1	—	—	—	24
−7	—	—	—	—	1	—	3	2½	3	4	10	8	5	2	6	1	—	1	—	—	1	—	47½
−5	—	—	—	1	1	1	5	5½	8½	9¼	23	28¾	16	8½	3½	1	2	—	—	—	—	—	114
−3	—	—	—	½	½	2	5	6½	16	28	64¼	91	27¼	12½	6½	—	1	—	—	1	—	1	263
−1	—	—	—	—	3	2½	4	12½	27½	86¾	232	236¾	78½	19	12½	1½	2	2	—	½	—	—	721
1	½	½	—	—	—	—	3½	10½	22	71	268½	207¼	82¼	29¾	6¼	3	1	—	1½	1	—	—	708½
3	—	—	—	—	—	—	1	5	13¾	30¾	82¼	89¼	47½	8¾	4¼	1½	½	—	—	—	—	—	284½
5	—	—	—	—	1	1	1½	3	7¾	16¼	18½	20½	11½	9	5	2	—	1	1	—	—	1	100
7	—	—	—	½	—	—	1	1	6	7½	11½	12½	6½	2½	—	1	—	—	—	—	—	—	50
9	—	—	—	½	1	—	—	1	2	2	1½	—	2½	1½	1½	2½	—	—	—	—	½	½	17
11	—	—	—	—	—	—	—	—	1	1	3½	—	½	1½	—	½	—	—	—	1	—	—	9
13	—	—	—	—	—	—	—	—	—	—	—	1	1	—	2	—	—	—	1	—	—	—	5
15	—	—	½	½	½	—	—	—	1	—	—	1	1	—	—	—	—	—	—	—	—	—	4½
17	—	—	—	—	½	—	—	—	1	—	—	3	—	—	—	—	—	—	—	—	—	—	4½
19	—	—	—	½	—	—	—	—	—	—	—	—	—	1	—	—	—	—	—	—	—	—	1½
21	—	—	—	½	—	1	—	—	—	1	—	1	—	—	—	—	—	—	—	—	—	—	3½
Totals	½	½	½	5	8½	7½	24	49½	114½	262½	721	707½	283½	100	50	17	8	5	4½	4½	1½	3½	2,379

Note.—There were also seven outlying pairs with values (−6·75, −53·0), (−53·0, 37·75), (37·75, −1·25), (−6·5, −30·75), (−30·75, 3·75), (−20·0. 53·0) and (53·0, 11·75). The last term of the series was taken with the first so that the total number of pairs is 2,386.

TABLE 2

Serial Correlations of First Differences of Wheat Prices (*Series* 20)

(Decimal points omitted)

Order of Correlation	*Series* 1883–1914 (n = 1,669)	*Series* 1921–1934 (n = 716)	*Whole Series* 1883–1934 (*Omitting* 1915–1920) (n = 2,385)
1	−078	−063	−071
2	+078	+051	+065
3	−000	+030	+014
4	−075	+042	−019
5	+007	−066	−028
6	−032	−071	−051
7	−016	−039	−027
8	−059	−064	−061
9	−047	+023	−014
10	−038	+034	−004
Large sample standard error	0·025	0·038	0·020

for the primary series and to calculate those of the differences from formulae such as

$$\sum_{i=1}^{n-2}(x_{i+2}-x_{i+1})(x_{i+1}-x_i)=2\sum_{i=1}^{n-1}x_ix_{i+1}-\sum_{i=1}^{n}x_i^2-\sum_{i=1}^{n-2}x_ix_{i+2}$$
$$+x_1^2+x_n^2-x_1x_2-x_{n-1}x_n \quad . \quad (1)$$

For long series when end effects can be neglected this is for all practical purposes equivalent to

$$\tau_k=-\tfrac{1}{2}\Delta^2\rho_{k-1}/(1-\rho_1) \quad . \quad . \quad . \quad . \quad . \quad . \quad (2)$$

where τ represents the autocorrelations of first differences and ρ those of the primary series. In this case it was not much extra trouble to take end-effects into account and the serial correlations for the constituent series in Table 2 were calculated from the formula

$$r_k=\frac{\sum_{i=1}^{n-k-1}u_iu_{i+k}-\frac{1}{n-k-1}\sum_{i=1}^{n-k-1}u_i\sum_{i=k+1}^{n}u_i}{\left[\left\{\sum_{i=1}^{n-k-1}u_i^2-\frac{1}{n-k-1}\left(\sum_{i=1}^{n-k-1}\right)^2\right\}\left\{\sum_{i=k+1}^{n}u_i^2-\frac{1}{n-k-1}\left(\sum_{i=k+1}^{n}u_i\right)^2\right\}\right]^{\frac{1}{2}}} \quad . \quad (3)$$

where u_i is the difference $x_{i+1}-x_i$.

In calculating the serials for the two series together a weighted average was taken, the formula being

$$r_k=\frac{(n_1-k-1)c_1+(n_2-k-1)c_2}{\{(n_1-k-1)v_{11}+(n_2-k-1)v_{21}\}^{\frac{1}{2}}\{(n_1-k-1)v_{12}+(n_2-k-1)v_{22}\}^{\frac{1}{2}}} \quad . \quad (4)$$

where n_1, n_2 are the number of terms, v_{11}, v_{21}, v_{12} and v_{22} refer to the variances of the first $n-k$ terms of the first and second series and the last $n-k$ terms of the first and second series respectively and c_1, c_2 are the covariances of the first and second series. Equation (4) has no very firm theoretical basis but it is obviously reasonable and will certainly not underestimate the magnitude of the serial correlations.

11. A comparison of the variances of the two parts of the series suggests that there has been an increase in variability since World War I. This is what one might perhaps have expected, but it is rather a nuisance from the point of view of analysis because it suggests that the series is not stationary. We have here an interesting and rather unusual case of a time-series for which the mean remains constant but the variance appears to be increasing. It is desirable to have a test of this type of departure from stationarity, distribution-free if possible. The problem is easy enough for a random series but not so easy for an autocorrelated series of unknown character, and as I do not require the test for present purposes I omit a discussion of the point.

12. A glance at Table 2 shows that the serial correlations are very small and, for most practical purposes, negligible. However, the first two serials are, in four cases out of six, more than three times their standard error as calculated from R. L. Anderson's formula

$$\text{var}\ r_k = \frac{n-1}{(n-2)^2} \quad . \quad . \quad . \quad . \quad . \quad . \quad . \qquad (5)$$

There are several possible explanations of this effect. Anderson's formula is based on normality and constant variance in the series. From general experience one would not expect the sampling variance to be increased by departure from parent normality in the direction of leptokurtosis, the distribution being symmetrical and n large. Nor, as it seems to me, can we appeal to non-stationarity to account for the difference, because the discrepancies occur most markedly in the first part of the series. I proceed to consider two more practical possibilities, one that the primary series was rounded up (or down) systematically, the other that for hour-to-hour or day-to-day changes in price there were systematic variations which have been lost to view in the weekly figures.

13. Suppose that a series of values $x_1, \ldots x_n$ is systematically rounded up by amounts $\varepsilon_1, \ldots \varepsilon_n$, ε thus having a positive mean but being serially independent. We shall then have for the differences $u_i (= x_{i+1} + \varepsilon_{i+1} - x_i - \varepsilon_i)$, in long series

$$\text{corr}\ (u_i, u_{i+k}) = \frac{-\Delta^2 \rho_{k-1}}{2(1-\rho_1) + 2\ \text{var}\ \varepsilon/\text{var}\ x} \qquad k > 1$$

$$= \frac{-\Delta^2 \rho_{k-1} - \text{var}\ \varepsilon/\text{var}\ x}{2(1-\rho_1) + 2\ \text{var}\ \varepsilon/\text{var}\ x} \qquad k = 1 \quad . \quad . \quad . \qquad (6)$$

In our present case var x is of the order of 10 and (the series being given to the nearest $\frac{1}{8}$th) var $\varepsilon = \frac{1}{12}\left(\frac{1}{8}\right)^2 = 0\cdot0^213$. The effect of factors in var ε/var x, of the order of $0\cdot0^31$, in (6) is therefore quite insufficient to account for the observed size of the first serial. In any case it would not account for the sign of the second. Even if we suppose that the rounding off was alternately up and down, which would induce autocorrelation in the ε's, the magnitude of the effect would not be large enough to explain the observations.

14. Consider then the attenuation in serial relationship which may be caused by averaging over time-periods. Suppose a market is open from Monday to Friday, that any Monday may be regarded as following immediately on the preceding Friday and that the scheme of generation from day to day is a simple Markoff process

$$u_{t+1} = \alpha u_t + \varepsilon_{t+1} \quad . \quad . \quad . \quad . \quad . \quad . \quad . \qquad (7)$$

It is fairly easy to show that the correlation between the average of the five daily prices in one week and the corresponding average in the succeeding week is given by

$$r = \frac{5\alpha^5 + 4(\alpha^6 + \alpha^4) + 3(\alpha^7 + \alpha^3) + 2(\alpha^8 + \alpha^2) + (\alpha^9 + \alpha)}{5 + 8\alpha + 6\alpha^2 + 4\alpha^3 + 2\alpha^4} \quad . \quad . \qquad (8)$$

For $\alpha = 0\cdot8$ $r = 0\cdot499$ and the correlation is reduced by 38 per cent. For $\alpha = 0\cdot5$ $r = 0\cdot178$ and the serial correlation between weekly averages is only a third of that between days. For $\alpha = 0\cdot1$ $r = \cdot02$. Evidently the averaging over successive terms may obliterate some quite substantial serial correlation in prices. The weekly wheat figures were not, in fact, averages over the days of the week, but it is evident that the choice of one day in the week as the representative point will stilll further attenuate any day-to-day correlations.

A simple Markoff process would still not account for the negative sign of r_1 in Table 3 unless the correlation between successive days was negative, which seems unlikely but is not impossible owing to the tendency of a sensitive market to swing too far and to correct itself. I ought perhaps to mention that the possibility of "significant" serials appearing owing to very occasional exceptional values was dismissed after examination of the series at such values.

15. It may well be, then, that the observed serials, small as they are, have some significance. But evidently the systematic element is slight compared with the random component and a very precise technique is going to be necessary in order to extract it, unless we know what to look for.

Some of my colleagues have pointed out that the operation of first differences could easily reduce quite a large cyclical effect below the threshold of detectability. For example, if there were a ten-year cycle in the data, corresponding to about 500 terms of the series, the differencing of a sine wave of amplitude comparable in magnitude to that of the observed fluctuations would be to reduce the amplitude in the differenced series by a factor of about $2\pi/500$ but to double the variance of random effects. We cannot say, then, that the analysis has ruled out the existence of periodic terms. There may well be traces of a seasonal effect present. What we may fairly say, I think, is that we do not need to invoke them to explain the observations.

16. Series 21 is a derivative of Series 20, being a monthly average instead of a weekly one. A frequency distribution of first differences, which I omit to save space, tells much the same story as Series 20. For the moments I find, with Sheppard's corrections

	Whole Series	*Omitting One Extreme Value at Each End*
$\mu_1' =$	0·0776	0·0504
$\mu_2 =$	44·0256	34·8271
$\mu_3 =$	208·2139	6·4153
$\mu_4 =$	32,721·3887	6,914·5455

For the series on the right

$$\beta_1 = 0{\cdot}00151 \qquad \beta_2 = 5{\cdot}701$$

The distribution is nearer normality, as one might expect under the central limit effect. Some of the serial correlations are given in the reply to the discussion.

17. A comparison of the variances and fourth moments of Series 20 and 21 raises no reason to suspect their wandering nature. If $u_1, \ldots u_{4n}$ is a series of $4n$ weekly terms, the series of monthly terms is $\frac{1}{4}(u_1 + u_2 + u_3 + u_4)$, $\frac{1}{4}(u_5 + u_6 + u_7 + u_8)$, etc. For the differences of these typified by η, we have (means being approximately zero)

$$\text{var}\ \eta = \frac{1}{16(n-1)} \sum_{j=0}^{n-2} (u_{4j+1} + u_{4j+2} + u_{4j+3} + u_{4j+4} - u_{4j+5} - u_{4j+6} - u_{4j+7} - u_{4j+8})^2$$

or, for large n,

$$\text{var}\ \eta = \frac{1}{16n} \sum_{=0}^{n-2} \{(u_{4j+1} - u_{4j+2}) + 2(u_{4j+2} - u_{4j+3}) + 3(u_{4j+3} - u_{4j+4}) + 4(u_{4j+4} - u_{4j+5}) + 3(u_{4j+5} - u_{4j+6}) + 2(u_{4j+6} - u_{4j+7}) + (u_{4j+7} - u_{4j+8})\}^2.$$

If the differences of u, typified by ε, are independent we then have

$$\text{var}\ \eta = \frac{1}{16}\ \text{var}\ \varepsilon\{1^2 + 2^2 + 3^2 + 4^2 + 3^2 + 2^2 + 1^2\}$$

$$= \frac{11}{4}\ \text{var}\ \varepsilon \quad . \quad . \quad . \quad . \quad . \quad . \quad . \quad . \quad . \quad . \quad (9)$$

The ratio of variances of the monthly to the weekly series is $34{\cdot}8271/12{\cdot}3294 = 2{\cdot}82$ against the theoretical value of $2{\cdot}75$. It may be shown similarly that $\kappa(\eta) = 113/64\ \kappa(\varepsilon)$. For the whole series the ratio $\kappa_4(\eta)/\kappa_4(\varepsilon)$ is $3{\cdot}12$ and for the series omitting extreme values $2{\cdot}12$ against the theoretical value $1{\cdot}77$.

18. There seems nothing to be gained by taking averages of the monthly figures to obtain an annual figure. Under the central limit effect the resulting series would be nearly normal; and successive values would almost certainly be nearly independent.

19. In the past I have often wondered whether annual figures were much use in the study of economic phenomena, except of course in relation to intermittent variables like crop-yields. So much can happen in a year that one feels the underlying causational system to require examination under a finer structure than an annual interval. But if this wheat-series is any guide, it seems that what we gain by observing the phenomenon at short intervals is lost, or at any rate deeply obscured, by the stochastic discontinuity of the process. It may be that the motion is genuinely random and that what looks like a purposive movement over a long period is merely a kind of economic Brownian motion. But economists—and I cannot help sympathizing with them—will doubtless resist any such conclusion very strongly. We can at this point suggest only a few conclusions:

(*a*) the interval of observation may be very important;

(*b*) it seems a waste of time to try to isolate a trend in data such as these;

(*c*) prediction in such a series, from internal behaviour alone, is subject to a wide margin of error and the best estimate of the change in price between now and next week is that there is no change.

British Industrial Share Prices (*Series* 1–19)

20. Series 1 to 19 are weekly index numbers based on 1930 = 100. (The price taken is that of Tuesday of each week, not an average of the week's quotations.) They are not all independently compiled, series 11 (Total industrial production) being an average of series 12–14, series 18 (Total industrial miscellaneous) being an average of series 16 and 17, and series 19 being an average of all classes. There are thus 15 independent series. During the period covered by the figures there were, I believe, some changes such as substitutions for quotations which dropped out. But this hardly affects my argument. Taken together, these are as complete and reliable a set of figures describing the inter-war movement of the U.K. industrial share market as one is likely to find.

21. The first 29 serial correlations were computed for each series and are given in Table 3. The method used was similar to that for series 20 except that corrections for means were omitted as being negligible and the variance of the whole series used in the denominator, the formula accordingly being

$$r_k = \frac{\dfrac{1}{n-k}\sum_{i=1}^{n-k} u_i u_{i+k}}{\dfrac{1}{n}\sum_{i=1}^{n} u_i^2} \quad . \quad . \quad . \quad . \quad . \quad (10)$$

It may be shown that for these series this approximation can affect the third decimal place at the most, except perhaps for series 3. To resolve any doubt the serials were worked out exactly for this series and the exact values were found to differ from the approximate ones only in the fourth decimal place.

For some of the series a bivariate distribution on the lines of Table 1 was constructed. In some cases the general picture of independence was presented, but in others there were signs of dependence as reflected in the serial correlations. I have space to reproduce only one, that for Series 3, where the dependence seems to be the highest (Table 4).

22. Such serial correlation as is present in these series is so weak as to dispose at once of any possibility of being able to use them for prediction. The Stock Exchange, it would appear, has a memory lasting less than a week except perhaps for Investment Trusts (Series 3), Stores and Catering (Series 14) and all series together. It may be, of course, that series of *individual* share prices would behave differently; the point remains open for inquiry. But the aggregates are very slightly correlated and some of them are virtually wandering. Investors can, perhaps, make money on the Stock Exchange, but not, apparently by watching price-moments and coming in on what looks like a good thing. Such success as investors have seems to be due (*a*) to chance,

TABLE 3

Serial Correlations of Series 1–19, *Lag* 1 *to* 29

(Decimal points omitted)

Number of Series (*see Para.* 4)

Lag	1	2	3	4	5	6	7	8	9	10
1	058	052	301	125	148	087	181	096	088	−013
2	061	−016	356	−001	075	121	055	185	−055	045
3	014	082	158	−025	061	061	053	049	011	−024
4	−050	−019	164	−079	−056	015	−016	086	043	−022
5	−087	−082	066	−025	−055	−030	−084	−005	−015	044
6	−028	−093	−101	−034	−056	−125	−085	−022	−091	−034
7	−015	−006	−030	001	−054	−044	−092	−036	−076	−029
8	−011	−042	−042	−013	−001	−058	−055	001	−020	004
9	004	−025	−030	−041	−084	−118	−067	056	−022	005
10	015	008	−033	010	−063	−067	−012	027	−015	−044
11	045	−050	−013	−022	042	−037	064	025	007	052
12	002	000	−094	048	040	004	000	043	046	004
13	−014	037	−052	045	021	048	040	049	094	009
14	−017	004	009	008	−037	−056	081	067	−060	−025
15	−062	032	027	050	−006	062	047	046	−039	−019
16	030	006	008	118	076	015	039	049	069	002
17	072	079	109	030	068	153	041	119	076	043
18	097	049	086	045	030	070	018	028	−028	036
19	098	−016	080	−016	031	−017	−009	−055	−036	−015
20	−037	−077	020	−026	−046	072	−023	016	−033	037
21	009	060	031	074	−031	−078	−056	052	004	−021
22	−027	011	−056	044	−067	−094	−029	074	−023	076
23	−044	−151	−047	056	−004	−084	−029	−041	−091	−048
24	034	−005	−086	−023	−020	−017	−001	043	−049	017
25	104	017	031	069	−039	015	016	133	076	−009
26	021	080	−048	038	−001	−012	029	067	044	−032
27	000	004	011	−017	082	−046	094	044	032	−052
28	045	130	053	071	097	−011	036	036	040	087
29	−011	−040	050	−014	060	−014	033	037	061	031

Number of Series

Lag	11	12	13	14	15	16	17	18	19
1	195	010	053	230	237	034	200	177	234
2	061	−057	029	054	076	087	103	106	105
3	053	044	−044	018	054	003	053	058	066
4	−034	039	−014	001	−040	−056	−013	−044	−046
5	−051	−015	−034	−084	−062	067	−080	−044	−055
6	−102	−081	058	−146	−093	−002	−115	−079	−094
7	−072	018	−013	−144	−072	019	−104	−087	−093
8	−043	139	036	−057	−044	−114	004	−033	−041
9	−047	033	037	−038	165	−018	−003	−003	−008
10	−047	−044	021	042	021	035	012	−091	−023
11	−003	079	046	−006	013	085	−014	013	005
12	039	069	044	086	089	017	024	142	050
13	082	−009	−009	−047	−006	077	037	064	045
14	−029	−041	−107	004	−008	−036	−037	−029	−025
15	036	−129	096	067	018	−084	045	001	033
16	069	030	090	064	077	055	158	122	114
17	145	005	168	029	086	001	124	120	148
18	047	072	063	047	082	029	089	079	084
19	−010	013	042	−001	057	047	010	022	038
20	007	005	008	035	014	−078	−001	−045	−017
21	−018	−012	−030	090	043	041	009	031	008
22	−035	009	−027	032	013	064	−015	033	−006
23	−082	−056	022	−039	−082	−005	−053	−045	−070
24	−041	−032	−037	−034	−021	028	−008	−008	−028
25	051	036	−000	002	−001	058	−001	017	026
26	025	037	086	−038	−005	046	036	049	036
27	047	055	092	−032	052	−019	030	012	038
28	168	048	016	−037	007	023	072	063	075
29	−069	039	−054	−004	032	−014	018	001	014

TABLE 4

Bivariate Distribution of First Differences of Series 3, *Difference between Weeks* t *and* t + 1 *against Difference between Weeks* t + 1 *and* t + 2; *Units, One-tenth of a Point*

Difference, t^{th} week, less $(t+1)^{th}$ week	*Difference* t + 1th *Week Less* t + 2th *Week*														
	< −12	−12, −11	−10, −9	−8, −7	−6, −5	−4, −3	−2, −1	0, 1	2, 3	4, 5	6, 7	8, 9	10, 11	⩾ 12	*Totals*
< −12	5	2	4	5	1	2	—	—	1	—	—	—	—	1	21
−12, −11	4	1	1	—	3	1	3	—	—	—	—	—	—	—	13
−10, −9	1	2	2	—	2	2	4	—	—	1	—	1	—	—	15
−8, −7	3	2	—	4	4	5	4	—	1	—	1	—	—	1	25
−6, −5	4	3	2	4	7	11	9	3	1	1	—	—	—	1	46
−4, −3	1	2	2	6	7	17	13	14	2	—	1	1	—	—	66
−2, −1	1	1	1	2	9	19	14	18	5	5	2	2	1	1	81
0, 1	—	—	—	1	6	8	22	26	13	9	3	2	1	—	91
2, 3	—	—	2	1	1	—	5	15	9	4	3	3	—	1	44
4, 5	1	—	—	1	3	—	3	7	7	5	1	1	2	1	32
6, 7	—	—	—	—	2	1	—	3	1	2	—	2	1	2	14
8, 9	—	—	1	—	1	—	1	2	1	3	2	—	—	4	15
10, 11	—	—	—	—	—	—	1	2	2	1	—	1	—	1	8
⩾ 12	—	—	—	—	—	—	1	1	1	1	—	2	3	1	14
Totals	21	13	15	25	46	66	81	91	44	32	14	15	8	14	485

(*b*) to the fact that at certain times all prices rise together so that they can't go wrong, (*c*) to having inside information so that they can anticipate a movement, (*d*) to their being able to act very quickly, (*e*) to their being able to operate on such a scale that profits are not expended in brokers' fees and stamp duties. But it is unlikely that anything I say or demonstrate will destroy the illusion that the outside investor can make money by playing the markets, so let us leave him to his own devices.

23. There are several factors of these series of correlations requiring explanation.

(*a*) If we arrange them according to the magnitude of the first serial correlation, which seems a fair summary measure of internal correlation, we get the following order:

Series	3	Investment Trusts.
	15	Total Distribution.
	19	All classes.
	14	Stores, etc.
	17	Miscellaneous.
	11	Total Production.
	7	Electric Light.
	18	Total Miscellaneous.
	5	Coal.
	4	Building Materials.
	8	Gas.
	9	Iron and Steel.
	6	Cotton.
	1	Banks.
	13	Shipping.
	2	Insurance.
	16	Breweries.
	12	Home rails.
	10	Oil.

We can understand why this list should be headed by investment trusts, which are almost an aggregative index in themselves, and why it should be tailed by shipping, breweries, home-rails and oil. But it is not clear why banks and insurance companies come so low.

(*b*) Apart from investment trusts and stores, the series showing the greatest internal correlations are the aggregative series 11, 15, 18 and 19. We get greater serial correlation in the averages than in the constituent series, which at first sight seems absurd and in any case is very misleading. One possibility is that this effect is generated by the method of construction of the series, e.g., the use of geometric means, the substitution of new quotations and so on. With the help of Mr.

S. T. David and Mr. Haycocks I went into this explanation fairly thoroughly but it does not seem to account for the phenomenon. The use of geometric means could raise or lower the serial correlations but need not bias them upwards; substitution of new quotations is carried out only in December and its effect, even if systematic, would be so diluted in the correlation of weekly figures as not to explain the magnitude of the aggregation effects. A more likely explanation lay in the existence of lag correlations between the series and I revert to the point below in paragraph 28.

24. Whatever the reason, the existence of these serial correlations in averaged series is rather disturbing. If the effect is a general one, it means that we must be very chary of drawing inferences from series of index numbers of the aggregative type, which is most unfortunate because such series are, in many instances, all that the statistician or the economist has to work on. The so-called "cycles" appearing in such series may not be due to endogenous elements or structural features at all, but to the correlations between disturbances acting on the constituent parts of the aggregative series.

25. I am led to infer that wherever possible the econometrician must study individual series rather than aggregates, just as he must study closely neighbouring points of time rather than observations at long intervals. This may be part of the reason why attempts to fit simple models to a whole economy have usually failed. It is as though a physicist set out to investigate the properties of light without being able to isolate a pure colour and with a diffraction grating ruled at intervals of an inch apart. Such conclusions would apply to price data, or to any nundinal data where there is a rapid adjustment of forces in a fluid market. They may not apply with equal force to production data.

Cross-correlation of the Share-prices

26. To study the series further we have to examine the correlations between them. It is possible to pick out pairs from 19 individuals in 171 ways and even with modern computing facilities it was too laborious to work out a number of lag correlations for them all. I chose 28 pairs, taking those for which the analysis seemed likely to be most rewarding, as follows:

(1, 4), (1, 5), (1, 6), (1, 9), (1, 12), (1, 13)
(4, 5), (4, 6), (4, 8), (4, 12), (4, 13)
(5, 6), (5, 8), (5, 9), (5, 12)
(6, 8), (6, 9), (6, 11), (6, 12)
(8, 11), (8, 18)
(9, 11), (9, 12), (9, 13)
(11, 12), (11, 13), (11, 18)
(12, 13)

The numbers correspond to the series set out in section 4. The lags even worked out as far as the sixth each way, e.g., (1, 4) means that lag correlations were worked out between first differences of series 1 and 4 (Banks, etc., and Building Materials) for lags of -6 to 6, thirteen coefficients in all. The results are given in Table 5.

27. The main feature of this set of correlations is the smallness of the magnitudes for the lags. There are traces of correlation in some instances but, however real, they are very slight. Thus no series acts as a "leader" for the others. Not only is it impossible to predict a series from its own internal behaviour but it seems equally impossible to predict it from the behaviour of the other price series.

28. We can now revert to the aggregation effect mentioned in paragraph 23 and discuss it in simplified terms. Suppose we have a series of consecutive pairs of terms $(\varepsilon_1, \eta_1)(\varepsilon_2, \eta_2) \ldots (\varepsilon_n, \eta_n)$ with zero means. Consider the correlation of sums $E \equiv \sum_{i=1}^{n} \varepsilon_i$ and $F \equiv \sum_{i=1}^{n} \eta_i$. If the variance of ε_i and η_i is σ_i^2 we find

$$\text{corr}\,(E, F) = \frac{\sum_{i,j} \sigma_i\,\sigma_j \,\text{corr}\,(\varepsilon_i\,\eta_j)}{\sum \sigma_i\,\sigma_j\,\text{corr}\,(\varepsilon_i, \varepsilon_j)} \qquad i, j = 1 \;\ldots\; n \quad . \quad . \quad . \quad (11)$$

TABLE 5

Lag Correlations of the Differences of Certain Pairs of Series from Series 1–19

Series	*Lag* −6	−5	−4	−3	−2	−1	0	1	2	3	4	5	6
(1, 4)	−008	−066	−077	−021	036	028	414	100	015	008	023	−080	−125
(1, 5)	−060	−134	−137	−068	015	030	255	077	102	−011	015	011	−066
(1, 6)	−080	−037	−074	−087	019	029	228	126	163	010	062	−010	−118
(1, 9)	−023	−102	−139	−014	−027	−021	295	053	050	028	−020	027	−028
(1, 12)	−037	−023	−046	−029	017	−044	275	154	073	0003	−001	022	−118
(1, 13)	−033	−068	−060	−124	113	051	258	121	030	−015	004	−063	−070
(4, 5)	−064	−108	−102	002	034	070	425	200	035	052	−044	−019	016
(4, 6)	−059	−036	−047	−003	059	058	424	201	076	054	−027	−003	−089
(4, 8)	−037	−067	−015	−024	049	085	310	−002	023	030	006	−027	−046
(4, 12)	022	−071	−049	026	−009	−006	367	187	−006	059	031	018	−066
(4, 13)	048	−052	−033	054	046	111	320	096	019	002	−068	001	010
(5, 6)	−080	009	051	083	048	182	454	161	081	054	−043	−108	−134
(5, 8)	−013	−013	026	00001	044	009	272	004	025	035	−056	−042	007
(5, 9)	−075	−016	−065	070	013	096	661	226	033	024	015	−065	−053
(5, 12)	026	−014	−036	074	037	063	392	167	024	−013	025	−048	−072
(6, 8)	−040	−033	037	−004	064	024	194	014	−031	012	−003	032	−072
(6, 9)	−122	083	004	078	038	085	482	228	030	013	071	024	−115
(6, 11)	−127	−073	004	092	104	156	750	140	080	033	018	011	−124
(6, 12)	−095	−034	−031	110	014	089	368	194	−027	011	−024	033	−114
(8, 11)	−020	−049	−027	069	002	006	389	066	075	−002	023	−046	−030
(8, 18)	−033	−064	−032	055	−005	−046	399	079	125	002	010	−051	−028
(9, 11)	−087	−004	042	022	022	228	795	071	−032	033	−038	−060	−101
(9, 12)	−056	−025	041	039	008	016	469	110	−070	060	042	−048	−149
(9, 13)	007	−031	026	133	092	144	279	102	−037	047	−111	−029	−002
(11, 12)	−053	−035	−060	072	010	054	522	224	−024	059	036	−004	−120
(11, 13)	066	−076	038	082	042	199	407	140	041	011	−056	−067	021
(11, 18)	−051	−004	007	036	094	196	824	166	069	065	−108	−084	−101
(12, 13)	−039	−022	044	041	039	146	281	045	031	001	−085	−008	−031

Note.—Decimal points omitted. Lag counted as interval of first series after second, e.g., lag 2 for series (1, 4) is correlation of series 1 at time $t+2$ and series 4 at time t.

or, if all the variances are equal,

$$\text{corr}\,(E, F) = \frac{\text{mean corr}\,(\varepsilon_i, \eta_j)}{\text{mean corr}\,(\varepsilon_i, \varepsilon_j)} \quad . \quad . \quad . \quad . \quad . \quad . \qquad (12)$$

For my series I do not possess all the lag correlations but as an example of the kind of thing which may happen let us suppose that all the correlations $(\varepsilon_i, \varepsilon_j)$ have a mean value of 0·4 (this being about the mean of those values we have in Table 5) and that the mean correlation of (ε_i, η_j) is 0·1. Then we find

$$\text{corr}\,(E, F) = \frac{225\,(0\cdot 1)}{15 + 210\,(0\cdot 4)} = 0\cdot 23$$

so that the first serial of the aggregate series is more than twice the mean serials of its constituents. It is a piece of good luck, I suppose, that the first serial of series 19 turns out to be 0·234.

29. The cross correlations of order zero suggest that there is a good deal of *simultaneous* sympathetic movement between the series; (simultaneous, that is, within the compass of a single week). They are more or less as one would expect but not always so big as a simple model of the economy would suggest. If we regard these random changes from week to week as due to exogenous elements, we must suppose that the elements are substantially different from one series to another, albeit not independent.

30. One could clearly take this analysis further, for instance by working out multiple regressions or attempting a component analysis to see whether there were any common factors. I have refrained from doing so at this stage for two reasons; one is that further experiments on other

price series and for different intervals of time are desirable before we refine the analysis; the other is that the existence of the aggregative effect would probably vitiate the findings and I think we shall have to go back to primary series before attempting to isolate components.

New York Cotton Prices

31. The New York spot cotton prices of series 22 are a salutary warning against undue generalization. They are monthly, not weekly, series and cover about 125 years. One might have expected them to behave like the wheat series, but they do not. Table 6 gives the serial correlations up to the sixteenth order. As there were obvious breaks in continuity during the

TABLE 6

Serial Correlations of Cotton Prices (Series 22)

(Decimal points omitted)

Order of Correlation	*Period* 1816–1860	*Period* 1868–1914	*Period* 1921–1950	*Total Series* (1816–1950)
1	346	393	227	313
2	096	046	−012	039
3	−049	−096	−027	−053
4	−045	−153	−172	−125
5	−075	−060	024	−032
6	−053	−075	095	−000
7	−013	−068	031	−010
8	015	−064	135	041
9	−037	−028	021	−012
10	001	071	122	068
11	029	088	−000	033
12	−051	080	−106	−038
13	−093	−006	024	−014
14	−143	032	047	012
15	−105	−011	025	016
16	−022	−050	042	008

American Civil War and during World War I, I have calculated the serials separately for the periods 1816–1860, 1868–1914 and 1921–1950.

The bivariate distributions of the type of Table 1 are regular, looking very much like a normal form, but the correlations are significant. We now find a pattern of behaviour rather like that of a simple Markoff series. The price-change from month t to month $t+1$ is now clearly correlated with that from month $t+1$ to month $t+2$, but we do not require a memory of more than a month in the market to account for the serial correlation.

There are several possible explanations of this effect, although it is difficult to be sure whether any of them is the right one. Wheat is a commodity which is grown in both hemispheres and a continual supply on the market may keep the price more fluid than for cotton, a product of the northern hemisphere only. Moreover, crop forecasts are better for wheat than for cotton and there is a tendency for the latter to be underestimated at the beginning of the season; the slow improvement in estimates as the growing season proceeds might induce some correlation in the prices. The slower rate of progress of cotton through the manufacturing process may also have some effect. But whatever the reason, it seems that we are not entitled to generalize from one agricultural commodity to another and a systematic survey of the major raw materials of commerce is necessary before we can lay down any general rules.

First Experiments in Trend Fitting

32. The typical Yule autoregressive scheme which has been fitted to economic data with a certain amount of success may be written

$$\alpha u_{t+2} + \beta u_{t+1} + \gamma u_t = \varepsilon_{t+2} \qquad (13)$$

It seems natural to generalize this by considering the case when the constants α are themselves slowly moving through time as the economy changes. I therefore examined the scheme

$$(\alpha_0 + \alpha_1 t + \alpha_2 t^2)\, u_{t+2} + (\beta_0 + \beta_1 t + \beta_2 t^2)\, u_{t+1} + (\gamma_0 + \gamma_1 + \gamma_2 t^2)\, u_t = \varepsilon_{t+2} \quad . \quad (14)$$

A second-order Yule scheme was chosen because general experience so far has indicated that nothing much is to be gained by taking higher orders into account; and second-degree polynomials were chosen because the general run of the series suggested that they would be sufficient and would keep the arithmetic within reasonable bounds. But a decision on these points is rather arbitrary and more extended schemes might be necessary in some cases. The fitting was carried out on the primary series not on the differences.

33. In equation (14), it will be noted, I have allowed the coefficient of u_{t+2} to vary with time. It would be more consonant with the idea of auto*regression* to take the coefficient as a constant, say unity. However, the more general expression has one advantage, namely that it allows for movements in the variance of the disturbance term ε, which may well vary in time, especially for price data. If the coefficients are small compared with the maximum observed value of t^2 equation (14) may be written

$$u_{t+2} + (\beta_0' + \beta_1' t + \beta_2' t^2)\, u_{t+1} + (\gamma_0' + \gamma_1' + \gamma_2' t^2) = (\delta_0 + \delta_1 + \delta_2 t^2)\, \varepsilon_{t+2} \quad . \quad (15)$$

which exhibits the scheme as an approximation to a generalized Yule process with trend in the variance of the disturbance.

34. It will also be noticed that (14) is not a stationary process. There is no reason why it should be. No economic system yet observed has been stationary over long periods. It is true that schemes of type (14) may be explosive, which all economic systems are not; but that only implies that in representing the slow movements by polynomials we are approximating to some non-explosive function, a familiar procedure to which no exception is taken in other fields of application.

35. I fitted a scheme of type (14) by least squares, that is to say, by minimising the square of the expression on the left for variations in the nine parameters. This leads to nine equations with coefficients of the type $\Sigma\, t^2 u^2_{t+2}$, $\Sigma\, t\, u_{t+1} u_{t+2}$, etc. They are troublesome to compute but there is no theoretical difficulty. For the experiment series 1, 6 and 8 (Banks, etc., Cotton and Gas) were taken. The following are the results. The origin in each case is at the 244th term so that t ranges from -243 to 242.

Series 1

$$\begin{aligned}(1 &+ 0{\cdot}021{,}279{,}208t - 0{\cdot}0^4 31{,}214t^2)\, u_{t+2} + (-\,1{\cdot}236{,}892{,}571 - 0{\cdot}0^2 2{,}127{,}249t \\ &+ 0{\cdot}0^4 40{,}297t^2)\, u_{t+1} + (0{\cdot}236{,}578{,}860 + 0{\cdot}0^3{,}848{,}707t - 0{\cdot}0^5 9{,}021t^2)\, u_t = \varepsilon_{t+2} \quad . \quad (16)\end{aligned}$$

Series 6

$$\begin{aligned}(1 &- 0{\cdot}0^2 3{,}435{,}032t + 0{\cdot}0^3{,}135{,}958t^2)\, u_{t+2} + (-\,0{\cdot}086{,}729{,}101 - 0{\cdot}0^4 95{,}266t \\ &- 0{\cdot}0^4 54{,}730t^2)\, u_{t+1} + (-\,0{\cdot}921{,}767{,}317 + 0{\cdot}0^2 3{,}562{,}777t - 0{\cdot}0^4 81{,}816t^2) \\ &= \varepsilon_{t+2} \quad . \quad (17)\end{aligned}$$

Series 8

$$\begin{aligned}(1 &+ 0{\cdot}0^2 1{,}273{,}056t - 0{\cdot}0^4 37{,}456t^2)\, u_{t+2} + (-\,1{\cdot}421{,}129{,}272 - 0{\cdot}0^2 2{,}055{,}170t \\ &+ 0{\cdot}0^4 56{,}754t^2)\, u_{t+1} + (0{\cdot}422{,}070{,}596 + 0{\cdot}0^3 822{,}663t - 0{\cdot}0^4 19{,}549)\, u_t = \varepsilon_{t+2} \quad . \quad (18)\end{aligned}$$

36. These equations were worked out before it had dawned on me that the parent series were behaving in a wandering way, and at that stage it was a surprise when Mr. K. H. Medin, working in my department, pointed out that there were identities in the coefficients. In fact, very approximately

$$\left.\begin{aligned}\alpha_0 + \beta_0 + \gamma_0 &= 0 \\ \alpha_1 + \beta_1 + \gamma_1 &= 0 \\ \alpha_2 + \beta_2 + \gamma_2 &= 0\end{aligned}\right\} \quad . \quad (19)$$

For instance in (16)

$$1 - 1{\cdot}236{,}892{,}571 + 0{\cdot}236{,}578{,}860 = 0{\cdot}0^3313{,}911$$

and in (18)

$$-0{\cdot}0^437{,}456 + 0{\cdot}0^456{,}754 - 0{\cdot}0^419{,}549 = -0{\cdot}0^6251$$

It follows, of course, that we can write (12) as

$$(\alpha_0 + \alpha_1 t + \alpha_2 t^2)(u_{t+2} - u_{t+1}) - (\gamma_0 + \gamma_1 t + \gamma_2 t^2)(u_{t+1} - u_t) = \varepsilon_{t+2} \quad . \quad . \quad (20)$$

so that first differences obey a generalized Markoff scheme. The ratio

$$(\gamma_0 + \gamma_1 t + \gamma_2 t^2)/(\alpha_0 + \alpha_1 t + \alpha_2 t^2)$$

varies to some extent over the range of t; for Series 1 from about 0·3 at one extreme to 0·2 at the other, being about 0·2 over the middle range; for Series 6 from about − 0·7 to − 0·5, about minus unity over the middle range; for Series 8 about 0·4 over the middle range.

On the whole I regard this experiment as a failure. It seems asking too much of quadratic polynomials to expect them to give a good representation over series of 500 terms; and in any case it is doubtful whether the series are sufficiently systematic to react significantly to this kind of treatment. The only reason I refer to the trend-fitting problem at all at this stage is to call attention to its difficulty in the hope that others may be led to study it.

37. I am well aware that this paper raises more difficulties than it resolves. Most papers on time-series do. The results concerning the time-interval of observation, the aggregative-effects and trend-fitting seemed to me, however, to be important enough to justify publication, if only to prevent my fellow-workers from spending time on profitless inquiries. More positive results must await further exploration of individual series, not only of prices but of production, and attempts to improve available statistical techniques for analysing them.

Acknowledgments

38. The computations involved have been severe, particularly in connection with the lag correlations of series 1 to 19. I am indebted to Mrs. Joan Humphries and Miss Julia Grahame for a great deal of the work, which they carried out with zeal and gratifying cheerfulness. I am also indebted to Mr. E. C. Fieller and the National Physical Laboratory for allowing me to use the electronic A.C.E. in determining serial covariances. My colleagues in the Division of Research Techniques at the London School of Economics, particularly Mr. S. T. David and Mr. K. H. Medin, have been very helpful in exploring lines of inquiry, many of which are referred to only very briefly or are not mentioned at all, that had to be followed up to decide incidental points; and my colleagues among the economists have endured much in discussions about the econometric implications of the results.

Discussion on Professor Kendall's Paper

Professor R. G. D. Allen: In moving this vote of thanks to Professor Kendall I should like to express the indebtedness of the Society not only to him but also to the Division of Research Techniques at the London School of Economics of which he is the Director. The Division, only a few years old, has been generously supported by the Nuffield Foundation. The publications of the Division are numerous, massive, and indicative of a long life for the Division. The work has not been at all concentrated; it has ranged over a surprisingly wide field. Some of the work on survey techniques has already been presented to this Society. Other work has been in the econometric field, as represented by the paper Professor Kendall has read tonight. This contribution is a generally understandable one. Not all econometric work is as intelligible to the layman or even to the expert. I hope that all economists as well as statisticians will read the paper and study it.

I want to take the point of view of the economist, not the statistician, who approaches the problem of handling time-series rather hopefully and at the same time apprehensively. Economists have learned much on this subject during the two decades or more since Yule's famous paper was read to the Society. They have come to look upon time-series with considerable suspicion. They have been told to beware of a fearsome devil—serial correlation—and to look to statisticians to cast the devil out for them. It seems now that the more statisticians work on the casting out of devils, the more they cast out everything else, including what the economists want.

If the economist is to design a model to relate to actual data, he must deal in averages and aggregates, e.g., such broad aggregates as national consumption, income, savings, investment. Moreover, the testing of models in the past has usually been in the terms of annual data. The question is how to break up aggregates and how to shorten the intervals of recording in the time-series. Short annual series of data are usually not good enough to test any model whatever, and what Professor Kendall now appears to show is that, with more frequent data, many different models fit the data very well. This is a serious dilemma.

In para. 19 of the paper which indicates Professor Kendall's conclusion on wheat prices he points out that what we gain by observing the phenomena at short intervals is lost or obscured by "the stochastic discontinuity of the process", that it may be that the motion is genuinely random, and that what looks like a purposive movement over a long period is "merely a kind of economic Brownian motion". That is a very depressing kind of conclusion to the economist. The conclusions on cotton prices (para. 31) are rather more encouraging, a "memory" of a month. Even so, this may arise because of defects in the cotton market as opposed to the wheat market. Cotton is practically confined to the northern hemisphere and the making of crop estimates is a rather peculiar process. Finally, on stock prices, apart from the question of how to "play the market", we would expect to have dependable aggregate series of prices of all stocks together. Paragraphs 24 and 25, however, indicate that the problem is not a simple one.

In the present state of knowledge the economist must work with simple models relating to broad aggregates. If he ventures further the variables and relations—and the problems—crowd in on him. Professor Kendall's conclusion is that any attempt to fit simple models to data must inevitably fail. The econometrician is far from attaining the position in which he can hope to determine the parameters in which he is interested. This paper must be regarded as the first dividend on a notable enterprise. Some "shareholders" may feel disappointed that the dividend is not larger than it is, but we hope to hear more from Professor Kendall and to have further, and larger, declarations of dividends.

Professor Champernowne (in seconding the vote of thanks): Professor Kendall is a pioneer in the study of time series by examining serial correlation coefficients, and a paper from such an authority gives us hope for further revolutionary advances. Every hint that old techniques are faulty and every suggestion of new methods of analysis will eagerly be followed by those who have time and resources to experiment with new ideas. Therefore the more conservative amongst us, especially amongst the economists, may be excused if we endeavour to pick holes in the argument before it is allowed to become gospel. As seconder of the vote of thanks I feel an obligation in this respect, and shall devote my attention almost entirely to the points on which I disagree.

Turning, however, to the list of conclusions given in para. 3, I do not find that I disagree violently with anything except perhaps the last sentence of (*d*), that the more systematic behaviour of aggregative index numbers may be due to chance, and that if so there will appear spurious time-correlations in aggregative series, and the use of index numbers in econometric work will need extensive reconsideration. I hope that those responsible for using or criticizing or constructing index numbers will not be led by these conclusions to suppose that there is something wrong about index numbers, and that they should use individual quotations or do something more drastic instead. Unless I misunderstand Professor Kendall's arguments, he is drawing a red herring across any discussion of the merits of index numbers for the purposes for which they are usually intended.

He has shown that if in a set of price-series the intercorrelations lagged by one time-unit are nearly all positive, and if the unlagged correlations are not large, then the serial correlation coefficient with unit lag for the index number may be much larger than for any individual series. If the phenomenon is due to a cancelling-out of random elements he and all of us would agree that this merely shows that the index number expresses the systematic part of the price-movements better than any individual series. But Professor Kendall suggests that the effect may be entirely spurious and due merely to chance. This is possible, but it seems to me that pure chance alone will only on very rare occasions lead to the intercorrelations of lag 1, nearly all having positive sign and an average nearly as high as that of 0·1 found in Table 5. This does not look like chance; it indicates that there is a tendency towards a positive correlation between one term in one series and the following term in another. If so, the higher serial correlation of the index number reflects its advantageous quality and is not spurious.

With reference to conclusion (*e*), I may perhaps be excused for agreeing with the moral drawn there. The professionals have to keep a sharp eye on the movement of prices, on the one hand stifling large shifts of price that are likely to be immediately reversed, and on the other anticipating any movements of prices that would otherwise continue over a period of weeks; and the low serial correlation coefficients found in this particular series may reflect the success with which the professionals are doing their job. It is reassuring that statisticians will not be able to earn vast sums by scientific gambling on the Stock Exchange, and will still find it worth while to continue their own work. I cannot resist saying that the analysis of prices of less speculative markets will have even greater interest for the economist.

Turning to conclusion (*c*), again it is difficult to disagree. It may be impossible to distinguish between different hypotheses which all fit the data, and the choice of possible hypotheses by prior analysis is, in the present state of knowledge, as essential as the testing and comparing of hypotheses by statistical methods. But there is a danger that people reading this conclusion will think that because Professor Kendall finds certain speculative price series have correlograms like those of wandering series, *ergo* it is futile to compare the trend, of say, real wages with the trend of other variables which, according to economic theory, determine real wages in the long run. Leaving aside the question whether seasonal and cyclical variations of prices have any precise meaning, there can be little doubt that the graphs of many price series exhibit such movements. If one is merely analysing the series by taking weekly changes in such prices and drawing up tables like Table 3 of the paper, I doubt whether the results would be so different from those of that table, my doubt being based on Professor Kendall's own admirable argument in the second part of para. 15, where he says that one does not need to invoke seasonal or cyclical changes to explain the observations. The danger is that this might be interpreted by some as meaning that cyclical and seasonal influences are not an important part of the movement of many economic time series. The fact that the correlograms of many economic series are like those of wandering series need not disturb us. If this effect is widespread it should give greater confidence in applying regression analysis to first differences of series. For example, even if the wholesale price and the retail price of a commodity, each regarded separately, behave in a manner indistinguishable from the wandering series, it might still be demonstrated that the relation between the two could not be best explained on the basis of chance. Here again I wish to defend Professor Kendall from those who might possibly misinterpret his own conclusions.

I fear that, after all, in what I have said I have failed to attack and have merely rushed in to defend the author against his more enthusiastic admirers, who may overstate some of his more spectacular claims and suggest that we should all study the correlograms of individual time series instead of the equations connecting them. This cannot be Professor Kendall's intention: indeed he has foreshadowed a further paper on the relations between price series and production series, to which we shall look forward with keen interest.

Mr. QUENOUILLE: Professor Kendall's paper is, as other speakers have indicated, a provocative one. Professor Kendall has dealt with such a wide diversity of subjects that, upon first reading this paper, I was at rather a loss as to where to start any investigation of his analysis.

As a first step, I decided to see what happened if the traditional autoregressive series were taken at a somewhat shorter interval than usual. I therefore assumed that the damping factor was $1-\varepsilon$ and that the period was θ, where both ε and θ are small. The first two serial correlations of the differences of this series would then be expected to be of the order

$$-\frac{\varepsilon}{2}+\frac{\theta^2}{2\varepsilon+\theta^2-2\varepsilon\theta}.$$

It is obvious from this formula that if ε and θ are both small, we should expect to have small serial correlations.

My next step was to substitute values for ε and θ in this formula. The first serial correlation coefficient for Kendall's series No. 21 calculated using the figures for the first week of every year was 0·66, and, since this was not far from the corresponding correlation coefficient of Kendall's artificial series $u_n - 1{\cdot}1\, u_{n-1} + 0{\cdot}5\, u_{n-2} = \varepsilon_n$, I decided to use the latter series in the further analysis. If this series were observed at monthly intervals, i.e., at one-twelfth the interval normally used, the damping factor would be 0·97 and θ would be 0·055 roughly. With these two values, the first serial correlation works out at 0·04. For narrower spacing, at weekly intervals, it works out at 0·01.

These results come out much as might be expected. Basically, the use of first differences of the original series gives rise to quantities which are proportional to the second differences of the correlogram. If the interval is small, these differences are correspondingly small, and so are the serial correlations. It is impossible to distinguish between different types of correlogram when we are working only with second differences of values from near the origin. It is not surprising that Professor Kendall concludes that the data behave almost like wandering series.

The above formula also throws some light on what happens when autoregressive schemes with the same random element are added together. It is fairly easy to demonstrate that this gives rise to an autoregressive scheme of higher order. Thus, the addition of two Markoff processes gives rise to a non-oscillatory second-order process (one for which θ^2 is negative). These higher processes will usually have constants such as θ, which will tend to increase the serial correlation coefficients of their differences. To put it simply, the increased complexity of the correlogram will increase the second differences. These results seem to indicate that Professor Kendall's conclusions are similar to those that would be reached by taking the usual autoregressive schemes at narrower intervals.

Concerning Professor Kendall's method of taking first differences, I remember when I first started at Rothamsted being shown a set of data of catches of insects in a light trap which was said to have been analysed by Professor Fisher. The method of analysis involved the use of differences between successive catches. It is interesting that this method is coming into use in the economic field to an increasing extent. Dr. Orcutt has been advocating it for some time, and he deserves some acknowledgment for his foresight. It is a very useful method, but it is necessary to beware of its limitations.

Obviously, it is restricted to series for which the serial correlations are high, so that if we carry out too complicated an analysis with the first differences there is a danger that we shall get to a position where we are, in effect, dealing with differences at a wider spacing, and our analysis may consequently be invalid. In general it will be valid, although there will be a limitation placed upon the narrowness of spacing by the errors of observation (in the simplest case by the errors of rounding-off) which will, if the observations are too close, swamp the normal random element. Also we shall run into problems of lag correlation for narrower intervals of spacing; for while it may be reasonable to assume that two events are simultaneous when our observations are yearly averages, it will often be altogether unreasonable when they are daily or weekly figures.

As to methods of dealing with trends in time series, I agree that it is difficult to distinguish between time trends and highly-correlated stochastic elements, but I do not think the position is as bad as Professor Kendall indicates. There are several rough and ready methods for use in this connection. One of these fits polynomials, examines the residuals in the light of the fitting, re-examines the fitting, and so on. A second method involves the use of moving averages (and includes the use of first differences as a special case) and the correction, if necessary, of the residuals to allow for the effect of the moving average. This is rather difficult, since initially it is desirable to maintain an open mind upon the form of average required to eliminate the trend. Yet another method employs the use of the spectrum of a series to discover its stochastic nature. This can be done, since trend is removed with the components of low periodicity.

We have had a stimulating paper, and the subject has benefited from the careful investigation carried out by Professor Kendall.

Mr. S. J. Prais: I wish to make three points. The first is that the reader must be grateful for the opportunity of examining analyses of really long economic time-series in which large sample theory can be used without the usual qualms. The use of an electronic computer to carry out the heavy numerical work is also noteworthy, and I wish to add my appreciation to that of earlier speakers.

Secondly, there is the question of the relevance of the type of autoregressive equation, the parameters of which Professor Kendall has been trying to estimate, to the type of dynamic theory economists usually deal with.

A dynamic economic system consists of a number of equations expressing relationships (generally demand or supply relationships) between economic variables, some of which at least involve time in an essential way in the form of time derivatives or time integrals. Since observations of economic phenomena are well separated in time, these relationships may without any considerable loss of generality be expressed in the form of difference equations. In a matrix notation such a complete set of equations may be written as

$$\mathbf{B}\,\mathbf{y} = \boldsymbol{\epsilon}, \qquad (1)$$

where **y** is a vector of n variables and **B** is a square matrix, the elements of which are polynomials in the delaying operator E^{-i}. The relationships are not supposed to be exact, so there is written on the right-hand side of the equation a vector of random variables ε with means zero. It is this type of economic relationship that economists usually investigate.

Now, in the usual way, if

$$\boldsymbol{\epsilon} = \mathbf{o}, \qquad (2)$$

the path in time of every economic variable in the system would be given by the same characteristic equation,

$$|\,\mathbf{B}\,|\; y = o, \qquad (3)$$

where $|\mathbf{B}|$ is obtained by expanding the coefficients of the system (1) as a determinant, and (3) may be written in the form of a difference equation

$$y_t = a_1 y_{t-1} + a_2 y_{t-2} + \ldots a_p y_{t-p}. \qquad (4)$$

Since in these circumstances the same difference equation will apply to all variables it is possible to investigate the dynamic properties of the system by examining the fluctuations of any one variable over time, no matter which variable is chosen. It is assumed that an argument of this kind lies at the base of Professor Kendall's investigations.

If, however, the equations of the system (1) are not exact, the autoregressive equation of the system will also not be exact. Further, each variable will have a different path over time; for (1) may then be "solved" for each variable in the form

$$|\mathbf{B}|\; y_i = \mathbf{B}'_i \boldsymbol{\epsilon} = \eta_i, \qquad (5)$$

say, where $\mathbf{B}_i$ is the vector of co-factors of the ith column of $\mathbf{B}$. While the variances of the residuals of the behaviour equations (1) may be small so that good dynamic theory is possible, the variance of the residual of (5) may be very large; for, if the variance matrix of the $\boldsymbol{\epsilon}$ is $\sigma^2 I$, the variance of η_i is given by

$$V(\eta_i) = \mathscr{E}\{\eta_i^2\} = \mathscr{E}\{\mathbf{B}'_i \boldsymbol{\epsilon}\boldsymbol{\epsilon}' \mathbf{B}_i\} = \sigma^2(\mathbf{B}'_i \mathbf{B}_i), \qquad (6)$$

and the increase in variance depends on the size of the cofactors $\mathbf{B}_i$ or, generally, on the degree of ill-conditioning of the system.

As a simple example a Keynesian-type system was considered:

$$C = \alpha E^{-1} Y + \varepsilon_1 \qquad (7a)$$

$$I = \beta E^{-1} Y + \gamma E^{-2} Y + \varepsilon_2 \qquad (7b)$$

$$Y = C + I, \qquad (7c)$$

where C, I and Y are consumption, investment and income; and α, β and γ are parameters. The autoregressive equation for consumption is

$$[1 - (\alpha + \beta) E^{-1} - \gamma E^{-2}]\, C = [1 - (\beta E^{-1} + \gamma E^{-2})]\, \varepsilon_1 + \alpha E^{-1} \varepsilon_2 . \qquad (8)$$

The variance of the expression on the right-hand side of (8) on the assumptions that $\mathscr{E}(\varepsilon_1^2) = \mathscr{E}(\varepsilon_2^2) = \sigma^2$ and $\mathscr{E}(\varepsilon_1\varepsilon_2) = 0$ is

$$V(\eta) = (1 + \alpha^2 + \beta^2 + \gamma^2)\, \sigma^2, \qquad (9)$$

which is always greater, and may be much greater, than σ^2. It may be noted that the values of η are not serially independent and not independent of lagged values of consumption.

It is, of course, true that the auto-correlations of an economic variable may be of negligible size if they are not generated by a dynamic system. But it follows from the above argument that, even if the dynamic behaviour relationships are well established, the auto-correlations of the variables will be weak on account of the increase in the residual variance of the transformed system. It may therefore be concluded that Professor Kendall's investigations of auto-correlations cannot in principle throw any light on the possibility of estimating the kind of dynamic economic relationships in which economists are usually interested.

The argument so far has been concerned with formal matters. My third and final point is concerned with the material nature of the markets investigated. These are share and commodity markets which are the best examples of markets that are dynamically perfect. That is, any expected future changes in the demand or supply conditions are already taken into account by the price ruling in the market as a result of the activities of hedgers and speculators. There is, therefore, no reason to expect changes in prices this week to be correlated with changes next week; the only reason why prices ever change is in response to unexpected changes in the rest of the economy; but when prices do change, as shown by Professor Kendall's calculations in Table 5, they all tend to change together.

In formal terms this could be expressed by saying that the system of behaviour relationships linking the stock market to the rest of the economy was ill-conditioned, and from the point of view of investigating the dynamic properties of the system it is therefore particularly unfortunate that Professor Kendall found it necessary to choose these markets for his investigations.

Professor F. Paish: I speak with some trepidation in such an assembly, but there are two things I should like to say. I am extremely grateful to Professor Kendall for his conclusion.

As to his arguments I am not competent to judge. Many people in this world have spent many years trying to forecast what the stock markets are going to do on the basis of what they have already done. I always thought it futile, and am glad Professor Kendall has confirmed my opinion. It seems inevitable that where prices are based on expectation markets are as likely to go down as up. If the markets thought they were more likely to go up they would have gone up already. For this reason when there is option dealing, one always finds that the price of a double option is twice that of a single. Nevertheless, I should have thought that there was a possibility that some rather more complicated patterns might be discernible. The market depends on two different sorts of people—speculators and investors—or, in produce markets, hedgers. The former come in quickly but cannot stay in for long. The latter come in more slowly but stay. If the investors do not come in or do not come in as much as expected, the speculators get tired and have to sell. I should have expected some slight negative correlation after a lapse of some weeks. When the market has gone up sharply the initial rise is often followed by a fall as the speculators get tired of waiting for the investors to come in and take over. There are possibly patterns of that sort which could be distinguished in the data by somebody competent to do so.

Professor BARTLETT: I think that all I wanted to say has effectively already been said, but I should like to reiterate what seems to me to be the fundamental point. Professor Kendall in stressing the apparent randomness of his series has raised a paradox, but perhaps it may largely be resolved as a matter of interpretation. Incidentally, even if something is *apparently* random it does not necessarily mean that it *is* so; it may depend on something of which one is not aware. I have heard, for example, of a cotton firm in Manchester which in the good old days used to base its prices of cotton goods largely on the previous year's rainfall in Morocco, because they had a big export to Morocco, and if the rainfall there had been good the people would have a good harvest and could afford to pay high prices. If the annual rainfall were random, so would the annual prices appear so.

But the main point that one asks is: what are the chief characteristics of the data which Professor Kendall has been discussing? One is really not in a position to say from his paper because, as he has said, he is not able to quote the whole of the data, and if one works merely with differences one may be missing much of importance. It is true that the particular problem of predicting prices on the Stock Exchange can be discussed on the basis of the correlations of weekly differences, but these do not complete the summary of the variability and characteristics of the data, as previous speakers, and Professor Kendall himself, are aware. Mr. Quenouille has referred to second-order autoregressive series, so I will content myself with even simpler examples. Thus the price might be rising steadily, and in one sense be perfectly predictable; or we might have a stationary Markoff series in which the random residual became a negligible fraction of the total variance. (I do not think the total variance of Professor Kendall's series is given.)

I had the impression from the size of the cross-correlations in the paper that there may be more of a pattern than appears from the autocorrelations. It may be worth noting that if two autoregressive series are cross-correlated through a cross-correlation between the contemporaneous residuals, this cross-correlation will be preserved between the differences, even if the auto-correlations have been drastically reduced by differencing.

On a minor point, I was not altogether clear whether the trend-fitting at the end was on first differences, or on the original series.

Professor KENDALL: They were on the original series.

Professor BARTLETT: Yes, but the same notation was, I think, earlier used for first differences. If so, this might be made clearer.

Dr. K. S. RAO: Professor Kendall's criticism of the economist and especially the trade-cyclist is perhaps not of very wide generality. Large Sample methods of treatment of economic time-series, however attractive from the point of view of statistical theory, are difficult to justify from the point of view of economic theory, which such methods are intended to help.

A frequency distribution of a time-series of large length and the smallness of the first few serial correlation coefficients may not necessarily rule out systematic models, as can be seen by the following example, which is possibly one of a set that can be constructed.

Let m be an odd prime number and l an integer not divisible by m. Let n be a "large" integer. Consider a "time-series" of $N = n(m - 1)$ terms defined as follows:
Let

$$\theta = 2\pi\, l/m \quad . \quad . \quad . \quad . \quad . \quad . \quad . \quad . \quad (1)$$

and

$$u_t = \frac{a}{\sqrt{m}} \sin \theta \left[\frac{n+t}{n}\right] + a \sin \theta \left[\frac{n+t}{n}\right] t, \quad . \quad . \quad . \quad . \quad . \quad (2)$$

for $t = 1, 2, \ldots N$, where $[x]$ denotes the greatest integer in x and a any given constant.

The Series in (2) is formed by using only the sine function. It can be noticed that, when we form a relative frequency table, we get a symmetric distribution with zero mean. Purely for purposes of simpler algebra, if we take n to be divisible by m, then

$$\bar{u} = \frac{1}{N} \sum_{t=1}^{N} u_t = 0. \quad . \quad . \quad . \quad . \quad . \quad . \quad . \quad . \quad (3)$$

$$\text{Var } u_t = \frac{1}{N} \sum_{t=1}^{N} u_t^2 = \frac{a^2}{2} \frac{m}{m-1}. \quad . \quad . \quad . \quad . \quad . \quad . \quad (4)$$

If P_k, for $1 \leqslant k < m$, denotes the kth serial product of the series defined by the expression

$$P_k = \sum_{t=1}^{N-k} u_t u_{t+k}, \quad . \quad . \quad . \quad . \quad . \quad . \quad . \quad (5)$$

after some simplification we get

$$P_k = na^2 \left\{ \sum_{s=1}^{m-1} \left(\frac{1}{m} \sin^2 \theta s + \frac{1}{2} \cos \theta sk \right) \right\} + O(1), \quad . \quad . \quad . \quad . \quad (6)$$

where the constant implied in O depends only on m, a^2 and k. It can be seen that the first term on the right-hand side of (6) vanishes.

Hence, if r_k is the kth serial correlation coefficient of the series in (2), then, for finite m, from (3), (4) and (6), it follows that

$$r_k = O\left(\frac{1}{N}\right) \text{ for } k = 1, 2, \ldots (m-1) \quad . \quad . \quad . \quad . \quad . \quad (7)$$

Professor JEVONS: I am not competent to deal with the mathematics of Professor Kendall's paper, but I feel that the economist is entitled to say something about the assumptions which are subjected to mathematical processes. At one time I paid some attention to time-series, and what became perfectly plain to me was that variations of the price of an individual commodity were perfectly random, and this was more particularly the case with highly organized markets such as the Chicago wheat market or the London Stock Exchange. If the day-to-day fluctuations of these markets are examined they will be found to be due to all sorts of causes—rumours of wars, changes of weather, forecasts from agricultural departments, or again the operations of bulls and bears; so I am not in the least surprised that Professor Kendall did not claim any very significant results from his analysis. I would suggest that if he had taken the time series of, say, agricultural commodities in some backward country, where the market is not properly organized, as, for instance, India or Burma, he would have found many more interesting things in the way of correlation.

Professor Kendall mentioned trade cycles in his paper. I did a little work myself on that subject from 1907 on. I was anxious to see whether statistics would support my father's theory that commercial crises depended on causes coinciding with the sunspot cycle. I found that there are two trade cycles—one a short period of about 3½ years, and the other a longer period which is usually ten years, or may be seven or eleven years. In order to put the idea to proof it seemed to me that we must take, not individual series of prices, but aggregate production—that is to say, if fluctuations in harvests produce any effect upon the activities of trade in general, it is not the result of fluctuations in one or more commodities, but of all the important commercial products. Therefore I took the production figures of the United States as far as available, and took weighted averages of these from year to year—weighted according to their average prices. I found quite remarkable evidence with regard to the shorter trade cycle. I later found, working on long series of statistics, both of prices and trade production, that the true period for the short cycle is 3·4 or 3·3 years. This period, practically one-third of ten years, has since been definitely established and accepted in the United States—the 40-month period, as they call it; and it is pretty obvious that the trade cycle is composed of two or three, or sometimes four, of these short periods, which are themselves fluctuating, probably, on the whole, with greater intensity than the longer periods.

This is a big subject. I should like to suggest that if you aggregate your figures (or sometimes you may take them individually, it depends on your purpose) you will find some interesting results. I only hope that Professor Kendall will continue his investigations along these lines.

The following contributions were received in writing after the meeting:

Mr. H. S. Houthakker: We are all indebted to the author for his work in the development and exposition of statistical methods, and his desire to apply some of these methods himself has much to commend it. Such applications, if based on a sound understanding of the material considered, can be of great benefit both to the subject under investigation and to the mathematical statistics. The qualification is important, for even the most refined methods, in fact especially those, will not lead to valid and interesting results unless they are applied to suitable data within an acceptable theoretical framework.

The evaluation of Professor Kendall's paper by the latter criterion is made difficult by the fact that there is no reference to a theoretical framework anywhere, nor indeed to work of others which the author may have had in mind. In his section 1, which is the only indication of the author's motives in undertaking this study, it is claimed that "it has been customary to analyse an economic time series by extracting from it a . . . trend . . . and then scrutinizing the residual portion for short term oscillatory movements and random fluctuations." This sentence would be correct if it began "it was customary twenty or thirty years ago." From about 1930 the development of econometrics has led to a general conviction among economists that little purpose is served by studying individual time series separately, and that they become interesting only when linked to each other in economically meaningful relations. The models thus constructed may well be of an autoregressive type, but trends or short-term oscillatory movements occupy only a minor place in them. A time series like the one for first differences of wheat prices may appear purely random if considered in isolation, yet regression analysis will soon reveal that the means of these apparently random differences are functions of supply and demand factors. And can there be any doubt that the movements of share prices are connected with changes in dividends and the rate of interest? To ignore these determinants is no more sensible or rewarding than to investigate the statistical properties of the entries in a railway time-table without recognizing that they refer to the arrivals and departures of trains.

It would nevertheless be wrong to dismiss this inquiry as merely the backwash of the Yule tide. Professor Kendall's iconoclastic efforts, even though directed against straw men, point to an interesting conclusion which is almost exactly the opposite of his own conclusion (*c*) in section 3. The series considered all refer to variables which in most econometric models would be regarded as endogenous. In recent statistical work much attention has been given to the complicated nature of the errors in these endogenous variables; it would make our work much easier if they were in fact random, for they could then be treated as if they were predetermined and the model accordingly simplified. However, the incompleteness of the author's analysis makes this conclusion so tentative that it would not yet seem safe to act on it.

Dr. D. R. Cox: The serial correlations in Professor Kendall's Table 3 are all small, but nevertheless the correlograms for the different series behave in a remarkably similar fashion. This can be seen in an elementary way by considering the number of negative coefficients:

Lag	1	2	3	4	5	6	7	8
Number of negative coefficients out of 19 .	1	4	3	13	16	18	16	14

9	10	11	12	13	14	15	16	17	18 . . .
13	10	7	1	6	13	6	0	0	1 . . .

Similarly for the cross-correlations in Table 5 we find

Lag	1	2	3	4	5	6
Number of negative coefficients out of 56	5	10	13	31	45	47

In both cases there is very strong evidence for the existence of a non-random component, presumably of approximately the same structure for all or most of the 19 series. This, of course, does not affect Professor Kendall's point that the correlations are small, and hence useless for short-term prediction.

It is worth pointing out explicitly that the presence of certain types of non-randomness is shown, not by the occurrence of high serial correlations, but by other features of the correlogram, such as the existence of a run of coefficients all with the same sign. For example if the successive differences of price were completely random, price itself would undergo a simple random walk and would not be stationary. If there were in addition some tendency to return to a stable level of price, the correlations of differences would be small and negative (see, for example, Table 2, last column.)

As another example, Mr. W. L. Smith and I, in some recent work to appear in *Biometrika*, have discussed a special time-series arising in neuro-physiology, and have shown that a certain simple completely deterministic time-series, relevant to the neurological problem, is indistinguishable from a random series by the conventional correlogram method.

The problem, raised by Professor Kendall in Section 1, of dealing with long-term and short-term movements in one model arises in an acute form in investigations of irregularity in textile processing, notably in drawing and spinning. Drawing is a process of successive attenuation, each stage introducing an irregularity. In the final time-series the irregularities from the last stage have a characteristic length of a few cm., while the irregularities from the initial stages spread over several hundreds of yards. A method that is useful here is analysis of variance, in which the total variation is divided into variation between and within lengths l, and the coefficient of variation between the means of different sections of length l is plotted against l. This method was suggested by Tippett, and later by Yule under the name of the lambdagram. It is roughly equivalent to integrating the correlogram twice, and so is sensitive to runs of correlations all of the same sign. A related method has been useful in the neurological problem mentioned above.

Mr. A. S. C. Ehrenberg: Whilst it is difficult to describe time-series fully, they seem to possess features of some relevance apart from the first differences discussed by Professor Kendall. For example, the two wheat price series (§§6–19) are measured in cents on a scale with a more or less absolute minimum, and the mean, the scatter and any general trends of the primary data might be of interest. "End-effects" would be expected if the mean is low, and "scaling effects" generally. Thus the threefold increase in variance of the first differences of the weekly wheat series (§§9–11) might be connected with an upward trend of the *primary* data. (The constant mean referred to in §11 is presumably that of the first differences.) Since heteroscedasticity greatly reduces the utility of correlation coefficients, any mathematical transformation to counteract it might simplify the description of the data. Supposing that a transformation suggested by the trend (if any) could be found which gives homoscedasticity, it would of course not necessarily be useful to think of the original change in variance as just a scaling effect. But this might be so if the same transformation were to simplify many different price series, or perhaps if it were closely related to the economists' ideas on "real" prices. Information on such points could, one feels, be given simply enough. The scaling argument, for example, could not be countenanced at all in the *absence* of any upward trend in the primary series.

Some of the relevant information is admittedly contained in the description of the first differences given in the paper, though extracting it may raise more problems than it settles. Thus, there does appear to be a trend, of 69 cents, in the primary weekly wheat prices—entirely after the Great War—whilst in the same data treated as *monthly* instead of *weekly* figures (series 21, §16) the trend is only 28 cents, omitting one value at each extreme (or 43 cents for the whole series). No other deductions can be made about the variability of the primary series, at least not without some wild assumptions about their serial correlations. For example, if one supposed that in fact only the linear trend was striking, and that the second order serial correlation was no more than two-thirds of the first order correlation, the standard deviation would seem to be about five cents or less (and the first order correlation already as high as about ·75). In as far as such a standard deviation would then be anywhere near the truth—this would be likely in view of the rejection of widely outlying values (footnote, §9) of the order of 40 cents—these trends and discrepancies seem striking enough at least to deserve being explained away.

Turning to Professor Kendall's lack of sympathy with aggregative series and their "spurious" serial correlations (e.g., §§3, 23, 28), it may avoid confusion to mention that the correlations are *not* spurious in the sense that they *do* describe such series. Difficulties arise when one wishes to compare different aggregative series, let alone propound any so-called "causal" arguments. In Professor Kendall's illustrative example (§28), the correlation between the aggregative series E and F, each made up of n primaries, depends not only on the intercorrelations, but also on the *number*, of primaries. (Equation 12 may be a little misleading here since the n cor $(\varepsilon_i\varepsilon_i)$ in the denominator are unity.) Thus the correlation between E and F varies from ·1 ($n = 1$) through ·14 ($n = 2$), ·17 ($n = 3$), ·23 ($n = 15$—Kendall's example) to a limit of ·25. The "spurious" element arises because by averaging similar readings the strength of the relationship is affected, and not only the accuracy of its determination. This phenomenon is of course one of the reasons why correlation coefficients are used less and less frequently elsewhere—nothing like it seems to occur with most models of the analysis of variance kind, for example.

It should perhaps be remembered that in aggregating time-series like this, we are not necessarily averaging similar readings in the usual sense (e.g., sampling from an infinite population), since the supply of relevant time-series satisfying the original requirements (§28), that cor $(\varepsilon_i\varepsilon_j) = \cdot 4\ (i \neq j)$

and cor $(\varepsilon_i\eta_j) = \cdot 1$, may be strictly limited. There can, however, be little doubt that when the primary series are available, their serial correlations will be more useful than those of some aggregative series. For example, suppose that one wishes to compare the behaviour of two sets of data and that one of the primary series is missing, or different, in one of the sets of data. Then if the two sets of serial correlations for the aggregative series are identical, the two sets of data cannot be the same, as judged by the serial correlations of the primaries, whilst if these latter are identical, the correlations of the aggregative series would not show it.

Professor KENDALL, in reply: I should like to thank the proposer and seconder for their remarks and all the contributors to a helpful discussion. I made this paper as little mathematical as possible in the expectation that the mathematicians would all agree with me and the economists perhaps disagree. But as it has turned out, the mathematicians are doubtful, and the economists in their conclusions are inclined to agree.

Professor KENDALL subsequently wrote as follows:

With much of the discussion I agree and will confine myself to points of difficulty or disagreement.

(*a*) As Professor Champernowne points out, I am not attacking the use of index numbers for customary purposes, and I agree with points made by him, Professor Bartlett and Mr. Ehrenberg that "spurious" is not a good word to describe the serial correlations found in series of index-numbers. At the same time, it seems to me necessary to proceed very cautiously in the interpretation of serials in aggregative series, especially if they are used to estimate autoregressive periods. The problem is similar to that of measuring correlation in divisible units, and one has to be sure that one is measuring a property of the system itself, not merely a property of aggregation.

(*b*) In his observations on rough and ready methods of trend-determination Mr. Quenouille has, I think, failed to see my point of view. I do not deny that there are methods. What troubles me is whether they have any purpose except a purely descriptive one.

(*c*) Dr. Rao is quite correct in pointing out that one can construct non-random series for which a set of serial correlations vanish. But if we find in practice a set of small serial correlations we are, I think, on safe ground in supposing that successive values are nearly independent, as against the alternative that the generating system consists of a set of harmonic oscillations with very special periods.

(*d*) I look forward to seeing Dr. Cox's work and agree with him except on one point. He calls attention to patterns of signs in the serial correlations which indicate that, although small, the correlations are not haphazard. I think he is probably right, but I do not regard the point as settled beyond all doubt. Certain runs of signs might occur from bias in the estimations or from the fact that they may be serially correlated (by chance) even in random series. I propose to settle this question by constructing correlograms of a series of normal independent variates.

(*e*) Professor Allen, Professor Champernowne, Mr. Prais and particularly Mr. Houthakker all raise the question how far can one profitably analyse time-series without a theoretical model in the background? The general question of model-building in statistics and economics is one which I propose to discuss at length on another occasion. All I need say at this stage is that I think it is futile to lay down any rules as to whether fact should precede theory or theory fact. Scientific discovery proceeds by an alternation of the two. I feel entitled, however, to protest at some of Mr. Houthakker's comments. I could perhaps overlook his description of my work as to some extent Yulean backwash because of the mordancy of the phrase. But his claim that "even the most refined methods . . . will not lead to valid and interesting results unless they are applied to suitable data *within an acceptable theoretical framework*" seems to me the negation of the scientific spirit. I have tried to elicit certain facts about economic series. They may be wrong, but if they are correct they are facts, irrespective of any theoretical framework.

As a result of the ballot taken during the meeting, the candidates named below were elected Fellows of the Society:

Harry Brierley.
Ronald Edgar Brook.
Thomas Peter Browell.
Peter Richard Browning.
Charles Frederick Evinton.
Derek John Finch.
Denys William Humphries.
Michael Harold Kreps.
John Kenneth Lambert.
George Gordon Lilley.
Denis Walter McAllister.
Ronald Macer.
Thomas Alma Perkins.
Robert Brian Sanderson.
Om Chandra Sharma.
Robert John Taylor.
Alan Arthur Walters.
John Wrigley.

Corporate Representative

Peter Kenneth Wiggs, *representing* The Morgan Crucible Co. Ltd.

From *Annals of Eugenics*, Vol. **10**, pp 106–111 (1940)

SOME PROPERTIES OF k-STATISTICS

BY M. G. KENDALL

1. k-statistics were introduced in a fundamental paper by R. A. Fisher (1928). The statistic of order p, written k_p, is defined as being the symmetric function of the independent variate values $x_1, x_2, \ldots, x_n$, whose mean value over all possible samples is κ_p, the pth semi-invariant (or, in Fisher's terminology, cumulant) of the parent distribution. The value of the k-statistics rests chiefly on the relative simplicity which their use imports into certain branches of the theory of sampling. In this note I shall not be concerned with this aspect, but with the demonstration of some mathematical properties which they possess.

2. As a preliminary I consider an operator ∂_1 which, when acting on a moment μ_r, multiplies by r and reduces the order of the moment by unity, i.e.

$$\partial_1 \mu_r = r\mu_{r-1}.$$

The moment here means the moment about any point, but to save complicating the appearance of the formulae I omit the dash usual in such cases.

More generally I define

$$\partial_p \mu_r = r(r-1)(\ldots)(r-p+1)\mu_{r-p},$$

regarding ∂_p as a single operator, not p successive repetitions of ∂_1.

We have

$$\partial_p \mu_p = p!,$$

$$\partial_{p+\alpha} \mu_p = 0, \quad \alpha > 0.$$

The operator ∂_p acting on a product is defined to follow the distributive law, e.g. we have

$$\partial_p(\mu_r \mu_s) = (\partial_p \mu_r)\mu_s + \mu_r(\partial_p \mu_s).$$

Thus

$$\partial_p(\mu_r)^m = m\mu_r^{m-1}\partial_p \mu_r$$

$$= \frac{\partial(\mu_r)^m}{\partial \mu_r}\partial_p \mu_r.$$

It follows that if f is a polynomial function in the μ's

$$\partial_p f = \frac{\partial f}{\partial \mu_1}\partial_p \mu_1 + \frac{\partial f}{\partial \mu_2}\partial_p \mu_2 + \ldots + \text{etc.},$$

and this also holds if f can be expanded in a series of polynomials in the μ's.

3. Now consider the expression defining the semi-invariants in terms of the moments

$$e^{\kappa_1 t + \ldots + \frac{\kappa_p t^p}{p!} + \ldots} = 1 + \mu_1 t + \ldots + \frac{\mu_p t^p}{p!} + \ldots.$$

On operating on both sides by ∂_p there results

$$\left(e^{\kappa_1 t+\ldots+\frac{\kappa_p t^p}{p!}}\right)\left(\partial_p \kappa_1 t+\ldots+\frac{\partial_p \kappa_p t^p}{p!}+\ldots\right)$$

$$= t^p+\frac{\mu_1}{1!}t^{p+1}+\ldots$$

$$= t^p\left(1+\frac{\mu_1 t}{1!}+\ldots\right),$$

and hence
$$\partial_p \kappa_1 t+\ldots+\frac{\partial_p \kappa_p t^p}{p!}+\ldots = t^p.$$

This is an identity in t, and hence

$$\partial_p \kappa_p = p!,$$

$$\partial_q \kappa_p = 0, \quad q \neq p.$$

Thus

Theorem 1. The operator ∂_p annihilates every semi-invariant except that of the same order, and $\partial_p \kappa_p = p!$.

For example,

$$\kappa_4 = \mu_4 - 4\mu_3\mu_1 - 3\mu_2^2 + 12\mu_2\mu_1^2 - 6\mu_1^4,$$

$$\partial_1\kappa_4 = 4\mu_3 - 4\mu_3 - 12\mu_2\mu_1 - 12\mu_2\mu_1 + 24\mu_1^3 + 24\mu_2\mu_1 - 24\mu_1^3$$
$$= 0,$$

$$\partial_2\kappa_4 = 12\mu_2 - 24\mu_1^2 - 12\mu_2 + 24\mu_1^2$$
$$= 0,$$

$$\partial_3\kappa_4 = 24\mu_1 - 24\mu_1$$
$$= 0,$$

$$\partial_4\kappa_4 = 24$$
$$= 4!$$

4. The explicit expressions for the k's in terms of the symmetric functions of $x_1, x_2, \ldots, x_n$ were given by Fisher as the solution of an operational equation. They may also be obtained directly as follows:

k_p is of degree p and may therefore be written

$$k_p = SS(x_1^{p_1} x_2^{p_1} \ldots x_{\pi_1}^{p_1} x_{\pi_1+1}^{p_2} \ldots x_{\pi_1+\pi_2}^{p_2} \ldots x_{\pi_1+\pi_2+\ldots\pi_s}^{p_s}) \times A(p_1^{\pi_1} \ldots p_s^{\pi_s}), \qquad \ldots\ldots(1)$$

where the second summation takes place over all possible ways (including permutations) of assigning the $(\pi_1+\pi_2+\ldots+\pi_s)$ subscripts from the n available, and the first summation extends over partitions of the number p, $(p_1^{\pi_1} p_2^{\pi_2} \ldots p_s^{\pi_s})$. $A(p_1^{\pi_1} \ldots p_s^{\pi_s})$ is a number depending on the partitions but not on the x's.

We have
$$p_1\pi_1+\ldots+p_s\pi_s = p, \quad \ldots\ldots(2)$$
and define ρ by
$$\pi_1+\ldots+\pi_s = \rho. \quad \ldots\ldots(3)$$

On taking mean values of (1) we get
$$\kappa_p = S\{(\mu_{p_1}^{\pi_1}\mu_{p_2}^{\pi_2}\ldots\mu_{p_s}^{\pi_s})\,AB\}, \quad \ldots\ldots(4)$$
where B is the number of ways of picking out the ρ subscripts from n, permutations being allowed, and is therefore equal to $n(n-1)\ldots(n-\rho+1)$.

Now, from the equation defining the κ's we have
$$\kappa_1 t+\ldots+\frac{\kappa_p t^p}{p!}+\ldots = \log\left(1+\mu_1 t+\ldots+\frac{\mu_p t^p}{p!}+\ldots\right),$$
and hence, picking out the coefficient of t^p, the well-known expression
$$\kappa_p = p!\,S\frac{(\mu_{p_1}^{\pi_1}\ldots\mu_{p_s}^{\pi_s})}{(p_1!)^{\pi_1}\ldots(p_s!)^{\pi_s}}\cdot\frac{(-1)^{\rho-1}(\rho-1)!}{\pi_1!\ldots\pi_s!}, \quad \ldots\ldots(5)$$
the summation extending over all partitions subject to (2) and (3). On identifying corresponding terms in (4) and (5) we find the values of the A's, and on substituting in (1) we get
$$k_p = p!\,S\frac{(-1)^{\rho-1}(\rho-1)!}{n(n-1)\ldots(n-\rho+1)(p_1!)^{\pi_1}\ldots(p_s!)^{\pi_s}\pi_1!\ldots\pi_s!}S(x_1^{p_1}\ldots x_\rho^{p_s}). \quad \ldots\ldots(6)$$

5. If the mean value of a rational integral symmetric function of the variables x is zero then the function must be identically zero; for otherwise the mean value would yield a syzygy between moments, or equate certain moments to zero, which is impossible, no moment being expressible in terms of other moments or a constant for all distributions.

Consider now
$$k_p(x_1+h, x_2+h, \ldots, x_n+h) = k_p+\frac{h}{1!}D_1 k_p+\frac{h^2}{2!}D_1^2 k_p+\ldots,$$
D_1 standing for the operator
$$\frac{\partial}{\partial x_1}+\frac{\partial}{\partial x_2}+\ldots+\frac{\partial}{\partial x_n}.$$

Taking mean values of both sides we have, since κ_p is independent of a change of origin,
$$\kappa_p = \kappa_p+E\left(\frac{h}{1!}D_1 k_p+\ldots\right),$$
E denoting the mean value, and thus the mean value of $D_1^r k_p$ is zero. Hence

Theorem 2. The operator D_1, and *a fortiori* the operator D_1^r annihilates every k_p except in the case $D_1 k_1 = 1$.

It follows that
$$k_p(x_1+h, x_2+h, \ldots, x_n+h) = k_p, \quad p>1,$$
and thus

Theorem 3. The k's are invariant under a change of origin, except k_1.

6. If s_p stands for the symmetric sum $x_1^p + x_2^p + \ldots + x_n^p$ we have the known theorem that rational integral algebraic symmetric functions of the x's can be expressed uniquely in terms of the s's. From (6) it follows that the k's can also be so expressed. Conversely the s's can be expressed uniquely as rational integral algebraic functions of the k's. For we have

$$k_1 = \frac{1}{n} s_1,$$

$$k_2 = \frac{1}{n-1}\left(s_2 - \frac{1}{n} s_1^2\right),$$

and generally $\qquad k_p =$ a term in $s_p +$ terms containing lower s's.

It is clear that s_p can be expressed as a rational function of the k's if s_{p-1} and lower order s's can; and since s_1 can, the result is generally true. Hence

Theorem 4. Any rational integral algebraic function of the x's can be expressed uniquely in terms of the k's.

The expression is unique because otherwise the difference of two different expressions would give a syzygy in the k's. The mean value of this would give a syzygy in the moments of the parent which did not vanish identically. For the same reason we have

Theorem 5. The k-statistics are asyzygetic.

7. In the paper under reference Fisher proved the following remarkable result: writing K_p for the same function of the differential operators $\frac{\partial}{\partial x_1}, \ldots, \frac{\partial}{\partial x_n}$, as k_p is of $x_1, \ldots, x_n$

$$K_p s_p = p!,$$

$$K_p(s_{p_1} s_{p_2} \ldots s_{p_m}) = 0,$$

where $(p_1 \ldots p_m)$ is any partition of p.

I now prove a theorem which bears a sort of reciprocal relation to this result. I write S_p for the same function of the operators $\partial/\partial x$ as s_p is of x, and generally

$$S_{p_1 \ldots p_m} = S\left(\frac{\partial^{p_1}}{\partial x_1^{p_1}} \cdots \frac{\partial^{p_m}}{\partial x_m^{p_m}}\right),$$

the summation extending over the possible ways of choosing the m suffixes of the x's. Consider the function $k_p(x_1 + h, x_2, \ldots, x_n)$ which may be written symbolically $e^{h\theta} k_p$, where $\theta = \partial/\partial x_1$. On looking at a typical term in equation (6) we see that the effect on $k_p(x_1, \ldots, x_n)$ of increasing x_1 by h is to introduce terms like $(x_1 + h)^{p_1} x_2^{p_2} \ldots x_\rho^{p_s}$, $n(n-1) \ldots (n - \rho + 1)$ in number.

When mean values are taken this results in terms like

$$\mu'_{p_1} \mu_{p_1}^{\pi_1 - 1} \mu_{p_2}^{\pi_2} \ldots \mu_{p_s}^{\pi_s}, \qquad \ldots\ldots(7)$$

where μ'_{p_1} is now the moment about a new origin h to the left of the old one.

We recall that
$$\mu_p' = \mu_p + h\binom{p}{1}\mu_{p-1} + \dots$$
$$= \left(1 + h\partial_1 + \frac{h^2}{2!}\partial_2 + \dots\right)\mu_p,$$

which we may write $e^{h\partial}\mu_p$, remembering that the coefficient of $h^p/p!$ is ∂_p, not $(\partial_1)^p$.

When we make substitutions of this kind in (7) and sum to get the mean value of $k_p(x_1 + h_1 x_2, \dots, x_n)$, the terms are multiplied by the number of ways of picking $(\rho - 1)$ from $(n-1)$, i.e. by $(n-1)(n-2)\dots(n-\rho+1)$; that is to say, terms containing h are $1/n$ of the original, except for the first, which is unchanged. Thus the results of operating by $e^{h\theta}$ and then taking mean values is equivalent to taking mean values and then operating by

$$1 + \frac{1}{n}(e^{h\partial} - 1).$$

Hence
$$E\{k_p(x_1 + h, x_2, \dots, x_n)\} = \left\{1 + \frac{1}{n}(e^{h\partial} - 1)\right\} E(k_p)$$
$$= \left\{1 + \frac{1}{n}(e^{h\partial} - 1)\right\} \kappa_p$$
$$= \kappa_p + \frac{h^p}{n},$$

in virtue of Theorem 1, and thus

Theorem 6. The mean value of $k_p(x_1 + h_1, x_2, \dots, x_n)$ is $\kappa_p + h^p/n$.

Now consider

$$k_p(x_1 + h, x_2, \dots, x_n) + k_p(x_1, x_2 + h, \dots, x_n) + \dots + k_p(x_1, x_2, \dots, x_n + h), \quad \dots\dots(8)$$

which may be written

$$nk_p + hS_1 k_p + \frac{h^2}{2!} S_2 k_p + \dots + \frac{h^p}{p!} S_p k_p. \quad \dots\dots(9)$$

Taking mean values we have, in virtue of the symmetry of k_p, that the mean value of (8) is n times the mean value of $k_p(x_1 + h, x_2, \dots, x_n)$ and is therefore $n\kappa_p + h^p$. Hence, from (9),

Theorem 7.
$$S_p k_p = p!,$$
$$S_q k_p = 0, \quad q \neq p.$$

Now any symmetric operator $S_{q_1 q_2 \dots q_m}$ of total degree q can be expressed as sum of products of the operators $s_1, s_2, \dots, s_q$ and hence such an operator will annihilate any k_p unless $q = p$. Thus we have

Theorem 8.
$$S_{q_1 \dots q_m} k_p = 0, \quad q_1 + \dots + q_m \neq p$$
$$= Ap!, \quad q_1 + \dots + q_m = p,$$

where A is the coefficient of S_p in $S_{q_1 \dots q_m}$.

8. The use of the S operators yields some interesting results about the coefficients of the expansion of k's in terms of s's and *vice versa*. For instance, in the expression of k_p in terms of the s's, operate by S_p. The effect on any term $(s_{p_1} s_{p_2}, \ldots, s_{p_m})$ is to yield np! Hence

Theorem 9. In the expression of k_p in terms of the sums s the sum of the coefficients is $1/n$.

Considering similarly the operation of S_p on the expression of s_p in terms of the k's, we have

Theorem 10. If

$$s_p = A_p k_p + A_{p-1,1} k_{p-1} k_1 + \ldots A_{p_1 p_2 \ldots p_m} k_{p_1} k_{p_2 \ldots} k_{p_m} + \ldots,$$

then

$$1 = \frac{A_p}{n} + \frac{A_{p-1,1}}{n^2} + \ldots \frac{A_{p_1 p_2 \ldots p_m}}{n^m} + \ldots.$$

Operation by the S's is also a useful check on the expression of symmetric functions in terms of the k's, which is usually the first stage in a sampling inquiry into the moments of distributions of those functions.

REFERENCE

R. A. Fisher (1928). "Moments and product-moments of sampling distributions." *Proc. Lond. Math. Soc.* **30**, 199.

From *Annals of Eugenics*, Vol. **10**, pp 215–222 (1940)

PROOF OF FISHER'S RULES FOR ASCERTAINING THE SAMPLING SEMI-INVARIANTS OF k-STATISTICS

By M. G. KENDALL

1. In the original paper (1928) in which he introduced k-statistics, R. A. Fisher enunciated the rules by which their sampling semi-invariants could be determined and briefly indicated the lines of a proof of the validity of these rules. Fisher's proof has been found difficult to understand, and there even seems to be doubt whether it is complete. For example, Sukhatme (1938) works through a particular case in great detail and concludes: "The whole analysis supplies an empirical proof [? confirmation] of Fisher's procedure.... It should materially help to produce a rigorous proof of his combinatorial methods."

In view of the importance of k-statistics in the present theory of sampling, a full and rigorous proof of the validity of the rules is an urgent necessity; and the purpose of this paper is to give one.

2. The k-statistic of order p, k_p, is defined as the symmetric (rational, integral, algebraic) function of the independent variate values $x_1, x_2, x_3, \ldots, x_n$, whose mean value over all possible samples is κ_p, the pth semi-invariant of the parent distribution. Explicitly,

$$k_p = p!\,S\frac{(-1)^{\rho-1}(\rho-1)!}{n(n-1)\ldots(n-\rho+1)}\,S\frac{(x_1^{p_1}x_2^{p_1}\ldots x_{\pi_1}^{p_1}\ldots x_{\pi_1+\pi_2}^{p_2}\ldots x_\rho^{p_s})}{(p_1!)^{\pi_1}(p_2!)^{\pi_2}\ldots(p_s!)^{\pi_s}\pi_1!\ldots\pi_s!}, \qquad \ldots\ldots(1)$$

where

$$p_1\pi_1+p_2\pi_2+\ldots+p_s\pi_s = p, \qquad \ldots\ldots(2)$$

$$\pi_1+\pi_2+\ldots+\pi_s = \rho, \qquad \ldots\ldots(3)$$

the first summation extending over all partitions $(p_1^{\pi_1}p_2^{\pi_2}\ldots p_s^{\pi_s})$ of the number p and the second summation extending over all the possible ways of assigning the ρ subscripts from n available, permutations included.

To any partition $(a_1^{\alpha_1}a_2^{\alpha_2}\ldots)$ of the number a there will correspond a moment $\mu(a_1^{\alpha_1}a_2^{\alpha_2}\ldots)$ which is the mean value of $(k_{a_1}^{\alpha_1}k_{a_2}^{\alpha_2}\ldots)$; and to this moment where will correspond a semi-invariant $\kappa(a_1^{\alpha_1}a_2^{\alpha_2}\ldots)$. The problem is to express this semi-invariant in terms of those of the parent population. Evidently the k-product $(k_{a_1}^{\alpha_1}k_{a_2}^{\alpha_2}\ldots)$ is homogeneous and of total degree a in the x's. Hence, when mean values are taken $\mu(a_1^{\alpha_1}a_2^{\alpha_2}\ldots)$ will be homogeneous and of total order a in the parent μ's (that is, the sum of the orders of the moments in each term will be a). Since the κ's are themselves of homogeneous order in the μ's, it follows that $\kappa(a_1^{\alpha_1}a_2^{\alpha_2}\ldots)$ is of homogeneous order in the κ's. Hence we get

Rule 1*. $\kappa(a_1^{\alpha_1}a_2^{\alpha_2}\ldots)$ consists of the sum of terms each of which, except for constants, is a product of parent κ's of total order a.

We may write
$$\mu(a_1^{\alpha_1}a_2^{\alpha_2}\ldots) = S\{A(\kappa_{b_1}^{\beta_1}\kappa_{b_2}^{\beta_2}\ldots)\}, \qquad \ldots\ldots(4)$$
where the A's are dependent on numerical factors and n.

* My numbering and arrangement of these rules is adapted to the present treatment, and differs from Fisher's.

3. In a previous paper (1940) I introduced an operator ∂_p defined by

$$\partial_p \mu_r = r(r-1)\dots(r-p+1)\mu_{r-p}, \quad r>p,$$
$$\partial_p \mu_p = p!,$$
$$\partial_p \mu_r = 0, \quad r<p,$$
$$\partial_p(AB) = (\partial_p A)B + A(\partial_p B),$$

and showed that

$$\partial_p \kappa_p = p!,$$
$$\partial_p \kappa_q = 0, \quad p \neq q.$$

Now consider the result of operating on both sides of equation (4) by $(\partial_{b_1}^{\beta_1} \partial_{b_2}^{\beta_2}\dots)$. Every term on the right is annihilated except that in $(\kappa_{b_1}^{\beta_1} \kappa_{b_2}^{\beta_2}\dots)$, and the result of the operation on this term is to yield $(b_1!)^{\beta_1}(b_2!)^{\beta_2}\dots\beta_1!\beta_2!\dots$ times the A-coefficient which it is our object to find. Thus we have

$$A = \frac{\partial_{b_1}^{\beta_1}\partial_{b_2}^{\beta_2}\dots}{(b_1!)^{\beta_1}(b_2!)^{\beta_2}\dots\beta_1!\beta_2!\dots}\,\mu(a_1^{\alpha_1}a_2^{\alpha_2}\dots). \qquad \dots\dots(5)$$

I now consider a corresponding operator θ_p which, when acting on a power of x of degree p, multiplies by $r(r-1)\dots(r-p)$ and reduces the exponent by p. More generally the operator acting on a product of x's is distributive, e.g.

$$\theta_p(x_1^l x_2^m) = l(l-1)\dots(l-p+1)\,x_1^{l-p}x_2^m + m(m-1)\dots(m-p+1)\,x_1^l x_2^{m-p}.$$

Considering now $\mu(a_1^{\alpha_1}a_2^{\alpha_2}\dots)$ as the mean value of $(k_{a_1}^{\alpha_1}k_{a_2}^{\alpha_2}\dots)$, we see that the result of operating by the ∂'s on the mean value is the same as that given by taking the mean value of

$$\frac{\theta_{b_1}^{\beta_1}\theta_{b_2}^{\beta_2}\dots}{(b_1!)^{\beta_1}(b_2!)^{\beta_2}\dots\beta_1!\beta_2!\dots}\,(k_{a_1}^{\alpha_1}k_{a_2}^{\alpha_2}\dots). \qquad \dots\dots(6)$$

But the result of (6) is a constant and is therefore equal to its mean value. Consequently A is equal to the result of the operation (6).

4. Now consider the two-way array

$$\begin{array}{cccc|c} & & & & b_1 \\ & & & & b_1 \\ & & & & \vdots \\ & & & & b_2 \\ & & & & \vdots \\ \hline a_1 & a_1 & \dots & a_2 \;\dots & a \end{array} \qquad \dots\dots(7)$$

where there is a row corresponding to every θ in the operator of (6) and a column corresponding to every k in the operand. Suppose we fill the body of the array by numbers whose sums in column and row are the relevant a and b numbers. A little reflexion will show that there is one such array for every term in (6) which does not vanish by operation, and that every term in (6) has its corresponding array (7). The numbers in the body of the array are the powers of x occurring in the k-product; added horizontally they compose the orders of the operators; added vertically they compose the orders of the corresponding k's. A completed array (7) is, so to speak, a chart of part of the operation; and the rules with which we are concerned are based on the use of this array to evaluate the operation (6).

5. The operation (6) gives us the coefficients in $\mu(a_1^{\alpha_1} a_2^{\alpha_2} \ldots)$ but we wish to find those in the corresponding κ. The simplicity imported into the sampling formulae is well illustrated by the rule which makes the necessary allowance.

Rule 2. The coefficient of $(\kappa_{b_1}^{\beta_1} \kappa_{b_2}^{\beta_2} \ldots)$ in $\kappa(a_1^{\alpha_1} a_2^{\alpha_2} \ldots)$ is given by the sum of all terms obtained by completing an array of type (7), ignoring those in which the non-zero terms in the body of the table are resolvable into two or more parts each confined to separate rows and columns.

The relation between multivariate moments and semi-invariants may be expressed by

$$S\left\{\kappa(a_1^{\alpha_1} a_2^{\alpha_2} \ldots) \frac{t_{a_1}^{\alpha_1}}{\alpha_1!} \frac{t_{a_2}^{\alpha_2}}{\alpha_2!} \ldots\right\} = \log S\left\{\mu(a_1^{\alpha_1} a_2^{\alpha_2} \ldots) \frac{t_{a_1}^{\alpha_1}}{\alpha_1!} \frac{t_{a_2}^{\alpha_2}}{\alpha_2!} \ldots\right\}. \qquad \ldots\ldots(8)$$

$\kappa(a_1^{\alpha_1} a_2^{\alpha_2} \ldots)$ is thus the sum of terms composed of the product of one, two, three... moments. The first term is $\mu(a_1^{\alpha_1} a_2^{\alpha_2} \ldots)$ itself. Consider a two-part term such as $\mu(a_1^{\alpha'_1} \ldots)$ $\mu(a_1^{\alpha''_1} \ldots)$ where $\alpha'_1 + \alpha''_1 = \alpha_1$, etc. Its coefficient in the expansion on the right-hand side of (8) is

$$-\frac{1}{2} \frac{2!}{1!\,1!} \frac{t_{a_1}^{\alpha_1}}{\alpha'_1!\,\alpha''_1!} \frac{t_{a_2}^{\alpha_2}}{\alpha'_2!\,\alpha''_2!} \cdots$$

and hence the coefficient with which it appears in the formula for $\kappa(a_1^{\alpha_1} a_2^{\alpha_2} \ldots)$ is

$$-\frac{\alpha_1!}{\alpha'_1!\,\alpha''_1!} \frac{\alpha_2!}{\alpha'_2!\,\alpha''_2!} \cdots. \qquad \ldots\ldots(9)$$

Now $\mu(a_1^{\alpha'_1} a_2^{\alpha'_2} \ldots)$ will have an array of type (7) with column totals $(a_1^{\alpha'_1} a_2^{\alpha'_2} \ldots)$ and row totals say $(b_1^{\beta'_1} b_2^{\beta'_2} \ldots)$; and similarly for $\mu(a_1^{\alpha''_1} a_2^{\alpha''_2} \ldots)$. Provided that $\beta'_1 + \beta''_1 = \beta_1$, these arrays will correspond to terms in the k's which, when multiplied, will give a term in $(\kappa_{b_1}^{\beta_1} \kappa_{b_2}^{\beta_2} \ldots)$. Thus the product of these terms may be considered as an array of type (7) with column totals $(a_1^{\alpha_1} a_2^{\alpha_2} \ldots)$ and row totals $(b_1^{\beta_1} b_2^{\beta_2} \ldots)$ and with the body of the table resolvable into two separate blocks. Since there are α_1 columns with the same total a_1, there will be

$$\frac{\alpha_1!}{\alpha'_1!\,\alpha''_1!} \frac{\alpha_2!}{\alpha'_2!\,\alpha''_2!} \cdots$$

products of this type in the expression (7) which gives $\mu(a_1^{\alpha_1} a_2^{\alpha_2} \ldots)$. This factor is the same as (9) but of opposite sign. Hence, if we ignore the separate two-part blocks in the array for μ, we shall have allowed for the products of two moments which, in accordance with (8), must be subtracted from μ to give κ.

Now some of these separate blocks will themselves be separable into two blocks and in subtracting them all from $\mu(a_1^{\alpha_1} a_2^{\alpha_2} \ldots)$ we subtract too much. For example, if there are three separate blocks, L, M, N, we shall, by considering L and $(M+N)$ as two blocks, have subtracted L, M, N. We shall have done the same by considering M and $(L+N)$, and N and $(L+M)$ as two blocks. That is, we have subtracted $2L$, $2M$, $2N$ too much. We must restore these blocks to the array for μ again. Such additions, summed over all blocks of three, will be found to equal the terms in the expansion of (8) which result from the product of three moments.

In restoring these blocks we restore too many of the cases where there are four separate blocks. These must be subtracted again, and correspond to the negative term in (8) involving the product of four moments. Proceeding in this way we establish rule 2.

6. We now come to the rules for evaluating the principal parts of the arrays (7).

Rule 3. Subject to the ignoration of terms enjoined by rule 2, to the coefficient of $(\kappa_{b_1}^{\beta_1}\kappa_{b_2}^{\beta_2}\ldots)$ in $\kappa(a_1^{\alpha_1}a_2^{\alpha_2}\ldots)$ there will be a contribution corresponding to each way of completing the array (7). Some of these contributions may vanish. The others consist of (i) a numerical coefficient multiplied by (ii) a function of n.

Rule 4. The numerical coefficient is the number of ways in which the column totals, considered as composed of separate units, can be allocated to form the array considered.

Rule 5. The function of n, called the pattern function, depends only on the configuration of zeros in the array, not on the actual numbers composing it or on the row and column totals. The function is given by considering the separations of the rows into distinct groups or separates.

(i) With one separate there is associated the number n, with two separates $n(n-1)\ldots$, with q separates $n(n-1)\ldots(n-q+1)$.

(ii) In each separation we count the number of separates in which a particular column is represented by a non-zero entry. If in ρ separates, we assign the factor $\dfrac{(-1)^{\rho-1}(\rho-1)!}{n(n-1)\ldots(n-\rho+1)}$.

(iii) This is done for each column.

(iv) The various factors given by (ii) and (iii) are multiplied together for each separation, multiplied by the factor appropriate under (i) and the results summed to give the pattern function.

These rules are rather difficult to grasp and I recall an example of Fisher's to illustrate them. Suppose we seek the coefficient of $\kappa_6\kappa_2^2$ in $\kappa(4^2\,2)$. One contribution will arise from the array

$$\begin{array}{ccc|c} 2 & 3 & 1 & 6 \\ 1 & 1 & . & 2 \\ 1 & . & 1 & 2 \\ \hline 4 & 4 & 2 & 10 \end{array} \qquad \ldots\ldots(10)$$

Under rule 4, the numerical coefficient is the number of ways of allocating 4, 4, and 2 individuals to form the array. The two units in the first column can be chosen in $4!/(2!\,1!\,1!) = 6$ ways, and that in the second in 4 ways. The units in the last column can be chosen in 2 ways, but it is important to note that either individual can be allocated in the first row, making 4 ways in all. The total coefficient is therefore $6\times4\times4 = 96$.*

* Fisher gives 192 because he considers the two arrays

$$\begin{array}{ccc} 2 & 3 & 1 \\ 1 & 1 & . \\ 1 & . & 1 \end{array} \quad \text{and} \quad \begin{array}{ccc} 3 & 2 & 1 \\ 1 & 1 & . \\ . & 1 & 1 \end{array}$$

as the same. For purposes of calculation this of course is the most convenient, the arrays having the same pattern function and the same numerical coefficient; but for the purposes of formal exposition it is better to keep them distinct.

Under rule 5 the pattern function will depend on the configuration

$$\begin{array}{ccc} \times & \times & \times \\ \times & \times & . \\ \times & . & \times \end{array}$$

where $\times$ stands for a non-zero entry and a period for a zero entry. There are five separations of this, one of one separate, three of two separates, and one of three separates. The contribution from the first is

$$n\frac{1}{n}\frac{1}{n}\frac{1}{n} = \frac{1}{n^2}$$

for each column has a non-zero entry in the separate. The contribution from the three separations given respectively by isolating the first, second and third row will be found to be

$$n(n-1)\left[\frac{-1}{n^3(n-1)^3}+\frac{1}{n^3(n-1)^2}+\frac{1}{n^3(n-1)^2}\right] = \frac{2n-3}{n^2(n-1)^2}.$$

The contribution from the separation of three separates is

$$n(n-1)(n-2)\left[\frac{2!}{n(n-1)(n-2)}\frac{-1}{n(n-1)}\frac{-1}{n(n-1)}\right] = \frac{2}{n^2(n-1)^2}.$$

The pattern function is the sum of these three contributions and is thus $1/(n-1)^2$.

7. Turning now to the proof of rules 3, 4 and 5, which are best taken together, consider the operation typified by an array (7) in which the body of the table has been filled in. To fix the ideas consider the example (10) just given. This array will represent a number of terms in the operation each of which consists of the operation of θ_6 on a term $x^2.x^3.x$ (the first row), θ_2 on $x.x$ (the second row) and so on. Provided that the suffixes of the x's in any row are alike, every suffix of the x's will provide a term, for k_p contains terms with every distribution of powers (adding to p) and suffixes. There will, for instance, be terms of the following kind

$$\begin{array}{ccc} x_1^2 & x_1^3 & x_1 \\ x_2 & x_2 & . \\ x_3 & . & x_3, \end{array} \qquad \begin{array}{ccc} x_1^2 & x_1^3 & x_1 \\ x_1 & x_1 & . \\ x_2 & . & x_2, \end{array} \qquad \begin{array}{ccc} x_1^2 & x_1^3 & x_1 \\ x_1 & x_1 & . \\ x_1 & . & x_1. \end{array}$$

In fact, for any completed array, we have terms in which

(i) all the x's have the same suffix (n in number, one for each suffix),

(ii) all the x's but one row have the same suffix ($n(n-1)$ in number),

(iii) all the x's but two rows have the same suffix and the remaining two are the same ($n(n-1)$ in number),

and so on. These cases correspond to the various separations dealt with in rule 5.

Now in case (i) the term in any column arises from the term in x^p in k_p and (apart from numerical factors which are considered presently) is n^{-1}, from equation (1). Hence any column which contains an entry contributes a factor n^{-1} and the total function of n arising from case (i) is the product of n and of (n^{-1}) to the power of the number of columns containing a non-zero entry.

Similarly in cases (ii) and (iii) the n-function for each separation is the product of $n(n-1)$ and, for each column, a factor in n^{-1} or $-1/n(n-1)$ according as the column contains non-zero entries in one or in both parts of the separation; and so on.

This explains the origin of the pattern function as described in rule 5. But in order to establish that rule completely (and incidentally to establish rules 3 and 4) we have to show that the numerical coefficients arising from each separation are the same. When this is done the validity of rule 5 is demonstrated, for the separate contributions in n may be added together to give the pattern function and the whole multiplied by the numerical coefficient.

8. In view of the simplicity of the result that the numerical coefficient is the same for all separations one suspects that there is some way of looking at this problem in which the equality of the coefficients is at once evident. The following is the simplest approach I can find.

Consider again equation (1) giving the k's explicitly in terms of the x's. Note that apart from the n-factors there is associated with each x^{p_1} a factorial $p_1!$, the remaining factors being $p!$ itself and the factorials $\pi!$, the latter of which appear because permutations of suffixes of the x's are permitted. Let us consider the x's as different individuals, whether their suffixes are the same or not. We may then write

$$k_p = S(x_{\gamma_1} x_{\gamma_2} \dots x_{\gamma_p}) \frac{(-1)^{\rho-1}(\rho-1)!}{n(n-1)\dots(n-\rho+1)}, \qquad \dots\dots(12)$$

where the summation takes place over all the possible ways of allotting suffixes; or in other words, a term will appear to correspond to each way of choosing suffixes from n (suffixes being permitted to appear any number of times in a term but being regarded as different individuals). This simple way of writing the k's brings out their essential structure very clearly.

Now θ_1 may be considered as the operation of picking out an x from the operand in all possible ways and replacing it by unity. Similarly $\theta_p/p!$ may be regarded as picking out p x's with the same suffix and replacing them by unity. It is thus evident that operating on a k product by a θ product $\theta_{b_1}\theta_{b_2}\dots/b_1!\,b_2!\dots$ of the same degree will yield a result which is the number of ways in which sets of x's can be picked out of the k product so that each set contains b_1 of one suffix, b_2 of a second suffix (which may be the same as the first) and so on.

Now consider the operation (6) in which the k's are expressed in the simplified form (12). The operations θ being distributive, we shall emerge from the operation with a sum of terms comprising all the possible ways in which the individual x's can be picked out of the k product such that the row and column totals of (7) are satisfied. Consider the sets corresponding to a particular array, such as (10). The contribution to the total will consist of the ways of picking out individuals such that

(i) from the individuals in the first k_4 are chosen four in the partition (2, 1, 1),

(ii) from the second k_4 are chosen four in the partition (3, 1),

(iii) from the k_2 are chosen two in the partition (1, 1),

(iv) these are associated in all possible ways such that individuals in a row arise from the same suffix.

On consideration it will be seen that the total number of ways of doing this is the number of ways of allocating the individuals from column totals as required by rule 4; *and this is true whether sets of rows have the same suffix or not.* There will be duplications in this process, every type of case occurring $\beta_1!\beta_2!\ldots$ times, i.e. just sufficient to cancel the factorial denominators in β of the operators.

Rules 4 and 5, and hence rule 3, follow at once. The remaining rules are useful ancillaries to these principal rules.

9. *Rule* 6. The expression of $\kappa(a_1^{\alpha_1} a_2^{\alpha_2}\ldots)$ in terms of $(\kappa_{b_1}^{\beta_1}\kappa_{b_2}^{\beta_2}\ldots)$ has no term containing κ_1.

This rule is seen most clearly from statistical considerations. All the k's are invariant under a change of origin* except k_1, and even that is invariable from sample to sample. The sampling semi-invariants $\kappa(a_1^{\alpha_1} a_2^{\alpha_2}\ldots)$ are therefore invariable, i.e. cannot contain the variable quantity κ_1.

10. *Rule* 7. Any pattern which contains a row consisting of a single non-zero element has a vanishing pattern function.

This rule follows from rule 6. In fact the pattern function is independent of the numbers composing the array, and the pattern with a row containing one element can therefore form the skeleton of an array in which that element is unity; and this would entail the appearance of κ_1 which by rule 6 is impossible.

11. *Rule* 8. Any pattern containing a column consisting of a single non-zero element has a pattern function n^{-1} times that of the pattern obtained by omitting that column.

This is an immediate consequence of rule 5. The column containing the single element appears in just one separate of all the separations and the contributions to the pattern function are thus all multiplied by n^{-1} owing to its presence.

12. *Rule* 9. The expression for any $\kappa(a_1^{\alpha_1} a_2^{\alpha_2}\ldots)$ which contains a unit part can be derived from that for the $\kappa(a_1^{\alpha_1} \alpha_2^{\alpha_2}\ldots)$ without the part by increasing the suffix of one of the κ's in every possible way and dividing by n.

For example
$$\kappa(2^2) = \frac{1}{n}\kappa_4 + \frac{2}{n-1}\kappa_2^2,$$

and hence
$$\kappa(2^2\,1) = \frac{1}{n^2}\kappa_5 + \frac{4}{n(n-1)}\kappa_2\kappa_3.$$

This rule follows from rule 8.

13. Fisher & Wishart (1930) have given processes for evaluating pattern functions in terms of simpler functions. Their methods are really generalizations of rule 8 and I need

* I have proved this result formally in the previous paper (1940). It was, of course, known to Fisher when his 1928 paper was written.

not repeat them here. For completeness, however, it may be as well to record a further rule concerning vanishing patterns.

Rule 10. If the non-vanishing elements of a pattern may be divided into two groups connected by a single column, its pattern function vanishes.

Any pattern function can be evaluated in terms of the function obtained by omitting one of the columns of the pattern. If the pattern consists of two groups connected only by a single column, we can reduce it step by step by omitting the other columns. We end up with this single column, and the pattern function of a single column must vanish. For the column total a corresponds to k_a, whose mean value the one column array expresses, and since by definition this mean value is κ_a no composite term such as would be given by two rows or more can appear.

REFERENCES

R. A. Fisher (1928). "Moments and product-moments of sampling distributions." *Proc. Lond. Math. Soc.* **30**, 199–238.

R. A. Fisher & J. Wishart (1930). "The derivation of the pattern formulae of two-way partitions from those of simpler patterns." *Proc. Lond. Math. Soc.* **32**, 195–208.

M. G. Kendall (1940). "Some properties of k-statistics." *Ann. Eugen., Lond.*, **10**, 106–111.

P. V. Sukhatme (1938). "On Fisher's combinatorial methods giving moments and cumulants of the distributions of k-statistics." *Sankhya*, **4**, 53–64.

From *Annals of Eugenics*, Vol. **10**, pp 392–402 (1940)

THE DERIVATION OF MULTIVARIATE SAMPLING FORMULAE FROM UNIVARIATE FORMULAE BY SYMBOLIC OPERATION

By M. G. KENDALL

1. Little attention has been given to the multivariate extension of the theory of univariate sampling moments of moments and similar statistics. The k-statistics introduced by Fisher (1929) may, as he himself pointed out, be generalized to the multivariate case without much difficulty, and the combinatorial method of determining sampling moments may be extended with comparatively little additional complication; but the subject has not been systematically investigated. Fisher showed that all the necessary formulae can be obtained by condensation from those for multivariate distributions of the most general kind depending on partitions of the multipartite number 1^h. These general forms, as Fisher (1929, p. 219) emphasized, are "extremely cumbrous"; it is therefore not a useful procedure to obtain the other formulae by condensation from them. In this paper I investigate the possibilities of proceeding in the reverse direction and deriving the multivariate formulae from the univariate by processes of symbolic evolution. It will appear that the univariate results already known may be used to generate multivariate results of any desired complexity, and in particular the bivariate results.

Definitions and nomenclature

2. The rth moment of a distribution, μ_r, is the mean value of the rth power of the variate; and the rth moment-statistic, m_r, of a set of observations $x_1, x_2, \ldots, x_n$, is $(\Sigma x^r)/n$. It is sometimes convenient to distinguish moments about the mean from moments about an arbitrary point by writing a prime against the latter, μ'_r. But the distinction is not important for present purposes, and to save printing I write all the moments without the prime.

Fisher (1929) has used the expression "moment-statistic" to denote any rational integral algebraic symmetric function of the observations. In view of the prevailing practice of confining "moment" and "moment-statistic" to the quantities just defined I shall not employ this convention, referring to the more general quantities as r.i.a.s. functions and r.i.a.s. statistics instead.

The quantities κ_r defined by

$$\exp\left\{\kappa_1 t+\frac{\kappa_2 t^2}{2!}+\ldots+\frac{\kappa_p t^p}{p!}+\ldots\right\} = 1+\mu_1 t+\frac{\mu_2 t^2}{2!}+\ldots+\frac{\mu_p t^p}{p!}+\ldots \tag{1}$$

have been called seminvariants, semi-invariants, half-invariants and cumulants. In previous papers I have used the second form, but henceforth I use the last, which is due to Fisher. Some recent work by Dressel (1940) shows the desirability of reserving the word "seminvariant" for more general r.i.a.s. functions. In fact, Dressel points out that, corresponding to the seminvariants occurring in the theory of algebraic forms, namely, the functions of coefficients of the binary quantic

$$a_0 x^n + \binom{n}{1} a_1 x^{n-1} y + \binom{n}{2} a_2 x^{n-2} y^2 + \ldots + a_n y^n$$

which are invariant under transformations of type

$$x = \xi + l\eta, \quad y = \eta,$$

there occur various sets of r.i.a.s. functions and statistics in statistical theory which are invariant under transformations of type

$$x = \xi + l,$$

i.e. transformations of the origin. The moments about the mean are one such set, the cumulants another. A third set is

$$\lambda_{2r} = \frac{1}{2} \sum_{j=0}^{2r} (-1)^j \binom{2r}{j} \mu_j \mu_{2r-j},$$

$$\lambda_{2r+1} = \sum_{j=0}^{r} (-1)^{j+r} \binom{2r}{j+r} \frac{2j+1}{j+r+1} \mu_{r-j} \mu_{r+j+1} + \sum_{j=0}^{2r} (-1)^{j+1} \binom{2r}{j} \mu_1 \mu_j \mu_{2r-j}.$$

Thus it seems best to use the word "seminvariant" to denote any r.i.a.s. functions of this kind, and so for "seminvariant statistics". The cumulants are a particular set of seminvariants.

The r.i.a.s. statistic k_r whose mean value is κ_r is called the "cumulant statistic" or "k-statistic". Whereas κ_r is the mean value of k_r, μ_r is not in general the mean value of m_r. From the point of view of complete consistency this is unfortunate, but it is not likely to cause any difficulty in practice.

Relations between moments and cumulants

3. From equation (1) it is easy to obtain the well-known relations expressing moments in terms of cumulants and vice versa:

$$\mu_p = \sum_{r=0}^{p} \sum \left(\frac{\kappa_{p_1}}{p_1!}\right)^{\pi_1} \left(\frac{\kappa_{p_2}}{p_2!}\right)^{\pi_2} \cdots \left(\frac{\kappa_{p_r}}{p_r!}\right)^{\pi_r} \frac{p!}{\pi_1!\,\pi_2! \ldots \pi_r!}, \tag{2}$$

the second summation extending over non-negative values of the π's such that

$$p_1 \pi_1 + p_2 \pi_2 + \ldots + p_r \pi_r = p, \tag{3}$$

and

$$\kappa_p = p! \sum_{r=0}^{p} \sum \left(\frac{\mu_{p_1}}{p_1!}\right)^{\pi_1} \left(\frac{\mu_{p_2}}{p_2!}\right)^{\pi_2} \cdots \left(\frac{\mu_{p_r}}{p_r!}\right)^{\pi_r} \frac{(-1)^{\rho-1}(\rho-1)!}{\pi_1!\,\pi_2! \ldots \pi_r!}, \tag{4}$$

the second summation extending over non-negative π's subject to (3) and to

$$\pi_1 + \pi_2 + \ldots + \pi_r = \rho.$$

It is worth noting that the rather tedious process of writing down the explicit relations for particular values of p may be shortened considerably. In fact, differentiating equation (1) by κ_r, we have

$$\frac{t^r}{r!}\left(1 + \mu_1 t + \ldots + \frac{\mu_p t^p}{p!} + \ldots\right) = \frac{\partial \mu_1}{\partial \kappa_p} + \ldots + \frac{t^p}{p!}\frac{\partial \mu_p}{\partial \kappa_p} + \ldots,$$

and hence, identifying powers of t,

$$\frac{\partial \mu_r}{\partial \kappa_p} = \binom{r}{p} \mu_{r-p}. \tag{5}$$

In particular
$$\frac{\partial \mu_r}{\partial \kappa_1} = r\mu_{r-1},$$
and thus, given the expression of any μ_r in terms of the κ's, we can write down successively those of lower orders. The first ten of these expressions are:

$$\left.\begin{aligned}
\mu_1 &= \kappa_1,\\
\mu_2 &= \kappa_2+\kappa_1^2,\\
\mu_3 &= \kappa_3+3\kappa_2\kappa_1+\kappa_1^3,\\
\mu_4 &= \kappa_4+4\kappa_3\kappa_1+3\kappa_2^2+6\kappa_2\kappa_1^2+\kappa_1^4,\\
\mu_5 &= \kappa_5+5\kappa_4\kappa_1+10\kappa_3\kappa_2+10\kappa_3\kappa_1^2+15\kappa_2^2\kappa_1+10\kappa_2\kappa_1^3+\kappa_1^5,\\
\mu_6 &= \kappa_6+6\kappa_5\kappa_1+15\kappa_4\kappa_2+15\kappa_4\kappa_1^2+10\kappa_3^2+60\kappa_3\kappa_2\kappa_1+20\kappa_3\kappa_1^3\\
&\quad+15\kappa_2^3+45\kappa_2^2\kappa_1^2+15\kappa_2\kappa_1^4+\kappa_1^6,\\
\mu_7 &= \kappa_7+7\kappa_6\kappa_1+21\kappa_5\kappa_2+21\kappa_5\kappa_1^2+35\kappa_4\kappa_3+105\kappa_4\kappa_2\kappa_1\\
&\quad+35\kappa_4\kappa_1^3+70\kappa_3^2\kappa_1+105\kappa_3\kappa_2^2+210\kappa_3\kappa_2\kappa_1^2+35\kappa_3\kappa_1^4\\
&\quad+105\kappa_2^3\kappa_1+105\kappa_2^2\kappa_1^3+21\kappa_2\kappa_1^5+\kappa_1^7,\\
\mu_8 &= \kappa_8+8\kappa_7\kappa_1+28\kappa_6\kappa_2+28\kappa_6\kappa_1^2+56\kappa_5\kappa_3+168\kappa_5\kappa_2\kappa_1\\
&\quad+56\kappa_5\kappa_1^3+35\kappa_4^2+280\kappa_4\kappa_3\kappa_1+210\kappa_4\kappa_2^2+420\kappa_4\kappa_2\kappa_1^2\\
&\quad+70\kappa_4\kappa_1^4+280\kappa_3^2\kappa_2+280\kappa_3^2\kappa_1^2+840\kappa_3\kappa_2^2\kappa_1\\
&\quad+560\kappa_3\kappa_2\kappa_1^3+56\kappa_3\kappa_1^5+105\kappa_2^4+420\kappa_2^3\kappa_1^2+210\kappa_2^2\kappa_1^4\\
&\quad+28\kappa_2\kappa_1^6+\kappa_1^8,\\
\mu_9 &= \kappa_9+9\kappa_8\kappa_1+36\kappa_7\kappa_2+36\kappa_7\kappa_1^2+84\kappa_6\kappa_3+252\kappa_6\kappa_2\kappa_1\\
&\quad+84\kappa_6\kappa_1^3+126\kappa_5\kappa_4+504\kappa_5\kappa_3\kappa_1+378\kappa_5\kappa_2^2+756\kappa_5\kappa_2\kappa_1^2\\
&\quad+126\kappa_5\kappa_1^4+315\kappa_4^2\kappa_1+1260\kappa_4\kappa_3\kappa_2+1260\kappa_4\kappa_3\kappa_1^2+1890\kappa_4\kappa_2^2\kappa_1\\
&\quad+1260\kappa_4\kappa_2\kappa_1^3+126\kappa_4\kappa_1^5+280\kappa_3^3+2520\kappa_3^2\kappa_2\kappa_1+840\kappa_3^2\kappa_1^3\\
&\quad+1260\kappa_3\kappa_2^3+3780\kappa_3\kappa_2^2\kappa_1^2+1260\kappa_3\kappa_2\kappa_1^4+84\kappa_3\kappa_1^6+945\kappa_2^4\kappa_1\\
&\quad+1260\kappa_2^3\kappa_1^3+378\kappa_2^2\kappa_1^5+36\kappa_2\kappa_1^7+\kappa_1^9,\\
\mu_{10} &= \kappa_{10}+10\kappa_9\kappa_1+45\kappa_8\kappa_2+45\kappa_8\kappa_1^2+120\kappa_7\kappa_3+360\kappa_7\kappa_2\kappa_1\\
&\quad+120\kappa_7\kappa_1^3+210\kappa_6\kappa_4+840\kappa_6\kappa_3\kappa_1+630\kappa_6\kappa_2^2+1260\kappa_6\kappa_2\kappa_1^2\\
&\quad+210\kappa_6\kappa_1^4+126\kappa_5^2+1260\kappa_5\kappa_4\kappa_1+2520\kappa_5\kappa_3\kappa_2+2520\kappa_5\kappa_3\kappa_1^2\\
&\quad+3780\kappa_5\kappa_2^2\kappa_1+2520\kappa_5\kappa_2\kappa_1^3+252\kappa_5\kappa_1^5+1575\kappa_4^2\kappa_2+1575\kappa_4^2\kappa_1^2\\
&\quad+2100\kappa_4\kappa_3^2+12600\kappa_4\kappa_3\kappa_2\kappa_1+4200\kappa_4\kappa_3\kappa_1^3+3150\kappa_4\kappa_2^3\\
&\quad+9450\kappa_4\kappa_2^2\kappa_1^2+3150\kappa_4\kappa_2\kappa_1^4+210\kappa_4\kappa_1^6+2800\kappa_3^3\kappa_1\\
&\quad+6300\kappa_3^2\kappa_2^2+12600\kappa_3^2\kappa_2\kappa_1^2+2100\kappa_3^2\kappa_1^4+12600\kappa_3\kappa_2^3\kappa_1\\
&\quad+12600\kappa_3\kappa_2^2\kappa_1^3+2520\kappa_3\kappa_2\kappa_1^5+120\kappa_3\kappa_1^7+945\kappa_2^5+4725\kappa_2^4\kappa_1^2\\
&\quad+3150\kappa_2^3\kappa_1^4+630\kappa_2^2\kappa_1^6+45\kappa_2\kappa_1^8+\kappa_1^{10}.
\end{aligned}\right\} \quad (6)$$

The expression of the κ's in terms of the μ's is not quite so straightforward. An analogous differentiation of equation (1) with respect to μ_r leads to the relations

$$\frac{\partial \kappa_r}{\partial \mu_p} + \mu_1\binom{r}{1}\frac{\partial \kappa_{r-1}}{\partial \mu_p} + \mu_2\binom{r}{2}\frac{\partial \kappa_{r-2}}{\partial \mu_p} + \ldots + \mu_{r-p}\binom{r}{p}\frac{\partial \kappa_p}{\partial \mu_p} = 0,$$

but these are not of much use for practical purposes. The simplest methods are to invert the formulae (6) or to work direct from equation (4); and useful checks are given by the annihilation of $\kappa_p(p+q)$ by the operator ∂_q, where

$$\partial_q \mu_r = r(r-1)\dots(r-q+1)\,\mu_{r-q}.$$

The first ten formulae for the κ's in terms of the μ's are:

$$\left.\begin{aligned}
\kappa_1 &= \mu_1, \\
\kappa_2 &= \mu_2 - \mu_1^2, \\
\kappa_3 &= \mu_3 - 3\mu_2\mu_1 + 2\mu_1^3, \\
\kappa_4 &= \mu_4 - 4\mu_3\mu_1 - 3\mu_2^2 + 12\mu_2\mu_1^2 - 6\mu_1^4, \\
\kappa_5 &= \mu_5 - 5\mu_4\mu_1 - 10\mu_3\mu_2 + 20\mu_3\mu_1^2 + 30\mu_2^2\mu_1 - 60\mu_2\mu_1^3 + 24\mu_1^5, \\
\kappa_6 &= \mu_6 - 6\mu_5\mu_1 - 15\mu_4\mu_2 + 30\mu_4\mu_1^2 - 10\mu_3^2 + 120\mu_3\mu_2\mu_1 - 120\mu_3\mu_1^3 \\
&\quad + 30\mu_2^3 - 270\mu_2^2\mu_1^2 + 360\mu_2\mu_1^4 - 120\mu_1^6, \\
\kappa_7 &= \mu_7 - 7\mu_6\mu_1 - 21\mu_5\mu_2 + 42\mu_5\mu_1^2 - 35\mu_4\mu_3 + 210\mu_4\mu_2\mu_1 \\
&\quad - 210\mu_4\mu_1^3 + 140\mu_3^2\mu_1 + 210\mu_3\mu_2^2 - 1260\mu_3\mu_2\mu_1^2 + 840\mu_3\mu_1^4 \\
&\quad - 630\mu_2^3\mu_1 + 2520\mu_2^2\mu_1^3 - 2520\mu_2\mu_1^5 + 720\mu_1^7, \\
\kappa_8 &= \mu_8 - 8\mu_7\mu_1 - 28\mu_6\mu_2 + 56\mu_6\mu_1^2 - 56\mu_5\mu_3 + 336\mu_5\mu_2\mu_1 \\
&\quad - 336\mu_5\mu_1^3 - 35\mu_4^2 + 560\mu_4\mu_3\mu_1 + 420\mu_4\mu_2^2 - 2520\mu_4\mu_2\mu_1^2 \\
&\quad + 1680\mu_4\mu_1^4 + 560\mu_3^2\mu_2 - 1680\mu_3^2\mu_1^2 - 5040\mu_3\mu_2^2\mu_1 \\
&\quad + 13440\mu_3\mu_2\mu_1^3 - 6720\mu_3\mu_1^5 - 630\mu_2^4 + 10080\mu_2^3\mu_1^2 \\
&\quad - 25200\mu_2^2\mu_1^4 + 20160\mu_2\mu_1^6 - 5040\mu_1^8, \\
\kappa_9 &= \mu_9 - 9\mu_8\mu_1 - 36\mu_7\mu_2 + 72\mu_7\mu_1^2 - 84\mu_6\mu_3 + 504\mu_6\mu_2\mu_1 \\
&\quad - 504\mu_6\mu_1^3 - 126\mu_5\mu_4 + 1008\mu_5\mu_3\mu_1 + 756\mu_5\mu_2^2 - 4536\mu_5\mu_2\mu_1^2 \\
&\quad + 3024\mu_5\mu_1^4 + 630\mu_4^2\mu_1 + 2520\mu_4\mu_3\mu_2 - 7560\mu_4\mu_3\mu_1^2 - 11340\mu_4\mu_2^2\mu_1 \\
&\quad + 30240\mu_4\mu_2\mu_1^3 - 15120\mu_4\mu_1^5 + 560\mu_3^3 - 15120\mu_3^2\mu_2\mu_1 + 20160\mu_3^2\mu_1^3 \\
&\quad - 7560\mu_3\mu_2^3 + 90720\mu_3\mu_2^2\mu_1^2 - 151200\mu_3\mu_2\mu_1^4 + 60480\mu_3\mu_1^6 \\
&\quad + 22680\mu_2^4\mu_1 - 151200\mu_2^3\mu_1^3 + 272160\mu_2^2\mu_1^5 - 181440\mu_2\mu_1^7 \\
&\quad + 40320\mu_1^9, \\
\kappa_{10} &= \mu_{10} - 10\mu_9\mu_1 - 45\mu_8\mu_2 + 90\mu_8\mu_1^2 - 120\mu_7\mu_3 + 720\mu_7\mu_2\mu_1 \\
&\quad - 720\mu_7\mu_1^3 - 210\mu_6\mu_4 + 1680\mu_6\mu_3\mu_1 + 1260\mu_6\mu_2^2 - 7560\mu_6\mu_2\mu_1^2 \\
&\quad + 5040\mu_6\mu_1^4 - 126\mu_5^2 + 2520\mu_5\mu_4\mu_1 + 5040\mu_5\mu_3\mu_2 - 15120\mu_5\mu_3\mu_1^2 \\
&\quad - 22680\mu_5\mu_2^2\mu_1 + 60480\mu_5\mu_2\mu_1^3 - 30240\mu_5\mu_1^5 + 3150\mu_4^2\mu_2 \\
&\quad - 9450\mu_4^2\mu_1^2 + 4200\mu_4\mu_3^2 - 75600\mu_4\mu_3\mu_2\mu_1 + 100800\mu_4\mu_3\mu_1^3 \\
&\quad - 18900\mu_4\mu_2^3 + 226800\mu_4\mu_2^2\mu_1^2 - 378000\mu_4\mu_2\mu_1^4 \\
&\quad + 151200\mu_4\mu_1^6 - 16800\mu_3^3\mu_1 - 37800\mu_3^2\mu_2^2 + 302400\mu_3^2\mu_2\mu_1^2 \\
&\quad - 252000\mu_3^2\mu_1^4 + 302400\mu_3\mu_2^3\mu_1 - 1512000\mu_3\mu_2^2\mu_1^3 \\
&\quad + 1814400\mu_3\mu_2\mu_1^5 - 604800\mu_3\mu_1^7 + 22680\mu_2^5 - 567000\mu_2^4\mu_1^2 \\
&\quad + 2268000\mu_2^3\mu_1^4 - 3175200\mu_2^2\mu_1^6 + 1814400\mu_2\mu_1^8 \\
&\quad - 362880\mu_1^{10}.
\end{aligned}\right\} \quad (7)$$

The results of equations (6) and (7) are obviously much simpler when the origin is transferred to the mean, but they are sometimes required in full.

Relations between product moments and cumulants in univariate distributions

4. If we have a set of r.i.a.s. statistics $f_1, f_2, \ldots$, we may define product moments

$$\mu(p_1^{\pi_1} \ldots p_r^{\pi_r}) = \text{mean value of } (f_{p_1}^{\pi_1} \ldots f_{p_r}^{\pi_r}), \tag{8}$$

and product cumulants by the relation

$$\exp\left\{\Sigma\kappa(p_1^{\pi_1} \ldots p_r^{\pi_r}) \frac{t_{p_1}^{\pi_1} \ldots t_{p_r}^{\pi_r}}{\pi_1! \ldots \pi_r!}\right\} = \Sigma\left\{\mu(p_1^{\pi_1} \ldots p_r^{\pi_r}) \frac{t_{p_1}^{\pi_1} \ldots t_{p_r}^{\pi_r}}{\pi_1! \ldots \pi_r!}\right\}. \tag{9}$$

The explicit relations for the product moments in terms of the product cumulants and vice versa may be derived from equations (6) and (7) as in the following examples:

It is evident from the definition that to every formula in (6) and (7) there will correspond a formula in (9) concerning one statistic, e.g. from the relation

$$\mu_3 = \kappa_3 + 3\kappa_2\kappa_1 + \kappa_1^3,$$

we have for the statistic f_r

$$\mu(r^3) = \kappa(r^3) + 3\kappa(r^2)\,\kappa(r) + \kappa^3(r).$$

Operating on this by $s(\partial/\partial r)$ we have

$$3\mu(r^2 s) = 3\kappa(r^2 s) + 3\{2\kappa(rs)\,\kappa(r) + \kappa(r^2)\,\kappa(s)\} + 3\kappa^2(r)\,\kappa(s),$$

or

$$\mu(r^2 s) = \kappa(r^2 s) + 2\kappa(rs)\,\kappa(r) + \kappa(r^2)\,\kappa(s) + \kappa^2(r)\,\kappa(s). \tag{10}$$

For example, if the functions f are the k-statistics, we have, since $\kappa(r) = \mu(r) = \kappa_r$, such relations as

$$\mu(3^2 2) = \kappa(3^2 2) + 2\kappa(32)\,\kappa_3 + \kappa(3^2)\,\kappa_2 + \kappa_3^2\kappa_2.$$

Operating on (10) by $t(\partial/\partial r)$ and cancelling a factor in 2, we have

$$\mu(rst) = \kappa(rst) + \kappa(r)\,\kappa(st) + \kappa(s)\,\kappa(rt) + \kappa(t)\,\kappa(rs) + \kappa(r)\,\kappa(s)\,\kappa(t). \tag{11}$$

In this way we may derive any of the appropriate formulae by operation on the expressions of equations (6) and (7).

The validity of the rule is perhaps seen most clearly from symbolic considerations. Suppose the expression

$$\kappa(p_1 t_{p_1} + \ldots + p_r t_{p_r})^p$$

expanded multinomially and $\kappa p_1^{\pi_1} \ldots p_r^{\pi_r}$ replaced by $\kappa(p_1^{\pi_1} \ldots p_r^{\pi_r})$. Then equation (9) may be written

$$\exp\left\{\Sigma\frac{1}{p!}\kappa(p_1 t_{p_1} + \ldots + p_r t_{p_r})^p\right\} = \Sigma\frac{1}{p!}\mu(p_1 t_{p_1} + \ldots + p_r t_{p_r})^p, \tag{12}$$

in which form the generalization of equation (1) is evident, the single parameter t therein being replaced by the symbolic parameter $p_1 t_{p_1} + \ldots + p_r t_{p_r}$. Thus to any equation derived from (1) there will correspond a generalized equation containing this parameter. This generalized equation may be expanded by Taylor's theorem, and the individual terms in the expansion are those given by the operations exemplified above; for example,

$$F(p_1 t_{p_1} + p_2 t_{p_2}) = F(p_1 t_{p_1}) + \frac{1}{1!} p_2 \frac{\partial}{\partial p_1} F(p_1 t_{p_1}) + \frac{1}{2!}\left(p_2 \frac{\partial}{\partial p_1}\right)^2 F(p_1 t_{p_1}) + \ldots,$$

the first term leading back to the case of a single parameter t, the second being a new type corresponding to one operation, the third to two operations and so on.

Relations between multivariate moments and cumulants

5. To simplify the exposition I will consider only the extension of the above results to the bivariate case; the theory is perfectly general, but it hardly seems worth while attempting to write down the multivariate formulae explicitly except for the bivariate case.

The bivariate analogue of equation (1) may be written

$$\exp\left\{\frac{\kappa_{10}t}{1!0!}+\frac{\kappa_{01}u}{0!1!}+\frac{\kappa_{20}t^2}{2!0!}+\frac{\kappa_{11}tu}{1!1!}+\frac{\kappa_{02}u^2}{0!2!}+\ldots\right\}$$

$$= 1+\frac{\mu_{10}t}{1!0!}+\frac{\mu_{01}u}{0!1!}+\frac{\mu_{20}t^2}{2!0!}+\frac{\mu_{11}tu}{1!1!}+\frac{\mu_{02}u^2}{0!2!}+\ldots, \qquad (13)$$

or symbolically
$$\exp\left\{\Sigma\frac{1}{p!}\kappa(t+u)^p\right\} = \Sigma\frac{1}{p!}\mu(t+u)^p, \qquad (14)$$

where $\kappa(t+u)^p$ is equivalent to

$$p!\left\{\frac{\kappa_{p0}\,t^p}{p!0!}+\frac{\kappa_{p-1,1}\,t^{p-1}u}{(p-1)!1!}+\ldots\right\}.$$

The generalization is somewhat similar to that of (12) but is distinct from it. The expressions connecting the multivariate κ's and μ's may also be derived from equations (6) and (7) by operation of a different kind. For example, we have

$$\kappa_4 = \mu_4-4\mu_3\mu_1-3\mu_2^2+12\mu_2\mu_1^2-6\mu_1^4.$$

Operate on this expression with the operator Δ_2 whose action "pushes the suffix along" in the manner exemplified as follows:

$$\begin{aligned}
\Delta_2\kappa_4 &= 4.3.\kappa_{22},\\
\Delta_2\mu_3\mu_1 &= (\Delta_2\mu_3)\,\mu_1+\mu_3(\Delta_2\mu_1)\\
&= 3.2.\mu_{12}\mu_{10}+0,\\
\Delta_2(\mu_2\mu_1^2) &= (\Delta_2\mu_2)\,\mu_1^2+\Delta_2(\mu_2\mu_1)\,\mu_1+\Delta_2(\mu_1^2)\,\mu_2\\
&= 2.1.\mu_{02}\mu_{10}^2+(2.1.\mu_{02}\mu_{10}+2\mu_{11}\mu_{01})\,\mu_{10}+\mu_{20}(2\mu_{01}^2).
\end{aligned}$$

We find, cancelling a factor in 12,

$$\kappa_{22} = \mu_{22}-2\mu_{12}\mu_{10}-2\mu_{21}\mu_{01}-2\mu_{11}^2-\mu_{20}\mu_{02}+2\mu_{02}\mu_{10}^2+2\mu_{20}\mu_{01}^2+8\mu_{11}\mu_{10}\mu_{01}-6\mu_{10}^2\mu_{01}^2,$$

giving an expression for the bivariate κ_{22} in terms of the bivariate moments.

There will be a Δ_p for any positive p, and operation by one of the Δ's on the appropriate univariate formula will give any desired bivariate formula. Since the operation of pushing along the suffix can change the first moment μ_1 only into one of the first moments μ_{01}, μ_{10}, we may derive formulae about the mean from formulae about the mean in the same manner, e.g.

$$\kappa_4 = \mu_4-3\mu_2^2,$$

and on operating by Δ_2
$$\kappa_{22} = \mu_{22}-\mu_{20}\mu_{02}-2\mu_{11}^2.$$

The operator Δ_p is distributive in action. When operating, say, on a product of μ's it will yield a term for each way in which a number of μ's of total order not less than p can be picked out of that product, and a term for each way in which the suffixes can be pushed along. If of a suffix α a part β is pushed along, the term is multiplied by $\alpha(\alpha-1)\ldots(\alpha-\beta+1)$.

A similar procedure applies to the formulae giving the μ's in terms of the κ's, e.g. about the mean

$$\mu_5 = \kappa_5 + 10\kappa_3\kappa_2,$$

and thus about the mean of the bivariate distribution we have, on operating by Δ_1,

$$\mu_{41} = \kappa_{41} + 6\kappa_{21}\kappa_{20} + 4\kappa_{30}\kappa_{11},$$

on operating by Δ_2

$$\mu_{32} = \kappa_{32} + 3\kappa_{12}\kappa_{20} + 6\kappa_{21}\kappa_{11} + \kappa_{30}\kappa_{02},$$

and so on. The last formula may be derived from that for μ_{41} by operating again by Δ_1; and generally, except for numerical factors, $\Delta_p\Delta_q = \Delta_{p+q}$.

The validity of the rule is seen from equations (1) and (14). Since the Δ-operators are distributive, we have

$$\frac{1}{q!}\frac{u^q}{t^q}\Delta_q \exp\left\{\Sigma\frac{\kappa_{p+q}t^{p+q}}{(p+q)!}\right\} = \exp\left\{\frac{1}{q!}\frac{u^q}{t^q}\Delta_q\Sigma\left(\frac{\kappa_{p+q}t^{p+q}}{(p+q)!}\right)\right\} = \exp\left\{\Sigma\left(\frac{\kappa_{pq}t^p u^q}{p!\,q!}\right)\right\},$$

and this

$$= \frac{1}{q!}\frac{u^q}{t^q}\Delta_q\Sigma\left(\frac{\mu_{p+q}t^{p+q}}{(p+q)!}\right)$$

$$= \Sigma\left(\frac{\mu_{pq}t^p u^q}{p!\,q!}\right),$$

i.e. equation (14), so far as the relevant powers of t and u are concerned. Thus operations on individual coefficients of t in equation (1) are legitimate and these coefficients are in fact those which give rise to (6) and (7).

Expression of multivariate k-statistics in terms of power sums

6. From equation (7) it is possible to write down at once the expressions for univariate k-statistics in terms of the product sums

$$(\pi_1 \ldots \pi_r) = \Sigma(x_{p_1}^{\pi_1} \ldots x_{p_r}^{\pi_r}).$$

For example, corresponding to $\kappa_3 = \mu_3 - 3\mu_2\mu_1 + 2\mu_1^3$,

we have

$$k_3 = \frac{(3)}{n} - \frac{3(21)}{n(n-1)} + \frac{2(1^3)}{n(n-1)(n-2)}.$$

For purposes of calculation it is more convenient to express the k-statistics in terms of the power sums s_r ($= \Sigma x^r = (r)$). We have, for example,

$$k_3 = \frac{n}{(n-1)(n-2)}\left\{s_3 - \frac{3s_2s_1}{n} + \frac{2s_1^3}{n^2}\right\}.$$

The first six expressions of this kind were given by Fisher (1929). Two more have been given by Dressel (1940).

Formulae for the multivariate k-statistics (defined as the r.i.a.s. statistics whose mean values are the multivariate cumulants) can be derived from the univariate formulae by pushing along the suffix in the manner of the preceding section. For instance,

$$k_4 = \frac{n}{(n-1)(n-2)(n-3)}\left\{(n+1)s_4 - \frac{4(n+1)}{n}s_3s_1 - \frac{3(n-1)}{n}s_2^2 + \frac{12s_2s_1^2}{n} - \frac{6s_1^4}{n^2}\right\}.$$

Operating by Δ_2 we get

$$k_{22} = \frac{n}{(n-1)(n-2)(n-3)}\left\{(n+1)s_{22} - \frac{2(n+1)}{n}s_{21}s_{01} - \frac{2(n+1)}{n}s_{12}s_{10} - \frac{n-1}{n}s_{02}s_{20} \right.$$
$$\left. - \frac{2(n-1)}{n}s_{11}^2 + \frac{2}{n}s_{02}s_{10}^2 + \frac{2}{n}s_{20}s_{01}^2 + \frac{8}{n}s_{11}s_{10}s_{01} - \frac{6}{n^2}s_{10}^2s_{01}^2\right\},$$

a result agreeing with that given by Fisher except that a factor 2 in his expression has been omitted from the term in $s_{12}s_{10}$.

The validity of this rule is evident when we consider that to the formulae in multivariate cumulants and moments there correspond formulae in k-statistics and product-sums, and that the formulae expressing multivariate product-sums in terms of multivariate power-sums are similar to those for the univariate case.

7. Results such as these suggest that the sampling formulae for bivariate statistics can be derived from the univariate formulae by symbolic processes; and this turns out to be true, though the processes are more complex than those described above.

As a preliminary I give a condensed proof of the validity of the combinatorial method of determining cumulants of k-statistics for the bivariate case. For details reference may be made to the similar proof for the univariate case given previously (Kendall, 1940).

The operator ∂_{pq} is defined by the equations

$$\partial_{pq}\mu_{rs} = r(r-1)(r-2)\ldots(r-p+1)(s)(s-1)\ldots(s-q+1)\mu_{r-p,\,s-q}, \quad (r \geqslant p,\, s \geqslant q),$$
$$= 0, \quad (r < p,\, s < q).$$

On operating on both sides of equation (13) and identifying coefficients of t and u, there results

$$\partial_{pq}\kappa_{rs} = 0,$$

unless $p = r$ and $q = s$, in which case

$$\partial_{rs}\kappa_{rs} = r!\,s!.$$

Now writing $\mu\begin{pmatrix}a_1a_2\ldots a_r\\ a_1'a_2'\ldots a_r'\end{pmatrix}$ for the mean value of $k_{a_1a_1'}\ldots k_{a_ra_r'}$, we have, analogously to the univariate case,

$$\kappa\begin{pmatrix}a_1\ldots a_r\\ a_1'\ldots a_r'\end{pmatrix} = \text{sum of terms of total order } \Sigma a + \Sigma a' \text{ in the } \kappa\text{'s}$$
$$= \Sigma A(\kappa_{b_1b_1'}\ldots\kappa_{b_sb_s'}).$$

On operating on both sides by $\partial_{b_1b_1'}\ldots\partial_{b_sb_s'}$, we get

$$A = \frac{\partial_{b_1b_1'}\ldots\partial_{b_sb_s'}}{b_1!\,b_1'!\ldots b_s!\,b_s'!}\kappa\begin{pmatrix}a_1\ldots a_r\\ a_1'\ldots a_r'\end{pmatrix}.$$

As in the univariate case, the coefficient A will then be the sum of contributions from the two-way array

$$\begin{array}{c|c} & (b_1 b_1') \\ & (b_2 b_2') \\ & \cdots \\ & (b_s b_s') \\ \hline (a_1 a_1')\,(a_2 a_2') \ldots (a_r a_r') & \end{array} \tag{15}$$

a term arising for each way in which the body of the array can be filled with bipartite numbers such that the row sums are the numbers (bb') and the column sums the numbers (aa'). Now the bivariate k-statistic can be written

$$k_{pq} = \Sigma(x_{\gamma_1\gamma_1'} \ldots x_{\gamma_p\gamma_q}) \frac{(-1)^{\rho-1}(\rho-1)!}{n(n-1)\ldots(n-\rho+1)},$$

and thus, in the contributions from the arrays (15) the pattern functions will be the same as for the univariate case with row and column totals $(b+b')$ and $(a+a')$; and the numerical coefficients will be the number of ways of allocating the column totals to form the body of the array, the units being considered as consisting of individuals of two kinds according to the two parts of the numbers.

When the corresponding univariate results are known in detail, by which I mean that the separate contribution from each array of type (15) is known, it is not difficult to extend the formulae to the bivariate case, for all that is necessary is to recalculate the numerical factors associated with particular patterns.

8. This recalculation may, however, be obviated by symbolic processes. To take an example: part of the coefficient of $\kappa_6\kappa_2^2$ in $\kappa(4^2\,2)$ is given by the following array:

$$\begin{array}{ccc|c} 2 & 3 & 1 & 6 \\ 1 & 1 & . & 2 \\ 1 & . & 1 & 2 \\ \hline 4 & 4 & 2 & 10 \end{array} \tag{16}$$

the numerical coefficient of which is 96 and the pattern function $1/(n-1)^2$. Suppose we want the corresponding terms in $\kappa\begin{pmatrix}2 & 2 & 1\\ 2 & 2 & 1\end{pmatrix}$, the cumulant corresponding to the mean value of $k_{22}k_{22}k_{11}$. This term is evidently the result of operating on the first k_4 in $k_4^2k_2$ by Δ_2, the second by Δ_2 and the third by Δ_1. Now write the term $\kappa_6\kappa_2^2$ as $\kappa_{2+3+1}\kappa_{1+1+0}\kappa_{1+0+1}$, the items of each suffix corresponding to the body of the contributory array. Operating on this term by a Δ_2 which is regarded as acting only on the first part of each suffix, we get

$$2\kappa_{02+30+10}\kappa_{10+10+00}\kappa_{10+00+10} + 4\kappa_{11+30+10}\kappa_{01+10+00}\kappa_{10+00+10}$$
$$+ 4\kappa_{11+30+10}\kappa_{10+10+00}\kappa_{01+00+10} + 2\kappa_{02+30+10}\kappa_{01+10+00}\kappa_{01+00+10}.$$

Here the operation on κ_{r0} results in $r(r-1)\,\kappa_{r-2,2}$, and operations on several suffixes are multiplied by the number of ways of distributing the parts of the operator over those suffixes, e.g.

$$\Delta_2 \kappa_{20}\kappa_{10} = (\Delta_2\kappa_{20})\kappa_{10} + \Delta_2(\kappa_{20}\kappa_{10})$$
$$= 2\kappa_{02}\kappa_{10} + 4\kappa_{11}\kappa_{01}.$$

We now operate by Δ_2 on the second parts in the composite suffixes and by Δ_1 on the third parts. Amalgamating the parts of the suffixes, we finally get

$$12(\kappa_{15}\kappa_{20}^2 + \kappa_{51}\kappa_{02}^2) + 72(\kappa_{24}\kappa_{20}\kappa_{11} + \kappa_{42}\kappa_{02}\kappa_{11}) + 72\kappa_{33}\kappa_{11}^2 + 48\kappa_{33}\kappa_{20}\kappa_{02}. \quad (17)$$

The sum of the coefficients in this expression is 288, and this, as it should be, is the coefficient in $\Delta_2 k_4 \Delta_2 k_4 \Delta_1 k_2$, namely, $12 \times 12 \times 2$.

Thus the bipartite terms corresponding to $\kappa_6\kappa_2^2$ from the array (16) are 96/288 times those of (17); and multiplying by 2 to take account of the similar array given by interchanging the first two columns of (16), we have, for the numerical parts of the coefficients in $\kappa\begin{pmatrix}2 & 2 & 1\\ 2 & 2 & 1\end{pmatrix}$,

$$8(\kappa_{15}\kappa_{20}^2 + \kappa_{51}\kappa_{02}^2) + 48(\kappa_{24}\kappa_{20}\kappa_{11} + \kappa_{42}\kappa_{02}\kappa_{11}) + 48\kappa_{33}\kappa_{11}^2 + 32\kappa_{33}\kappa_{20}\kappa_{02}. \quad (18)$$

Apart from (16) the only arrays making a contribution to $\kappa_6\kappa_2^2$ in $\kappa(4^2 2)$ are

2	2	2	6
1	1	.	2
1	1	.	2
4	4	2	10

3	3	.	6
1	.	1	2
.	1	1	2
4	4	2	10

which have numerical coefficients 72 and 32 and pattern functions $1/(n-1)(n-2)$ and $1/(n-1)^2$ respectively. The bipartite analogues will be found to be

$$2(\kappa_{15}\kappa_{20}^2 + \kappa_{51}\kappa_{02}^2) + 16(\kappa_{24}\kappa_{20}\kappa_{11} + \kappa_{42}\kappa_{02}\kappa_{11}) + 20\kappa_{33}\kappa_{11}^2 + 16\kappa_{33}\kappa_{02}\kappa_{20}, \quad (19)$$

and

$$8(\kappa_{24}\kappa_{20}\kappa_{11} + \kappa_{42}\kappa_{02}\kappa_{11}) + 8\kappa_{33}\kappa_{11}^2 + 8\kappa_{33}\kappa_{02}\kappa_{20}. \quad (20)$$

Adding (18), (19) and (20), each divided by its pattern function, we get, for the coefficients in $\kappa\begin{pmatrix}2 & 2 & 1\\ 2 & 2 & 1\end{pmatrix}$ corresponding to $\kappa_6\kappa_2^2$,

$$\frac{1}{(n-1)^2(n-2)}\{2(5n-9)(\kappa_{15}\kappa_{20}^2 + \kappa_{51}\kappa_{02}^2) + 8(9n-16)(\kappa_{24}\kappa_{11}\kappa_{20} + \kappa_{42}\kappa_{11}\kappa_{02})$$
$$+ 4(19n-33)\kappa_{33}\kappa_{11}^2 + 8(7n-12)\kappa_{33}\kappa_{20}\kappa_{02}\}. \quad (21)$$

An expression of this kind was given by Fisher (1929, p. 223), but apart from a misprint ($\kappa_{31}\kappa_{20}^2$ for $\kappa_{51}\kappa_{02}^2$) his result seems to be in error in the terms in $\kappa_{33}\kappa_{20}\kappa_{02}$ and $\kappa_{42}\kappa_{11}\kappa_{02}$.

The validity of the above rule is evident from the consideration that the operations merely split the univariate array into the contributory bivariate arrays and ensure the correct numerical proportion of each constituent.

9. When it is a matter of determining a single term in the expression for a bivariate cumulant the combinatorial method is the most direct and is also the quickest. However, that method, though the best, is itself not entirely free from difficulty in that it is easy to

overlook numerical factors, those in 2 being particularly elusive on occasion. The systematic development of the bivariate formulae from the univariate results has the advantage of giving simultaneously all the terms generated from the original univariate term and is less liable to errors of omission. Furthermore, as the example in the previous section shows, it provides a useful check on the combinatorial method.

REFERENCES

Paul L. Dressel (1940). "Statistical seminvariants and their estimates with particular emphasis on their relation to algebraic invariants." *Ann. Math. Statist.* **11**, 33–57.

R. A. Fisher (1929). "Moments and product-moments of sampling distributions." *Proc. Lond. Math. Soc.* **30**, 199–238.

M. G. Kendall (1940). "Proof of Fisher's rules for ascertaining the sampling semi-invariants of k-statistics." *Ann. Eugen., Lond.*, **10**, 215–22.

From *Annals of Eugenics*, Vol. **11**, pp 300–305 (1941-2)

ON SEMINVARIANT STATISTICS

By M. G. KENDALL

1. A seminvariant statistic is a function of the sample values $x_1, x_2, \ldots, x_n$ which is invariant under a change of origin. This general definition would include certain classes of statistics, such as the range, which are not symmetric functions of the sample values, but this paper deals only with statistics which are rational, integral, algebraic and symmetric functions of the observations.

There are evident advantages in using seminvariants as compared with statistics which are dependent on the origin of measurement. Once the distribution has been located we cease to be interested in the origin, and statistics which are independent of locatory measures are nearer to representing intrinsic properties of the parent form than those which are not. The seminvariants are, in fact, as their name implies, partly independent of the method of measurement. If it were possible to devise tractable statistics which were entirely independent of the measuring system we should have reached that degree of absoluteness conferred on some branches of physics by the tensor calculus. The seminvariants themselves can be used to produce statistics which are independent of both location and scale, and it does not seem impossible to look forward to the discovery of measures which are invariant under more general conditions. In the meantime the seminvariants are a step in the desired direction.

2. The n sample values may be considered as the roots in x/y of a binary n-ic

$$f = (x-x_1 y)(x-x_2 y)\ldots(x-x_n y) = b_0 x^n + \binom{n}{1} b_1 x^{n-1} y + \binom{n}{2} b_2 x^{n-2} y^2 + \ldots + y^n, \qquad (1)$$

and a seminvariant statistic is thus a seminvariant, in the sense of the algebra of quantics, of the form f. The functions b are symmetric sums of the products of the roots $x_1 \ldots x_n$, and we have, for example,

$$b_r = \frac{(-1)^r}{\binom{n}{r}} \Sigma(x_i x_j \ldots) \quad (i \neq j \neq \ldots), \qquad (2)$$

there being r factors in each term on the right.

In accordance with known results in the theory of binary forms, if a seminvariant consist of the sum of several parts each of different degree in the x's, then each of the parts is seminvariant. Furthermore, it is a necessary and sufficient condition for a homogeneous function of the x's (an isobaric function of the b's) to be a seminvariant that it should be annihilated by the operator

$$D = b_0 \frac{\partial}{\partial b_1} + 2b_1 \frac{\partial}{\partial b_2} + 3b_2 \frac{\partial}{\partial b_3} + \ldots + n b_{n-1} \frac{\partial}{\partial b_n} \qquad (3)$$

or the equivalent operator

$$D' = s_0\frac{\partial}{\partial s_1}+2s_1\frac{\partial}{\partial s_2}+3s_2\frac{\partial}{\partial s_3}+\ldots+ns_{n-1}\frac{\partial}{\partial s_n}, \tag{4}$$

where

$$s_r = \sum_{j=1}^{n}(x_j^r). \tag{5}$$

There is, in fact, a duality between the coefficients b and the power sums s.

3. Consider now the most general homogeneous function of the x's of degree p. Evidently this will contain a term of type $(x_i^{p_1}\,x_j^{p_2}\ldots x_m^{p_s})$ corresponding to every partition $(p_1p_2\ldots p_s)$ of the number p. It will thus contain $\varpi(p)$ terms, where $\varpi(p)$ is the number of partitions of p, (p) included, and hence $\varpi(p)$ assignable constants. If we operate on this expression by D the vanishing of the resultant is equivalent to $\varpi(p-1)$ linear conditions on these constants. Thus the most general seminvariant has $\varpi(p)-\varpi(p-1)$ assignable constants, namely, $\rho(p)$, where $\rho(p)$ is the number of non-unitary partitions of p. Hence there are $\rho(p)$ independent seminvariants of degree p, a result due to Macmahon (1884).

For example, there is no seminvariant of the first degree, only one of the second and third degrees, two of the fourth and fifth degrees, and three of the sixth degree. The moments about the sample mean are seminvariant except the first, which is identically zero. Consequently, any seminvariant of degree two or three is equal, except for constants, to the second or third mean-moment. Similarly, any seminvariant of degree four or five is a linear function of the fourth moment and the fourth k-statistic or the fifth moment and the fifth k-statistic as the case may be. It is not until we reach the seminvariant of the sixth degree that any function can appear which is not expressible linearly in terms of members of the same degree of two known systems of seminvariants, the mean-moments and the k-statistics.

4. There is clearly an infinite number of families of seminvariants, a family comprising one statistic of each degree from 2 onwards. All such families will have certain statistics of lower degree in common. Moreover, in any given family there will be $(p-1)$ statistics of degree p or less, and from these statistics we can build up the most general statistic of degree p provided that the members of the family are asyzygetic. For example, in the family of mean-moments there are five moments of degree six or less, and with them we can construct the most general statistic of degree six in the form $c_0m^6+c_1m_4m_2+c_2m_3^2+c_3m_2^3$. Thus all families are equivalent in the sense that a member of any one of degree p can be expressed rationally, integrally and algebraically in terms of the statistics of degree p and lower in any other family.

From the purely algebraic standpoint, therefore, there does not seem to be anything to choose between the possible families. Nor is any one clearly marked out from the others by ease of calculation. There is only one convenient way of calculating the statistics in practice, namely, by expressing them in terms of the symmetric sums s (or, in special cases, of factorial sums). It would thus appear that for a guide in choosing one family in preference to others we must look to purely statistical considerations.

5. The primary desideratum in most inquiries is simplicity of sampling formulae. Here Fisher's k-statistics have very decided advantages over the mean-moments. But it is natural to wonder whether there may not be other families with even simpler properties. In short, do the k-statistics represent the limit of possible achievement in sampling simplicity? This question can be discussed and, I think, answered affirmatively by considering the rules giving the sampling cumulants of k-statistics enunciated by Fisher (1928) and proved in a previous paper of mine (1940b).

Suppose that we have any family of statistics, not necessarily seminvariant, ω_p, where ω_p is homogeneous and of degree p in the observations. The mean value of $\omega_{p_1}^{\pi_1} \dots \omega_{p_s}^{\pi_s}$, which is of total order $p_1\pi_1 + \dots + p_s\pi_s = p$ in the x's, will, when expressed in terms of moments, consist of the sum of a number of terms each of order p. Hence the corresponding cumulant $\kappa(p_1^{\pi_1} \dots p_s^{\pi_s})$ will consist of the sum of terms each of order p in the cumulants. Thus Rule 1 (the references are to my 1940b paper) remains valid for any homogeneous statistics.

The argument of the 1940b paper then applies unchanged to show that the coefficient of any term $\kappa_{\beta_1}^{b_1} \dots \kappa_{\beta_m}^{b_m}$ is the result of the operation

$$\frac{\theta_{\beta_1}^{b_1} \dots \theta_{\beta_m}^{b_m}}{(\beta_1!)^{b_1} \dots (\beta_m!)^{b_m} b_1! \dots b_m!} \omega_{p_1}^{\pi_1} \dots \omega_{p_s}^{\pi_s}, \tag{6}$$

where θ_j is the operator which picks out j x's from the l available and replaces them by $l(l-1) \dots (l-j+1)\, x^{l-j}$.

Thus the two-way partition giving the term-by-term expansion of the k-statistics is generally applicable.

Furthermore, the term-by-term expansion of the cumulant $\kappa(p_1^{\pi_1} \dots p_s^{\pi_s})$ is obtained by omitting all contributions to equation (6) which, in the two-way array, divide into two separate parts each confined to distinct rows and columns. Rule 2 is thus generally true.

6. This seems to be as far as we can go without imposing some restriction on the statistics. Suppose, then, that they are seminvariant. Then Rule 6 will hold, namely, there is no term in $\kappa(p_1^{\pi_1} \dots p_s^{\pi_s})$ involving κ_1; for the statistics are independent of the origin and therefore of the parent mean, and hence κ_1 cannot appear in their sampling distribution and in particular in their sampling cumulants.

7. It may be noted in passing that the operator θ_j is equivalent to

$$\frac{\partial^j}{\partial x_1^j} + \frac{\partial^j}{\partial x_2^j} + \dots + \frac{\partial^j}{\partial x_n^j},$$

which may be written S_j to indicate that S_j is the same function of the operators $\partial/\partial x$ as s_j is of the variables x. Thus, from equation (6), the moment $\mu(a_1^{\alpha_1} \dots a_s^{\alpha_s})$ is the sum of terms such as

$$\frac{(\kappa_{\beta_1} S_{\beta_1})^{b_1} \dots (\kappa_{\beta_m} S_{\beta_m})^{b_m}}{(\beta_1!)^{b_1} \dots (\beta_m!)^{b_m} b_1! \dots b_m!} (\omega_{a_1}^{\alpha_1} \dots \omega_{a_s}^{\alpha_s}), \tag{7}$$

and hence the moment generator of the ω-statistics is

$$\Sigma\left\{\mu(a_1^{\alpha_1}\dots a_s^{\alpha_s})\frac{t_{a_1}^{\alpha_1}\dots t_{a_s}^{\alpha_s}}{\alpha_1!\dots\alpha_s!}\right\}$$

$$= \exp\left\{\frac{\kappa_1 S_1}{1!}+\frac{\kappa_2 S_2}{2!}+\dots\right\}\Sigma\left\{\omega_{b_1}^{\beta_1}\frac{t_{b_1}^{\beta_1}}{\beta_1!}\dots\omega_{b_m}^{\beta_m}\frac{t_{b_m}^{\beta_m}}{\beta_m!}\right\}$$

$$= \exp\left\{\frac{\kappa_1 S_1}{1!}+\frac{\kappa_2 S_2}{2!}+\dots\right\}\exp(\omega_1 t_1+\omega_2 t_2+\dots), \tag{8}$$

where non-constant terms after operation are ignored. This presents an interesting reciprocal relationship with a result proved by Fisher (1928), namely, that the generating function is equal to

$$\left[\exp(\Omega_1 t_1+\Omega_2 t_2+\dots)\exp\left(\frac{\kappa_1 s_1}{1!}+\frac{\kappa_2 s_2}{2!}+\dots\right)\right]_{x_1=\dots x_n=0}, \tag{9}$$

where Ω is the same function of the operators $\partial/\partial x$ as ω is of the variables x. From equation (8) we have an alternative proof of a result already given (1940*a*)

$$\left.\begin{aligned} S_p k_p &= p!, \\ S_q k_p &= 0, \quad q\neq p, \end{aligned}\right\} \tag{10}$$

for if the ω's are k-statistics $\mu(p)=\kappa(p)=\kappa_p$, and the coefficient of κ_p in equation (8) will be found to lead to this relationship.

8. The same equation also provides a rigorous proof of Fisher's derivation of the moments of the statistics

$$q_3=\frac{k_3}{k_2^{3/2}},\quad q_4=\frac{k_4}{k_2^2},\quad \dots,\quad q_r=\frac{k_r}{k_2^{r/2}},$$

in samples from a normal population (1930). Fisher used equation (9) to derive the expression for the moment generating function of the q's in terms of that of the k's in the form

$$M'(\tau_2,\tau_3,\dots)=\left[\exp\{\tau_2 K_2+\tau_3 Q_3+\dots\}M(t_2,t_3,\dots)\right]_{t=0}, \tag{11}$$

where Q is the same function of the operators $\partial/\partial t$ as q is of the k's. Observing that k_2 is independent of the q's and that its generating function is $\left(1-\frac{2\kappa_2 t_2}{n-1}\right)^{-\frac{1}{2}(n-1)}$, we find, for $2j=a+b+c+\dots$,

$$\mu(\dots 5^c 4^b 3^a)=\mu(\dots 5^c 4^b 3^a 2^{-j})\left[\frac{d^j}{dt^j}\left(1-\frac{2\kappa_2 t}{n-1}\right)^{-\frac{1}{2}(n-1)}\right]_{t=0}, \tag{12}$$

where 5 refers to k_5 and so on. The moment on the right is that of $(\dots q_5^c q_4^b q_3^a)$, and hence we obtain the moments of the q's from those of the k's.

But this process involves the use of the symbolic operators Q, which are fractional in $\partial/\partial t$. Though perhaps this could be justified (e.g. as in the case of Heaviside's operational calculus, by contour integration), it remains at present only a piece of suggestive formalism, leading in this instance to what is undoubtedly the right answer.

The use of equation (8) avoids the difficulty, the operators being integral. In fact we have that $(\dots q_5^c q_4^b q_3^a k_2^j)$ is a rational integral algebraic function of the observations and hence that equation (8) holds with the substitution of the q's for the ω's. Thus $\mu(\dots 5^c 4^b 3^a)$ is the result of operating by

$$\exp\left(\frac{\kappa_1 S_1}{1!}+\frac{\kappa_2 S_2}{2!}+\dots\right) \quad \text{on} \quad \exp(k_2 t_2 + q_3 t_3 + \dots),$$

non-constant terms being ignored. But all the κ's vanish except κ_1 and κ_2, and the former contributes nothing to the operation because S_1 annihilates every q and k_2. The operator is then simply $\exp(\frac{1}{2}\kappa_2 S_2)$. Further, since k_2 is independent of the q's the moment generating function of the whole is the product of the separate moment generating functions, and the operation becomes

$$\{\exp(\tfrac{1}{2}\kappa_2 S_2)\exp(k_2 t_2)\}\{\exp(\tfrac{1}{2}\kappa_2 S_2)\exp(q_3 t_3+\dots)\}. \tag{13}$$

The second function is the moment generating function of the q's. The first is that of k_2 and is thus $\left(1-\frac{2\kappa_2 t_2}{n-1}\right)^{-\frac{1}{2}(n-1)}$. Moreover, $\frac{1}{2}S_2 k_2 = 1$ and hence, identifying coefficients, we get equation (12).

9. Returning now to the problem of the sampling cumulants of the ω's we have to consider the crucial rules 3, 4 and 5, which permit of the evaluation of the operation (6), when the ω's are k-statistics, in two stages, the first consisting of the finding of a numerical coefficient from the number of ways in which the two-way array can be set up, the second of finding a pattern function in n depending only on the configuration of zeros in the array.

It is here that the superiority of the k-statistics manifests itself, for so far as I can see they are the only seminvariants possessing this property. It was shown in the 1940b paper that the basis of these rules is the fact that k_p can be written in the form

$$k_p = \Sigma(x_{\gamma_1} x_{\gamma_2} \dots x_{\gamma_p}) \frac{(-1)^{\rho-1}(\rho-1)!}{n(n-1)\dots(n-\rho+1)}, \tag{14}$$

where the summation extends over all ways of allotting the suffixes γ, repetitions included, from the n available and ρ is the number of different suffixes. The significance of this way of looking at the k-statistics is that there are no numerical factors in equation (14) depending on the numbers γ other than ρ itself. It would seem that this does not hold for any other seminvariant, for if l_p were such a one it could be expressed as $k_p + c_1 k_{p-2} k_2 + \text{etc.}$, and the products of the k's, themselves expressed in the form (14), would involve factors depending on the γ's, which would complicate either the determination of the numerical coefficient or that of the pattern function or both.

Although, therefore, the combinatorial method may be of general application, I doubt whether it is possible to improve materially on the k-statistics so far as sampling simplicity is concerned. The above remarks, however, cannot be regarded as a conclusive proof of this surmise. There may exist seminvariant parameters λ and seminvariants statistics l such that the sampling λ's of the l's can be expressed more simply in terms of parent λ's than is the case for cumulants and k-statistics. Such λ's would at least be restricted by the desideratum that $\lambda_p = 0$, $p > 2$, so that the important case of the normal parent would reduce in the usual way to the evaluation of a single term. This being so λ_p must, except for constants, be identical with κ_p as far as $p = 4$.

Moreover, it is not to be inferred that no further work remains to be done in this field. The evaluation of the numerical coefficient in two-way arrays still presents some practical difficulties owing to the possibility of overlooking factors in symmetrical arrangements. It would, for example, be useful if this process could be reduced to an automatic algebraic one by the devising of generating functions to give the numbers required.

REFERENCES

R. A. Fisher (1928). Moments and product moments of sampling distributions. *Proc. Lond. Math. Soc.* **30**, 199.

R. A. Fisher (1930). The moments of the distribution for normal samples of measures of departure from normality. *Proc. Roy. Soc.* A, **130**, 16.

M. G. Kendall (1940*a*). Some properties of *k*-statistics. *Ann. Eugen., Lond.*, **10**, 106.

M. G. Kendall (1940*b*). Proof of Fisher's rules for ascertaining the sampling semi-invariants of *k*-statistics. *Ann. Eugen., Lond.*, **10**, 215.

P. A. MacMahon (1884). Seminvariants and symmetric functions. *Amer. J. Math.* **6**, 131.

From *Annals of Eugenics*, Vol. 12, pp 138–142 (1943-5)

SAMPLING MOMENTS OF MOMENTS FOR A FINITE POPULATION

BY J. O. IRWIN AND M. G. KENDALL

INTRODUCTION

1. The problem of determining the moments of moment-statistics in samples from a finite population has been considered by several authors, particularly by Tschuprow (1923) and Isserlis (1931), who gives references. All these writers worked with sample-moments, and the formulae so derived are very cumbrous except for lower moments. Isserlis made a substantial advance by introducing new symmetric functions of the parent-moments and abbreviated forms for some of the coefficients which arise as functions of n (the sample number) and N (the population number). Even in his notation, however, the fourth moment of the variance takes a full page of the *Proceedings of the Royal Society* to write down.

In this paper we have approached the problem on lines which are novel in two respects: (*a*) we use k-statistics instead of moment-statistics and hence simplify the expressions to some extent; and (*b*) we use a method by which the formulae for a finite population can be derived directly from the known results for an infinite population. The results already obtained by other writers are confirmed and certain new results given for product-moments.

2. The k-statistics of the rth order in the sample of n will be written k_r. The corresponding quantity for the population of N will be written K_r. We shall imagine this population itself as a random sample from an infinite population such that the mean value of K_r in samples of N is the rth cumulant κ_r. It will then follow that the mean value of k_r in samples of n from this infinite population is κ_r and that the mean value of k_r in samples of n from the finite population is K_r. Denoting mean values for the infinite population by E and those for the finite population by E_N we have

$$E_N(k_r) = K_r, \tag{1}$$

$$E(k_r) = E(K_r) = \kappa_r. \tag{2}$$

To see that (1) is true we have only to appreciate that the mean value of k_r in samples of n from the finite population is some function (necessarily symmetric), say L_r, of the population values. Then the mean values of L_r in samples of N from the infinite population is $E(k_r) = \kappa_r$ in accordance with (2). This involves the identity of K_r and L_r, for the expectation of their difference is zero, and if they were different this in turn would involve a syzygy between the parent cumulants, which is impossible. The result of equation (1) has already been given by Anderson (1935, p. 217), who says that it was communicated to him by R. A. Fisher.

The same principle may be extended, and forms the basis of the method used below. If at any point we arrive at a function f of the sample values, such that

$$E(f) = \Sigma a_r \kappa_r, \tag{3}$$

where the a's are constants, then we have immediately

$$E_N(f) = \Sigma a_r K_r, \tag{4}$$

and hence can derive the expectations for the finite population from those for the infinite population.

Moments of the mean

3. We proceed to derive the first four moments of the mean-statistic m_1, which is equal to k_1. We have

$$E_N(k_1) = K_1, \tag{5}$$

$$E(k_1^2) = \frac{\kappa_2}{n} + \kappa_1^2, \quad E(K_1^2) = \frac{\kappa_2}{N} + \kappa_1^2.$$

Hence

$$E(k_1 - K_1)^2 = E(k_1^2) - E(K_1^2)$$
$$= \kappa_2\left(\frac{1}{n} - \frac{1}{N}\right).$$

In virtue of the remark at the end of § 2 it follows that

$$E_N(k_1 - K_1)^2 = K_2\left(\frac{1}{n} - \frac{1}{N}\right), \tag{6}$$

giving the sampling variance of the mean.

We shall find it convenient to write

$$\alpha_r = \frac{1}{n^r} - \frac{1}{N^r}. \tag{7}$$

Since

$$K_2 = \frac{N}{N-1} M_2,$$

where M_r is the rth moment about the mean of the finite population, we have from (6)

$$E_N(k_1 - K_1)^2 = \frac{\alpha_1 N}{N-1} M_2$$
$$= \frac{N-n}{N-1}\frac{M_2}{n}, \tag{8}$$

a familiar form.

4. For the third moment we have

$$E(k_1 - \kappa_1)^3 = E(k_1 - K_1 + K_1 - \kappa_1)^3$$
$$= E(k_1 - K_1)^3 + 3E(k_1 - K_1)^2 (K_1 - \kappa_1) + 3E(k_1 - K_1)(K_1 - \kappa_1)^2 + E(K_1 - \kappa_1)^3. \tag{9}$$

Now the third term on the right vanishes, since the expectation of $k_1 - K_1$ in the finite population vanishes. We have further

$$E(k_1 - \kappa_1)^3 = \frac{\kappa_3}{n^2}, \quad E(K_1 - \kappa_1)^3 = \frac{\kappa_3}{N^2}.$$

Hence, from (9),

$$E(k_1 - K_1)^3 = \kappa_3\left(\frac{1}{n^2} - \frac{1}{N^2}\right) - 3E(K_1 - \kappa_1)(k_1 - K_1)^2$$
$$= \alpha_2\kappa_3 - 3E(K_1 - \kappa_1)\alpha_1 K_2$$
$$= \alpha_2\kappa_3 - 3\alpha_1 E(K_1 - \kappa_1)(K_2 - \kappa_2)$$
$$= \alpha_2\kappa_3 - \frac{3\alpha_1}{N}\kappa_3$$
$$= \left(\alpha_2 - \frac{3\alpha_1}{N}\right)\kappa_3.$$

Hence, for the third moment,

$$E_N(k_1 - K_1)^3 = \left(\alpha_2 - \frac{3\alpha_1}{N}\right) K_3. \tag{10}$$

Substituting for α_2, α_1 and
$$K_3 = \frac{N^2}{(N-1)(N-2)} M_3,$$
we find
$$E_N(m_1 - M_1)^3 = \frac{(N-n)(N-2n)}{n^2(N-1)(N-2)} M_3. \tag{11}$$

5. The derivation of the fourth moment of the mean proceeds analogously:
$$E(k_1-\kappa_1)^4 = E(k_1-K_1)^4 + 4E(k_1-K_1)^3(K_1-\kappa_1) + 6E(k_1-K_1)^2(K_1-\kappa_1)^2 + E(K_1-\kappa_1)^4.$$

Hence, since
$$E(k_1-\kappa_1)^4 = \frac{\kappa_4}{n^3} + \frac{3\kappa_2^2}{n^2},$$
we have
$$E(k_1-K_1)^4 = \alpha_3\kappa_4 + 3\alpha_2\kappa_2^2 - 4\left(\alpha_2 - \frac{3\alpha_1}{N}\right) E(K_1-\kappa_1) K_3 - 6\alpha_1 E(K_1-\kappa_1)^2 K_2$$
$$= \kappa_4\left(\alpha_3 - \frac{4}{N}\alpha_2 + \frac{6}{N^2}\alpha_1\right) + 3\left(\alpha_2 - \frac{2}{N}\alpha_1\right)\kappa_2^2. \tag{12}$$

Now
$$E(K_2^2) = \frac{\kappa_4}{N} + \frac{N+1}{N-1}\kappa_2^2,$$
and hence, substituting in (12) for κ_2^2, we have
$$E(k_1-K_1)^4 = \kappa_4\left\{\alpha_3 - \frac{4}{N}\alpha_2 + \frac{6}{N^2}\alpha_1\right\} + 3\left(\alpha_2 - \frac{2}{N}\alpha_1\right)\frac{N-1}{N+1}\left\{E(K_2^2) - \frac{\kappa_4}{N}\right\},$$
from which, immediately,
$$E_N(k_1-K_1)^4 = K_4\left\{\alpha_3 - \frac{4}{N}\alpha_2 + \frac{6}{N^2}\alpha_1 - \frac{3(N-1)}{N(N+1)}\left(\alpha_2 - \frac{2}{N}\alpha_1\right)\right\} + K_2^2\frac{3(N-1)}{N+1}\left(\alpha_2 - \frac{2}{N}\alpha_1\right). \tag{13}$$
This is the required expression. Making the usual substitutions and in particular
$$K_4 = \frac{N^2}{(N-1)(N-2)(N-3)}\{(N+1)M_4 - 3(N-1)M_2^2\},$$
we find the known result
$$E_N(m_1-M_1)^4 = \frac{N-n}{n^3(N-1)(N-2)(N-3)}\{(N^2 - 6nN + N + 6n^2)M_4 + 3(n-1)N(N-n-1)M_2^2\}. \tag{14}$$

Moments of the variance

6. We now turn to the second statistic k_2. We have
$$E_N(k_2) = K_2, \tag{15}$$
which is equivalent to
$$E_N(m_2) = \frac{n-1}{n}\frac{N}{N-1}M_2. \tag{16}$$
For the variance we have
$$E(k_2^2) = \frac{\kappa_4}{n} + \frac{n+1}{n-1}\kappa_2^2, \quad E(K_2^2) = \frac{\kappa_4}{N} + \frac{N+1}{N-1}\kappa_2^2.$$
Hence
$$E(k_2^2) = \frac{\kappa_4}{n} + \frac{N-1}{N+1}\frac{n+1}{n-1}\left\{E(K_2^2) - \frac{\kappa_4}{N}\right\}.$$
Thus
$$E(k_2-K_2)^2 = E(k_2^2) - E(K_2^2)$$
$$= \kappa_4\left\{\frac{1}{n} - \frac{n+1}{n-1}\frac{N-1}{N(N+1)}\right\} + E(K_2^2)\left\{\frac{n+1}{n-1}\frac{N-1}{N+1} - 1\right\},$$
whence
$$E_N(k_2-K_2)^2 = \frac{(N-n)(Nn-n-N-1)}{n(n-1)N(N+1)}K_4 + \frac{2(N-n)}{(n-1)(N+1)}K_2^2, \tag{17}$$

which is equivalent to

$$E_N\left(m_2-\frac{n-1}{n}\frac{N}{N-1}M_2\right)^2 = \frac{(n-1)N(N-n)}{n^3(N-1)^2(N-2)(N-3)}\{(nN-N-n-1)(N-1)M_4 - (nN^2-3N^2+6N-3n-3)M_2^2\}. \quad (18)$$

Product-moments involving mean or variance

7. Proceeding now to product-moments we have

$$E(k_r-\kappa_r)(k_1-\kappa_1) = E(k_r-K_r)(k_1-K_1)+E(K_r-\kappa_r)(K_1-\kappa_1),$$

whence

$$E(k_r-K_r)(k_1-K_1) = \kappa_{r+1}\left(\frac{1}{n}-\frac{1}{N}\right) = \alpha_1\kappa_{r+1}.$$

Hence

$$E_N(k_r-K_r)(k_1-K_1) = \alpha_1 K_{r+1}, \quad (19)$$

a general expression for the covariance of k_r and k_1 in finite samples which we think is new. We observe in particular that if the finite population is symmetrical k_1 is uncorrelated with any k-statistic of even order.

In a similar way it appears that

$$E(k_r-K_r)(k_2-K_2) = \kappa(r2)_n-\kappa(r2)_N, \quad (20)$$

where $\kappa(r2)$ has the usual meaning and the suffixes n and N refer to the numbers to be substituted as sample numbers in the usual formulae. The results when $r = 1$ and $r = 2$ have already been obtained. When $r = 3$ equation (20) yields

$$E(k_3-K_3)(k_2-K_2) = \kappa_5\left(\frac{1}{n}-\frac{1}{N}-\frac{6(N-n)}{(n-1)N(N+5)}\right)+\frac{6(N-n)}{(n-1)(N+5)}E(K_3K_2),$$

whence

$$E_N(k_3-K_3)(k_2-K_2) = K_5\frac{(N-n)\{(n-1)(N-1)-6\}}{n(n-1)N(N+5)}+\frac{6(N-n)}{(n-1)(N+5)}K_3K_2. \quad (21)$$

Since $\kappa(r2)$ vanishes for odd r when the population is symmetrical (since then all K's of odd order vanish) it is clear that in this case k_2 is uncorrelated with any k-statistic of odd order.

Approximations to order $1/N$

8. The foregoing methods will yield sampling moments of higher orders but the formulae soon become extremely complex and of very little interest. We have therefore not thought it worth while working them out in full. We give a method of evaluating the formulae to order $1/N$ which would doubtless provide adequate approximations in the majority of cases where statistics of high order were worth calculating. Consider, for example, the variance of the third k-statistic. We have

$$E(k_3-K_3)^2 = E(k_3-\kappa_3)^2-E(K_3-\kappa_3)^2.$$

Now

$$E(k_3-\kappa_3)^2 = \frac{1}{n}\kappa_6+\frac{9}{n-1}\kappa_4\kappa_2+\frac{9}{n-1}\kappa_3^2+\frac{6n}{(n-1)(n-2)}\kappa_2^3,$$

and, to order $1/N$,

$$E(K_3-\kappa_3)^2 = \frac{1}{N}\kappa_6+\frac{9}{N}\kappa_4\kappa_2+\frac{9}{N}\kappa_3^2+\frac{6}{N}\kappa_2^3.$$

Hence

$$E(k_3-K_3)^2-\alpha_1\kappa_6 = \left(\frac{9}{n-1}-\frac{9}{N}\right)\kappa_4\kappa_2+\left(\frac{9}{n-1}-\frac{9}{N}\right)\kappa_3^2+\left(\frac{6n}{(n-1)(n-2)}-\frac{6}{N}\right)\kappa_2^3. \quad (22)$$

We also have the known formulae, to order $1/N$

$$E(K_4K_2)-\frac{1}{N}\kappa_6=\left(1+\frac{8}{N}\right)\kappa_4\kappa_2+\frac{6}{N}\kappa_3^2, \tag{23}$$

$$E(K_3^2)-\frac{1}{N}\kappa_6=\frac{9}{N}\kappa_4\kappa_2+\left(1+\frac{9}{N}\right)\kappa_3^2+\frac{6}{N}\kappa_2^3, \tag{24}$$

$$E(K_2^3)=\frac{3}{N}\kappa_4\kappa_2+\left(1+\frac{6}{N}\right)\kappa_2^3. \tag{25}$$

We have to eliminate $\kappa_4\kappa_2$, κ_3^2 and κ_2^3 from (22), (23), (24) and (25) and find, for the variance of k_3,

$$\begin{vmatrix} E_N(k_3-K_3)^2-\alpha_1K_6 & \frac{9}{n-1}-\frac{9}{N} & \frac{9}{n-1}-\frac{9}{N} & \frac{6n}{(n-1)(n-2)}-\frac{6}{N} \\ K_4K_2-\frac{1}{N}K_6 & 1+\frac{8}{N} & \frac{6}{N} & 0 \\ K_3^2-\frac{1}{N}K_6 & \frac{9}{N} & 1+\frac{9}{N} & \frac{6}{N} \\ K_2^3 & \frac{3}{N} & 0 & 1+\frac{6}{N} \end{vmatrix}=0, \tag{26}$$

which reduces to

$$\begin{aligned}\left(1+\frac{23}{N}\right)E_N(k_3-K_3)^2 = {} & K_6\left\{\frac{1}{n}-\frac{n^2+28n-11}{n(n-1)}\frac{1}{N}\right\} \\ & +K_4K_2\left\{\frac{9}{n-1}-\frac{9(n^2-7n+14)}{(n-1)(n-2)}\frac{1}{N}\right\} \\ & +K_3^2\left\{\frac{9}{n-1}-\frac{9(n-9)}{n-1}\frac{1}{N}\right\} \\ & +K_2^3\left\{\frac{6n}{(n-1)(n-2)}-\frac{6(n^2-11n-16)}{(n-1)(n-2)}\frac{1}{N}\right\}.\end{aligned} \tag{27}$$

REFERENCES

L. Isserlis (1931). On the moment distributions of moments in the case of samples drawn from a limited universe. *Proc. Roy. Soc.* A, **132**, 586.

A. A. Tschuprow (1923). On the mathematical expectation of the moments of frequency distributions. *Metron*, **2**, 461, 646.

O. Anderson (1935). *Einführung in die Mathematische Statistik*. Vienna: Springer.

From *Biometrika*, Vol. **39**, pp 14–16 (1952)

MOMENT-STATISTICS IN SAMPLES FROM A FINITE POPULATION

By M. G. KENDALL

Division of Research Techniques, London School of Economics

1. The problem of deriving the moment-statistics of sample moment-statistics is, in essence, extremely simple. With the usual notation for an augmented symmetric function of order ϖ we have, for samples from an infinite population,

$$E[p_1^{\pi_1}p_2^{\pi_2}\dots p_s^{\pi_s}] = n^{(\varpi)}(\mu'_{p_1})^{\pi_1}(\mu'_{p_2})^{\pi_2}\dots(\mu'_{p_s})^{\pi_s}, \tag{1}$$

and for samples from a finite population

$$E[p_1^{\pi_1}p_2^{\pi_2}\dots p_s^{\pi_s}]_n = [p_1^{\pi_1}p_2^{\pi_2}\dots p_s^{\pi_s}]_N n^{(\varpi)}/N^{(\varpi)}, \tag{2}$$

where the subscripts n and N in equation (2) mean that the symmetric functions relate respectively to the sample of n and the population of N members.

These two formulae contain all that is necessary to deal with the univariate case and can easily be extended to the multivariate case. The reasons that, in practice, this branch of the subject becomes algebraically complex are

(*a*) that the symmetric-function statistics with which we are usually concerned are not given in terms of augmented symmetrics and therefore have to be converted into them; and likewise the parent augmented symmetrics have to be converted back to the types we usually require;

(*b*) that powers and products of statistics, even those composed of augmented symmetrics, are not augmented symmetrics and have to be converted into them;

(*c*) that we may require the answer in terms of particular kinds of functions, particularly the cumulants, to deal more easily with special populations (e.g. the normal) or cases where the cumulants after a certain point can be neglected (e.g. the Edgeworth series);

(*d*) that we may require for conciseness and concinnity to use particular kinds of statistics and to express our results in terms of particular kinds of parameters.

2. Difficulties arising from (*a*) and (*b*) are now removed by the tables of symmetric functions. Difficulties under (*c*) were to some extent resolved by Fisher's combinatorial method of obtaining sampling cumulants, which is essentially a powerful short-cut. Since the publication of the paper by Irwin & Kendall (1944) it has been known how to derive the formulae for finite populations from those for infinite populations, this again being essentially a short-cut method of circumventing the straightforward use of equation (2). But hitherto no combinatorial method has been given for use in the finite case, and as Dr Wishart points out in the foregoing paper, the key lies in the generalized k-statistics suggested by Tukey (1950). The purpose of this note is to demonstrate the validity of the method developed by Wishart.

3. The generalized k-statistic $k_{pqr\dots}$ is most easily defined as having for its mean value the product of cumulants $\kappa_p\kappa_q\kappa_r\dots$. It then follows by the argument used by Irwin & Kendall that the mean value in samples from a finite population is $K_{pqr\dots}$.

Consider now the operators

$$S_p = \sum_{j=1}^{p} \frac{\partial^p}{\partial x_j^p} \quad (p = 1, 2, \dots), \tag{3}$$

and the operators δ_p (see, for example, Kendall, 1952, p. 275) such that

$$\left.\begin{aligned}\delta_p \mu'_r &= r^{(p)}\mu'_{r-p} \quad (r > p)\\ &= p! \qquad\qquad (r = p)\\ &= 0 \qquad\qquad (r < p).\end{aligned}\right\} \tag{4}$$

It is known that δ_p obliterates every cumulant except κ_p, i.e.

$$\left.\begin{aligned}\delta_p \kappa_q &= 0 \quad (q \neq p)\\ &= p! \ \ (q = p).\end{aligned}\right\} \tag{5}$$

The expectation of a symmetric function after operation by a set of S's is, if a constant, the same as the operation by corresponding δ's on the expectation. Thus for any combination of operators S of total order ϖ, say
$$S = S_{p_1}^{\pi_1} S_{p_2}^{\pi_2} \dots S_{p_s}^{\pi_s},$$

we find that
$$S\, k_{q_1 q_1 (\chi_1 \text{ factors})\, q_2 q_2 (\chi_2 \text{ factors}) \dots q_s q_s (\chi_s \text{ factors})} = 0, \tag{6}$$

unless $p_j = q_j$ and $\chi_j = \pi_j$ for $j = 1, \dots, s$, in which case

$$S\, k_{p_1 p_1 \dots (\pi_1 \text{ factors})\, p_2 p_2 \dots (\pi_2 \text{ factors}) \dots} = (p_1!)^{\pi_1} \dots (p_s!)^{\pi_s}. \tag{7}$$

This is a generalization of a result (Kendall, 1940*a*) for the ordinary k-statistics.

If now a function of k-statistics on the left, say, is expressed as a linear function of the generalized k-statistics on the right, including a term $k_{p_1 p_1 \dots (\pi_1 \text{ factors}) \text{ etc.}}$, and we operate on both sides by S, everything is obliterated on the right except the coefficient of this statistic and on the left we have the result of operating on the function by S. The evaluation of coefficients is then the same as in the combinatorial method evolved by Fisher. It can be demonstrated exactly in the manner given by myself (see Kendall, 1940*b*, 1952, pp. 276–9). The only difference is that, as Wishart points out, we cannot use rule 3 (which enjoins ignoration of patterns which fall into separate blocks) because we are here evaluating moments, not cumulants.

4. This proof, the simplest I can find, depends on the introduction of parent cumulants, just as the Irwin-Kendall method depends notionally on formulae for the infinite population. This is a perfectly valid method of reasoning, but we may note that an alternative proof could be obtained without introducing the infinite population. Equations (6) and (7), in fact, concern only the obliterating properties of the operators S on the k-statistics and could be established directly after the manner (Kendall, 1940*a*) in which I originally proved the less general results for ordinary k-statistics.

5. It may also be noted that the operational methods developed (Kendall, 1940*c*) for proceeding to the multivariate case are applicable here for developing formulae for the finite population. It hardly seems worth while displaying the resulting formulae, which would have a very limited usefulness in the present stage of development, but the methods are ready if they are ever required.

6. It is also interesting to note that the above methods solve another outstanding problem: given a function of moments or cumulants, to write down an unbiased estimator of it. For example, the variance of the second moment about the mean is

$$\operatorname{var} m_2 = \frac{(n-1)^2}{n^2}\left\{\frac{\kappa_4}{n} + \frac{2\kappa_2^2}{n-1}\right\}, \tag{8}$$

but if we wish to use this to test significance, the expression on the right usually has to be estimated from the sample and we require an unbiased estimator. This is written down at sight as

$$\frac{(n-1)^2}{n^2}\left\{\frac{k_4}{n}+\frac{2k_{22}}{n-1}\right\},$$

and can then be easily converted to ordinary k-statistics or sample-moments by the use of Wishart's tables. In this case

$$k_{22}=\frac{n-1}{n+1}\left\{k_2^2-\frac{k_4}{n}\right\},$$

and hence the unbiased estimator of $\operatorname{var} m_2$ is

$$\frac{(n-1)^2}{n^2(n+1)}\left\{\frac{n-1}{n}k_4+2k_2^2\right\}.$$

Parallel results hold for the case of sampling from a finite population.

REFERENCES

IRWIN, J. O. & KENDALL, M. G. (1944). *Ann. Eugen., Lond.*, **12**, 138.
KENDALL, M. G. (1940*a*). *Ann. Eugen., Lond.*, **10**, 106.
KENDALL, M. G. (1940*b*). *Ann. Eugen., Lond.*, **10**, 215.
KENDALL, M. G. (1940*c*). *Ann. Eugen., Lond.*, **10**, 392.
KENDALL, M. G. (1952). *The Advanced Theory of Statistics*, vol. **1**, 5th ed. London: Charles Griffin and Co.
TUKEY, J. W. (1950). *J. Amer. Statist. Ass.* **45**, 501.

From *Biometrika*, Vol. **30**, pp 81–93 (1938)

A NEW MEASURE OF RANK CORRELATION

By M. G. KENDALL

1. In psychological work the problem of comparing two different rankings of the same set of individuals may be divided into two types. In the first type the individuals have a given order A which is objectively defined with reference to some quality, and a characteristic question is: if an observer ranks the individuals in an order B, does a comparison of B with A suggest that he possesses a reliable judgment of the quality, or, alternatively, is it probable that B could have arisen by chance? In the second type no objective order is given. Two observers consider the individuals and rank them in orders A and B. The question now is, are these orders sufficiently alike to indicate similarity of taste in the observers, or, on the other hand, are A and B incompatible within assigned limits of probability? An example of the first type occurs in the familiar experiments wherein an observer has to arrange a known set of weights in ascending order of weight; the second type would arise if two observers had to rank a set of musical compositions in order of preference.

The measure of rank correlation proposed in this paper is capable of being applied to both problems, which are, in fact, formally very much the same. For purposes of simplicity in the exposition it has, however, been thought convenient to preserve a distinction between them.

Definition of τ

2. Consider a set of individuals, numbered from 1 to 10, whose objective order is that of the natural sequence 1, 2, 3, ..., 10, and consider an arbitrary ranking such as the following:

$$4 \quad 7 \quad 2 \quad 10 \quad 3 \quad 6 \quad 8 \quad 1 \quad 5 \quad 9$$

Consider the order of the nine pairs of numbers obtained by taking the first number 4, with each succeeding number. The first pair, 4 7, is in the correct order (in the sequence of 1, 2, ..., 10), and we therefore allot it a score $+1$. The second pair, 4 2, is in the wrong order and we score -1. The third pair, 4 10, scores $+1$, and so on, the nine scores being

$$+1-1+1-1+1+1-1+1+1, \text{ totalling } +3.$$

Consider also the scores of the second number, 7, with its eight succeeding numbers. They are

$$-1+1-1-1+1-1-1+1, \text{ totalling } -2.$$

The scores of the third number are

$$+1+1+1+1-1+1+1, \text{ totalling } +5.$$

Proceeding thus with each number, we have 9 scores, as follows

$$+3, \quad -2, \quad +5, \quad -6, \quad +3, \quad 0, \quad -1, \quad +2, \quad +1.$$

The total of these scores is $+5$.

Now the maximum score, obtained if the numbers are all in the objective order (1, 2, ..., 10), is 45. I therefore define a rank correlation coefficient between a variable ranked in the objective order (1, 2, ..., 10) and the variable ranked in the order above as

$$\tau = \frac{\text{actual score}}{\text{maximum possible score}} = \frac{5}{45} = +0{\cdot}11.$$

Generally, if there are n individuals, the maximum score, obtained if and only if they are all in objective order is $(n-1)+(n-2)+\ldots+1 = \frac{n(n-1)}{2}$. Denoting the actual score for any given ranking by Σ, we may calculate a measure of the rank correlation between this ranking and the objective ranking by putting

$$\tau = \frac{2\Sigma}{n(n-1)}. \qquad \ldots\ldots(1)$$

Two short methods for the calculation of τ

3. τ is calculable more easily than might appear at first sight from the above approach. Consider for example the order given above, viz.

4 7 2 10 3 6 8 1 5 9

We see that the number 1 has two numbers on its right and 7 on its left. We therefore score $+2-7=-5$, and then strike out the 1, being left with

4 7 2 10 3 6 8 5 9

The number 2 has 6 numbers on its right and two on its left and hence we score $6-2=+4$. We then strike out the 2 and proceed with the 3 and so on. It will be found that the scores obtained are

$$-5, \quad +4, \quad +1, \quad +6, \quad -3, \quad 0, \quad +3, \quad 0, \quad -1.$$

The total of these scores is $+5$, and is equal to Σ.

The above rule is quite general. Its validity will be evident when it is noted that instead of taking the first number with each succeeding number and so on, as in § 2, we consider the pairs contributing to Σ in a different way. Taking the number 1 first, and remembering that all the other numbers are greater than 1, we see that any number on the left must contribute -1 to Σ and any number on the right contributes $+1$. When 1 is struck out the procedure remains valid for 2, and so on.

4. Alternatively, the following procedure may be adopted:
Considering once again the order

$$4 \quad 7 \quad 2 \quad 10 \quad 3 \quad 6 \quad 8 \quad 1 \quad 5 \quad 9$$

we see that the first number, 4, has on its right 6 numbers which are greater. The second number, 7, has on its right 3 numbers which are greater. The third number, 2, has on its right 6 numbers which are greater; and so on. The numbers so obtained are

$$6, \quad 3, \quad 6, \quad 0, \quad 4, \quad 2, \quad 1, \quad 2, \quad 1$$

totalling 25.

There must, therefore, be $45-25 = 20$ numbers lying to the right of successive numbers in the order which are less than those numbers, and hence

$$\begin{aligned} \Sigma &= 25-20 \\ &= +5, \text{ as before.} \end{aligned}$$

Generally, if the number obtained by the above method of counting greater numbers is k

$$\Sigma = 2k - \frac{n(n-1)}{2}.$$

In practice, I find this method convenient and rapid. It has, moreover, the advantage of providing an independent check; for if the process is repeated counting greater numbers which lie *to the left*, giving a total of, say, l,

$$\Sigma = \frac{n(n-1)}{2} - 2l.$$

5. The use of τ can now be extended to the case where no objective order exists. In fact, given two rankings, A and B, of the same set of individuals τ may be defined as the coefficient obtained by regarding one order, A, as an objective order. If, for example, the orders are as follows:

A	6	9	4	3	5	10	2	1	8	7
B	6	5	10	2	3	9	7	4	1	8

τ is given by first rearranging A as an objective order, writing below it the corresponding member in B, thus

A'	1	2	3	4	5	6	7	8	9	10
B'	4	7	2	10	3	6	8	1	5	9

and then calculating τ in the manner of preceding paragraphs. Actually, as will be seen below, it is not necessary in any practical calculations to rewrite the orders in this way.

6. It is a notable fact that the same coefficient τ is reached whichever of the two orders, A and B, is rearranged as an objective order.

Consider again the orders given in the preceding paragraph, namely,

A'	1	2	3	4	5	6	7	8	9	10
B'	4	7	2	10	3	6	8	1	5	9

Rearranging B as an objective order we have

A''	8	3	5	1	9	6	2	7	10	4
B''	1	2	3	4	5	6	7	8	9	10

If we repeat this operation on the A'' and B'' we shall get back to A' and B'. A', B' and A'', B'' are thus reciprocally related and the permutations B' and A'' may be said to be *conjugate*.

We have to show that τ is the same when calculated from B' when A' is the objective ranking as when calculated from A'' when B'' is the objective ranking, i.e. that Σ is the same for two conjugate permutations with regard to an objective order 1, 2, ..., n.

In § 2, the value of Σ for B' was ascertained directly, the various items entering into the sum being

$$+3, \quad -2, \quad +5, \quad -6, \quad +3, \quad 0, \quad -1, \quad +2, \quad +1.$$

Consider now the value of Σ for A'' *obtained by the short method of* § 3.

The sums entering into Σ will be found to be

$$+3, \quad -2, \quad +5, \quad -6, \quad +3, \quad 0, \quad -1, \quad +2, \quad +1,$$

i.e. exactly the same as those for B' obtained by the more direct method; and hence Σ and τ are the same in the two cases.

This result is true in general. If the permutation B' begins with a number a_0 the contribution to $\Sigma_{B'}$ from pairs involving a_0 will be $(n-a_0)-(a_0-1)$. In A'' the a_0th number will be 1 and the contribution to $\Sigma_{A''}$ will also be $(n-a_0)-(a_0-1)$, in the manner of § 3. If the second number in B' is a_1 the contribution to $\Sigma_{B'}$ will be $(n-a_1)-(a_1-1)\pm 1$ according to whether a_1 is greater than a_0 or not. In A'' the a_1th number will be 2, and the contribution to $\Sigma_{A''}$ is also $(n-a_1)-(a_1-1)\pm 1$ according to whether 1 lies to the left or the right of 2 in A'', i.e. whether a_1 is greater than a_0 or not; and so on.

Thus Σ and τ are the same for two conjugate permutations with regard to the objective order 1, 2, ..., n.

7. In practical cases, the value of τ may be found as follows:

Write down above the given rankings the objective ranking. In the example already considered this would give

	1	2	3	4	5	6	7	8	9	10
A	6	9	4	3	5	10	2	1	8	7
B	6	5	10	2	3	9	7	4	1	8

The number 1 in B has an 8 above it in A. In the objective ranking 8 has two numbers to the right and seven to the left. Score, therefore, -5 and strike out the 8 in the objective ranking. The number 2 in B has a 3 above it in A, and 3 in the objective ranking has six numbers to its right (ignoring the number struck out) and two to its left, score $+4$; and so on, the scores being

$$-5, \quad +4, \quad +1, \quad +6, \quad -3, \quad 0, \quad +3, \quad 0, \quad -1,$$

totalling $+5$, which is equal to Σ.

8. τ satisfies certain elementary requirements of a measure of rank correlation. It is $+1$ if and only if correspondence between the rankings of A and B is perfect. It is -1 if and only if the rankings are exactly inverted. For intermediate values it appears to provide a satisfactory measure of the correspondence between the two rankings. A few examples for $n = 10$ will give some idea of the scale of measurement which it provides (an objective order 1, 2, ..., 10 is taken in each case):

Order	τ	ρ*
4 7 2 10 3 6 8 1 5 9	+0·11	+0·14
1 6 2 7 3 8 4 9 5 10	+0·56	+0·64
7 10 4 1 6 8 9 5 2 3	−0·24	−0·37
6 5 4 7 3 8 2 9 10 1	+0·02	+0·03
10 1 2 3 4 5 6 7 8 9	+0·60	+0·45
10 9 8 7 6 1 2 3 4 5	−0·56	−0·76

In the case where no objective ranking exists τ measures the closeness of correspondence between two given rankings in the sense that it measures how accurate either ranking would be if the other were objective. In other words it measures the *compatibility* of two rankings.

9. For the purpose of measuring correlation between ranks, therefore, τ appears to compare favourably with ρ. It is admitted that ρ can take $\frac{n^3-n}{6}$ values between -1 and $+1$, whereas τ can take only $\frac{n^2-n}{2}$ values in the range. This does not, however, appear to constitute a serious disadvantage to the sensitivity of τ.

On the other hand, τ possesses one marked advantage over ρ, in that it is not difficult to find the distribution of values obtained by correlating a given ranking with the members of a universe in which all possible rankings occur equally

* Throughout this paper ρ means the Spearman coefficient of rank correlation defined by

$$\rho = 1 - \frac{6S(d^2)}{n^3-n},$$

where d is a difference in ranks.

frequently. It is shown below that the distribution of τ tends to normality for large n, resembling ρ in this respect; but in fact τ is surprisingly close to normality even for low values of n, whereas the distribution for ρ has not yet been given, and appears to present peculiar features.*

The sampling distribution of Σ

10. To judge the significance of an observed value of τ or of Σ in the case where an objective order is given, we wish to know whether the value could have arisen by chance from a universe in which all the possible rankings of the n objects occur an equal number of times. It is, therefore, necessary to consider the distribution of Σ in such a universe. The distribution of τ may be found at once from that of Σ by dividing the variate values of Σ by $\frac{n(n-1)}{2}$.

The same distribution may be used to judge the significance of a value of τ expressing the compatibility of two rankings. A significantly negative τ, for example, would mean that if one ranking is taken to be objective the other has not, as judged by the τ-distribution, arisen by chance from the universe in which all possible rankings occur equally frequently; in other words that the two rankings are significantly incompatible.

Consider then the universe of values of Σ obtained from an objective order 1, 2, 3, ..., n and the $n!$ possible permutations of the first n integers. Let the number of values of a given Σ be denoted by $u_{n,\Sigma}$. Consider a given ranking of the numbers 1, ..., n, and the effect of inserting an additional number $(n+1)$ in the various possible places in this ranking, from the first place (preceding the first member of the rank of n) to the last place (following the last member of the rank of n).

Inserting the number $(n+1)$ at the beginning will add $-n$ to the value of Σ. Inserting it between the first and second members will add $-(n-2)$ to Σ. Inserting it between the second and third will add $-(n-4)$ to Σ, and so on. Adding the number $(n+1)$ at the end will add $+n$ to Σ.

It follows that

$$u_{n+1,\Sigma} = u_{n,\Sigma-n} + u_{n,\Sigma-n+2} + u_{n,\Sigma-n+4} + \dots + u_{n,\Sigma+n-4} + u_{n,\Sigma+n-2} + u_{n,\Sigma+n}. \qquad \dots\dots(2)$$

This recursion formula permits of the calculation of the frequency array of Σ.

11. If $n = 2$, there are two values of Σ, $+1$ and -1, i.e. $u_{2,-1} = u_{2,1} = 1$, $u_{2,0} = 0$. From (2) we have

$$u_{3,\Sigma} = u_{2,\Sigma-4} + u_{2,\Sigma-2} + u_{2,\Sigma} + u_{2,\Sigma+2} + u_{2,\Sigma+4},$$

* The fact that ρ tends to normality for large n has recently been proved by Hotelling & Pabst (1936). The remarks above on the behaviour of ρ for low values of n are founded on an expression for the sampling distribution of ρ which will be discussed in a further communication shortly to be published. This communication will also deal with the relation between τ and ρ.

the possible values of Σ ranging from -3 to $+3$. By substituting $\Sigma = -3, \dots +3$ in the above, we find

$$u_{3,3} = 1, \quad u_{3,2} = 0, \quad u_{3,1} = 2, \quad u_{3,0} = 0,$$

and similar values for the negative values of Σ.

Applying equation (2) again we find

$$u_{4,6} = 1, \quad u_{4,5} = 0, \quad u_{4,4} = 3, \quad u_{4,3} = 0,$$

$$u_{4,2} = 5, \quad u_{4,1} = 0, \quad u_{4,0} = 6, \quad \text{etc.}$$

The successive arrays of Σ may in fact be built up by the following process:

```
1  1
   1  1
      1  1
--------
1  2  2  1
   1  2  2  1
      1  2  2  1
         1  2  2  1
--------------------
1  3  5  6  5  3  1
etc.
```

At each stage, to find the array for $(n+1)$ we write down the n-array $(n+1)$ times, one under the other and moving one place to the right each time, and then sum the $(n+1)$ arrays. If the total array has a central value, that value is the frequency for $\Sigma = 0$, and all values of Σ must be even. If the total array has two central values, these values are the frequencies for $\Sigma = \pm 1$, and all values of Σ must be odd.

12. The above procedure may be condensed by forming a kind of figurate triangle as follows:

```
1     1
2     1  1
3     1  2  2   1
4     1  3  5   6   5   3   1
5     1  4  9  15  20  22  20  15  9  4  1
etc.  etc.
```

In this array, a number in the rth row is the sum of the number immediately above it and the $(r-1)$ numbers to the left of that number. The formation of the array is quite simple and several devices shorten the arithmetic. For instance, in part of the array towards the left a number in the rth row is the sum of the number immediately above it and the number immediately to the left. A check is provided by the fact that the total in the rth row is $r!$.

The following table shows the frequency distribution of Σ for values of n from 1 to 10.

TABLE I

Distribution of Σ for values of n from 1 to 10 (only the positive half of the symmetrical distribution shown)

Values of Σ

Values of n	0	1	2	3	4	5	6	7	8	9	10	11	12	13	14
1	1														
2	0	1													
3	0	2	0	1											
4	6	0	5	0	3	0	1								
5	22	0	20	0	15	0	9	0	4	0	1				
6	0	101	0	90	0	71	0	49	0	29	0	14	0	5	0
7	0	573	0	531	0	455	0	359	0	259	0	169	0	98	0
8	3836	0	3736	0	3450	0	3017	0	2493	0	1940	0	1415	0	961
9	29228	0	28675	0	27073	0	24584	0	21450	0	17957	0	14395	0	11021
10	0	250749	0	243694	0	230131	0	211089	0	187959	0	162337	0	135853	0

Values of Σ

Values of n	15	16	17	18	19	20	21	22	23	24	25	26	27	28
6	1													
7	49	0	20	0	6	0	1							
8	0	602	0	343	0	174	0	76	0	27	0	7	0	1
9	0	8031	0	5545	0	3606	0	2191	0	1230	0	628	0	285
10	110010	0	86054	0	64889	0	47043	0	32683	0	21670	0	13640	0

Values of Σ

Values of n	29	30	31	32	33	34	35	36	37	38	39	40	41	42	43	44	45
9	0	111	0	35	0	8	0	1									
10	8095	0	4489	0	2298	0	1068	0	440	0	155	0	44	0	9	0	1

The frequency polygon of the distribution is quite close to normality even for $n = 6$. For $n = 10$ the correspondence is very good over the material part of the range, as may be judged roughly by drawing the frequency polygon to Σ and the normal curve with the same area and standard deviation. On an ordinary scale the two curves are hardly distinguishable by the eye above $\Sigma = 5$.

STANDARD ERROR OF τ

13. A little consideration of the above method of obtaining the frequency distribution of Σ will show that the distribution may be arrayed by the function:

$$f = (x^{-1}+x)(x^{-2}+1+x^2)(x^{-3}+x^{-1}+x+x^3)\ldots$$
$$(x^{-(n-1)}+x^{-(n-3)}+\ldots+x^{(n-3)}+x^{(n-1)}). \qquad \ldots\ldots(3)$$

The coefficient of x^Σ in f is the frequency of Σ in the distribution.

If we differentiate f with respect to x and then multiply by x the coefficient of x^Σ is multiplied by Σ. Writing then θ for the operator $x\dfrac{\partial}{\partial x}$ we have

$$\mu_0\mu_1 = (\theta f)_{x=1},$$

and generally

$$\mu_0\mu_r = (\theta^r f)_{x=1}. \qquad \ldots\ldots(4)$$

Applying equation (4) when $r = 2$, I find

$$\mu_2 = \frac{n(n-1)(2n+5)}{18}, \qquad \ldots\ldots(5)$$

and hence the standard error of τ is given by

$$\sigma_\tau = \frac{1}{3}\sqrt{\frac{2(2n+5)}{n(n-1)}}, \qquad \ldots\ldots(6)$$

which, as n becomes large, gives

$$\sigma_\tau \sim \frac{2}{3}\cdot\frac{1}{\sqrt{n}}. \qquad \ldots\ldots(7)$$

Table II shows the proportion of the total frequencies falling outside ranges $\pm\sigma$, $\pm 2\sigma$, $\pm 3\sigma$ for some of the distributions of Table I.

The expected values on the hypothesis of a normal distribution are 0·3173, 0·0455, 0·0027 and it is clear that for most practical purposes in testing the significance of an observed τ for $n = 10$ or greater, the standard error may be used in the ordinary way.

14. Applying equation (4) when $r = 4$, I find

$$\mu_4 = \frac{n(n-1)}{2}\left\{1+\frac{74}{9}(n-2)+\frac{37}{6}(n-2)(n-3)+\frac{32}{25}(n-2)(n-3)(n-4)\right.$$
$$\left.+\frac{2}{27}(n-2)(n-3)(n-4)(n-5)\right\}. \qquad \ldots\ldots(8)$$

TABLE II

Proportion of frequencies of the distribution of Σ falling in certain ranges

n	σ_Σ	Proportion falling outside range		
		$\pm\sigma$	$\pm 2\sigma$	$\pm 3\sigma$
6	5·32	0·272	0·056	0·0000
7	6·66	0·381	0·030	0·0004
8	8·08	0·275	0·031	0·0004
9	9·59	0·359	0·045	0·0009
10	11·18	0·291	0·047	0·0009

From this β_2 may be obtained and it is evident that as n becomes large β_2 tends to the value 3. In fact it remains below that value, so that the distribution of Σ and therefore of τ is slightly platykurtic. The following table shows the values of β_2 for some values of n. The corresponding values of β_2 for the distribution of ρ are also given and it will be observed that, as judged by β_2, the approach of τ to normality is appreciably quicker than that of ρ.

TABLE III

Values of β_2 in the distribution of Σ and of ρ for certain values of n

n	$\beta_2(\Sigma)$	$\beta_2(\rho)$
5	2·53	2·07
10	2·78	2·54
20	2·89	2·77
30	2·93	2·85

In general, as will be seen below, the moment of order $2s$ is a polynomial of degree n^{3s}.

Proof of the normality of τ for large n

15. We shall prove that as $n \to \infty$

$$\mu_{2s} \sim \frac{(2s)!}{2^s s!} (\mu_2)^s,$$

where μ_{2s} is the $2s$th moment of the distribution of Σ. In virtue of the symmetry of the distribution moments of odd order vanish and it follows from the Second Limit Theorem of Probability (see Fréchet & Shohat, 1931) that the distribution of Σ, and hence that of τ, tends to normality in the sense that the frequency between τ_1 and τ_2 tends to

$$\frac{1}{\sigma\sqrt{(2\pi)}}\int_{\tau_1}^{\tau_2} e^{-\frac{x^2}{2\sigma^2}}dx.$$

16. Consider the effect of operating on the product f (equation (3)) by $\theta \equiv x\frac{\partial}{\partial x}$. The first operation will result in a sum of terms of type

$$\{-rx^{-r}-(r-2)\,x^{-(r-2)}-\ldots+(r-2)\,x^{(r-2)}+rx^r\}$$

multiplied by the remaining terms of f unchanged. When x is put equal to unity we may write this as the sum of terms

$$\frac{-r-(r-2)-\ldots+(r-2)+r}{1+1+\ldots+1+1}n! = \frac{-r-(r-2)-\ldots+(r-2)+r}{r}n!.$$

Similarly the second operation will bring out terms like

$$\frac{r^2+(r-2)^2+\ldots+(r-2)^2+r^2}{r}n!$$

and $$n!\left\{\frac{-r-(r-2)-\ldots+(r-2)+r}{r}\right\}\left\{\frac{-t-(t-2)-\ldots+(t-2)+t}{t}\right\}.$$

Generally, operating $2s$ times will bring out terms like

$$n!\left\{\frac{r^{2s}+(r-2)^{2s}+\ldots+(r-2)^{2s}+r^{2s}}{r}\right\},$$

$$n!\left\{\frac{-r^{2s-1}-(r-2)^{2s-1}-\ldots+(r-2)^{2s-1}+r^{2s-1}}{r}\right\}$$

$$\times\left\{\frac{-t-(t-2)-\ldots+(t-2)+t}{t}\right\},$$

etc.

When x is put equal to unity any term beginning with an odd superscript in the powers will vanish. Consider now the sum of terms like

$$n!\left\{\frac{r^2+\ldots+r^2}{r}\right\}\left\{\frac{t^2+\ldots+t^2}{t}\right\}\ldots\left\{\frac{u^2+\ldots+u^2}{u}\right\}, \qquad \ldots\ldots(9)$$

containing s factors.

It will be proved below that this term contributes the greatest power of n to the total sum giving $\mu_0\mu_{2s}$.

Further, in virtue of the multinomial form of Leibniz' theorem, the factor by which this term is multiplied in the expansion of $(\theta^{2s}f)$ is

$$\frac{(2s)!}{2!\,2!\ldots 2!} = \frac{(2s)!}{2^s}.$$

Hence, since $\mu_0 = n!$ we have

$$\mu_{2s} \sim \frac{(2s)!}{2^s}\{\text{sum of terms like (9)}\}. \qquad \ldots\ldots(10)$$

Each term in (9) is of type

$$\frac{1}{r}\{r^2+(r-2)^2+\ldots+(r-2)^2+r^2\},$$

i.e. is of order $\frac{r^2}{3}$. The summation will therefore tend to the sum of terms like $\frac{1}{3^s}\{1^2.2^2.\ldots.s^2\}$, each term containing s squares of the numbers 1, 2, ..., $(n-1)$. Call this Π_s.

Then Π_s is $\frac{1}{s!}$ times the sum of terms in

$$\frac{1}{3^s}\{1^2+2^2+\ldots+(n-1)^2\}^s, \qquad \ldots\ldots(11)$$

which contain s different factors.

Now (11) is of order $\frac{n^{3s}}{9^s} \sim (\mu_2)^s$. Hence if the product term Π_s tends to the sum (11)

$$\Pi_s \sim \frac{(\mu_2)^s}{s!},$$

and in virtue of (10)

$$\mu_{2s} \sim \frac{(2s)!}{2^s}\frac{(\mu_2)^s}{s!}.$$

To complete the demonstration, we have therefore to show that (11) tends asymptotically to the sum of its terms $s!\,\Pi_s$, i.e. that sums of terms like

$$1^4.2^2.\ldots.(s-1)^2, \quad 1^6.2^2.\ldots.(s-2)^2$$

tend in comparison to zero.

This may be shown inductively.

Consider first of all

$$\{1^2+2^2+\ldots+(n-1)^2\}^2 = 2\Pi_2+1^4+2^4+\ldots+(n-1)^4.$$

The expression on the left $\sim\frac{n^6}{9}$. But the sum of fourth powers on the right $\sim\frac{n^5}{5}$, which is of lower order. Hence the sum on the right $\sim 2\Pi_2$. Multiplying by $\{1^2+2^2+\ldots+(n-1)^2\}$ we have

$$\{1^2+2^2+\ldots+(n-1)^2\}^3 \sim 2\Pi_2\{1^2+2^2+\ldots+(n-1)^2\}$$
$$\sim 6\Pi_3 + \text{terms of type } 1^4 2^2.$$

These terms will be less in sum than

$$2\{1^2+2^2+\ldots+(n-1)^2\}\{1^4+2^4+\ldots+(n-1)^4\},$$

which $\sim 2 . \frac{n^3}{3} . \frac{n^5}{5}$, of order 8. But the expression on the left is of order 9. Hence $\{1^2 + 2^2 + \ldots + (n-1)^2\}^3 \sim 6\Pi_3$ and so on.

We can now justify the assertion that the maximum power of n arises from terms like $(1^2 . 2^2 . \ldots . s^2)$. In fact, by a similar line of reasoning to that just given it will be seen that sums of terms of type $\{1^4 . 2^2 . \ldots . (s-1)^2\}$, etc. are of lower order.

The demonstration is complete.

17. It appears therefore that the coefficient τ has a good claim to serious consideration as a measure of rank correlation. It is easily calculable. In the important case of the distribution wherein all possible rankings occur equally frequently its standard error is known; for the values of n likely to be required in practice it may be taken to be normally distributed; and where there is doubt the distribution can be obtained in an exact form.

It should also be remarked that τ has a natural significance. An observer who is given a set of objects (such as coloured discs) to rank appears to follow a process something like this: First of all he searches for the beginning of the series, say the disc of lightest shade. Having selected a disc, he compares it with each of the remainder to verify the propriety of his choice. The coefficient τ gives him one mark for each comparison which is made correctly, and subtracts a mark for each error.* When the first disc is selected, he proceeds as before with a second; and so on. τ follows this process exactly. It appears to be a logical measure of ranking carried out by the process and should therefore prove useful in psychological work.

REFERENCES

FRÉCHET, M. & SHOHAT, J. (1931). "A proof of the generalised second limit theorem in the theory of probability." *Trans. Amer. Math. Soc.* **33**, 533.

HOTELLING & PABST (1936). "Rank correlation and tests of significance involving no assumptions of normality." *Ann. Math. Statist.* **7**, 29.

* Inasmuch as comparisons between extremes in the series will generally be easier than comparisons between neighbouring members it might in some cases be preferable to weight the marking given to different comparisons according to some selected scale. The determination of such a scale, however, would depend to some extent on the circumstances of individual cases and would present considerable difficulty where no objective order is known to exist, apart from adding greatly to the complexity of the distribution of the measure so obtained.

From *Annals of Mathematical Statistics*, Vol. **10**,
pp 275–287 (1939)

THE PROBLEM OF m RANKINGS

BY M. G. KENDALL AND B. BABINGTON SMITH

1. **Introduction.** If n objects are ranked by m persons according to some quality of the objects there arises the problem: does the set of m rankings of n show any evidence of community of judgment among the m individuals? For example, if a number of pieces of poetry are ranked by students in order of preference, do the rankings support the supposition that the students have poetical tastes in common, and if so is there any strong degree of unanimity or only a faint degree?

The problem in its full generality permits of no assumption about the nature of the quality according to which the objects are ranked, other than that ranking is possible. No hypothesis is made that the quality is measurable, still less that there is some underlying frequency distribution to the quantiles of which the rankings correspond. The quality is to be thought of as linear in the sense that any two objects possessing it are either coincident or may be put in the relation "before and after." A metric may, of course, be imposed on this linear space by convention; but the relationship between objects is invariant under any transformation which stretches the scale of measurement. In particular, it is not a condition of the problem that the ranking shall be based on a distribution according to a normal variate.

It is permissible to denote the rankings by the *ordinal* numbers $1, 2, \cdots n$; but it is not permissible, without further discussion, to operate on these numbers as if they were cardinals. This point seems to have been inadequately appreciated. For instance, when $m = 2$ we have the familiar case of rank correlation between a pair of rankings; and this is frequently treated by subtracting corresponding ranks, squaring, and forming the Spearman coefficient

$$\rho = 1 - \frac{6S(d^2)}{n^3 - n}. \tag{1}$$

To justify this procedure it is necessary to explain what is meant, for example, by such a process as (4th minus 8th), or what the square of this difference of ordinal numbers represents.

It is worth stressing that the necessary transition from ordinals to cardinals can be made without invoking a scale of measurement. When we rank an object as first we mean, in effect, that no member of the set of n is preferred to it; when we rank it as the rth we mean that $(r - 1)$ objects are preferred to it. The ordinals of the ranking are then biunivocally related to the cardinals expressing the number of objects which are preferred. It is thus legitimate to apply the laws of cardinal arithmetic to them. For example, if an object A is ranked r_1 by Brown, r_2 by Jones and r_3 by Robinson we may form the sum $(r_1 + r_2 + r_3)$, which is to be interpreted as meaning that, taking the preferences of the three persons together, there were $(r_1 + r_2 + r_3 - 3)$ cases in which some other object was preferred to A. The point is of some importance, in view of the prevailing practice of replacing ranks by quantiles of the normal distribution—a practice which cannot always be regarded as justifiable and is sometimes little short of desperate.

To fix the ideas, consider the following three rankings of six objects

Object:	A	B	C	D	E	F
	5	4	1	6	3	2
	2	3	1	5	6	4
	4	1	6	3	2	5
Sum of ranks	11	8	8	14	11	11

We may sum the ranks for each object, as shown. These sums (which must add to 63, and in general to $mn(n + 1)/2$) reflect the degree of resemblance among the rankings. If the resemblance were perfect the sums would be 3, 6, 9, 12, 15, 18 (though not necessarily, of course in that order) and in such a case would be as different as possible. On the other hand, when there is little or no resemblance, as in the example given, the sums are approximately equal. It is thus natural to take the variance of these sums as providing some measure of the concordance in the rankings. If S is the observed sum of squares of the deviations of sums of ranks from the mean value $m(n + 1)/2$ (i.e. is n times the variance) we may write

$$W = \frac{12S}{m^2(n^3 - n)} \tag{2}$$

and call W the coefficient of concordance. Here $m^2(n^3 - n)/12$ is the maximum possible value of S, occurring if there is complete unanimity in the rankings, so that W may vary from 0 to 1. In the example given, $S = 25.5$, $W = 0.16$.

The coefficient W has arisen in several ways.

(a) W is simply related to the average of the $\binom{m}{2}$ Spearman rank correlation coefficients between pairs of the m rankings. It is easy to show that the average ρ is given by

$$\rho_{av} = \frac{\dfrac{12S}{n^3 - n} - m}{m^2 - m} \tag{3}$$

$$= \frac{mW - 1}{m - 1} \tag{4}$$

ρ_{av} has been considered by Kelley [3] as a measure of average intercorrelation in rankings, but he gives no results for testing the significance of observed values.

It is to be noted that whereas W may vary from 0 to 1, ρ_{av} may vary from $-1/(m-1)$ to 1, i.e. it is asymmetrical like the coefficient of intraclass correlation, to which it bears some resemblance.[1]

(b) Friedman [1] has considered a quantity χ_r^2 related to W by the equation

$$\chi_r^2 = m(n-1)W. \tag{5}$$

(c) Welch [6] and Pitman [5] have considered the problem of the distribution of variance in an array

$$a_1, a_2, \cdots a_n$$

$$b_1, b_2, \cdots b_n$$

etc., for permutations of the numbers a, b, etc. in rows.

This is more general than the ranking case, in which $a_1 \cdots a_n$, $b_1 \cdots b_n$ etc. reduce to permutations of the numbers $1 \cdots n$. Since S', the total sum of squares in an array of m rankings of n, is $m^2(n^3 - n)/12$, we have

$$W = \frac{S}{S'} \tag{6}$$

i.e. the ratio of variance between columns to the total variance.

2. **Significance of W.** To test whether an observed value of W is significant it is necessary to consider the distribution of W (or, more conveniently, of S) in the universe observed by permuting the n ranks in all possible ways. No generality is lost by supposing one ranking fixed, and the others will then give rise to $(n!)^{m-1}$ values of S.

The actual distribution of W (or S), as will be seen below, is irregular for low values of m and n, and likely to be quite irregular for moderate values. It may, however, be shown that the first four moments of W are

$$\mu_1' \text{ (about 0)} = \frac{1}{m} \tag{7}$$

$$\mu_2 = \frac{2(m-1)}{m^3(n-1)} \tag{8}$$

$$\mu_3 = \frac{8(m-1)(m-2)}{m^5(n-1)^2} \tag{9}$$

$$\mu_4 = \frac{24(m-1)}{m^7(n-1)^2}\left\{\frac{25n^3 - 38n^2 - 35n + 72}{25(n^3-n)} + 2(n-1)(m-2) + \frac{n+3}{2}(m-2)(m-3)\right\}. \tag{10}$$

---

[1] The Spearman rank correlation coefficient is the product-moment coefficient of correlation between the ranks considered as ordinary variate values. ρ_{av} is the intraclass correlation coefficient for the m sets of ranks, also considered as variate values.

Results equivalent to these for the first three moments were given by Friedman [1]; and for the first four moments by Pitman [5].

In a valuable contribution to the subject Friedman showed that the distribution of χ_r^2 tends to that of χ^2 with $(n - 1)$ degrees of freedom as m tends to infinity and suggested the use of χ_r^2 (equation (5)) for an ordinary test of significance in the χ^2 distribution. This is satisfactory for moderately large values, but for small values it is subject to the disadvantage inherent in any attempt to represent a distribution of finite range by one of infinite range—the fit near the tails is not likely to be very good. An improvement is obtained by noting that the first four moments of the Type I distribution,

$$df = \frac{1}{B(p, q)} W^{p-1}(1 - W)^{q-1} \tag{11}$$

are approximately those of W if m and n are moderately large, and

$$p = \frac{n-1}{2} - \frac{1}{m} \tag{12}$$

$$q = (m - 1)\left\{\frac{n-1}{2} - \frac{1}{m}\right\}. \tag{13}$$

For practical purposes it is most convenient to put

$$z = \tfrac{1}{2} \log_e \frac{(m-1)W}{1 - W} \tag{14}$$

so that z can be tested in Fisher's distribution with $(n - 1) - \frac{2}{m}$ $(= n_1)$ and $(m - 1)\left\{(n - 1) - \frac{2}{m}\right\}$ $(= n_2)$ degrees of freedom.

There can be little doubt that this test is quite reliable for moderate values of m and n; but it has hitherto been far from clear how reliable it is for low values of m and n. This point we attempt to clear up in the present paper.

3. **Distribution of S.** For the case $m = 2$ the distribution of S is the same as the distribution of the $S(d^2)$ used in calculating Spearman's rank correlation coefficient. A table showing the distribution up to and including $n = 8$ has already been given (Kendall and others, [4]). Tables giving probabilities that specified values of χ_r^2 would be attained or exceeded were given by Friedman [1] for $n = 3$, $m = 2$–9; and $n = 4$, $m = 2$–4. We have taken this work somewhat further and obtained the distributions for $n = 3$, $m = 2$–10; $n = 4$, $m = 2$–6; and $n = 5$, $m = 3$. Tables 1–4 give the probabilities based on these distributions.

These distributions were obtained by two methods. The first consisted of building up the distribution for $(m + 1)$ and n from that of m and n. For

TABLE 1

Probability that a given value of S will be attained or exceeded, for $n = 3$ and values of m from 2 to 10

Values of m

S	2	3	4	5	6	7	8	9	10
0	1.000	1.000	1.000	1.000	1.000	1.000	1.000	1.000	1.000
2	.866	.944	.931	.954	.956	.964	.967	.971	.974
6	.500	.528	.653	.691	.740	.768	.794	.814	.830
8	.167	.361	.431	.522	.570	.620	.654	.685	.710
14		.194	.273	.367	.430	.486	.531	.569	.601
18		.028	.125	.182	.252	.305	.355	.398	.436
24			.069	.124	.184	.237	.285	.328	.368
26			.042	.093	.142	.192	.236	.278	.316
32			.0046	.039	.072	.112	.149	.187	.222
38				.024	.052	.085	.120	.154	.187
42				.0085	.029	.051	.079	.107	.135
50				$.0^{3}77$	.012	.027	.047	.069	.092
54					.0081	.021	.038	.057	.078
56					.0055	.016	.030	.048	.066
62					.0017	.0084	.018	.031	.046
72					$.0^{3}13$	.0036	.0099	.019	.030
74						.0027	.0080	.016	.026
78						.0012	.0048	.010	.018
86						$.0^{3}32$	.0024	.0060	.012
96						$.0^{3}32$	.0011	.0035	.0075
98						$.0^{4}21$	$.0^{3}86$	.0029	.0063
104							$.0^{3}26$	.0013	.0034
114							$.0^{4}61$	$.0^{3}66$	.0020
122							$.0^{4}61$	$.0^{3}35$	.0013
126							$.0^{4}61$	$.0^{3}20$	$.0^{3}83$
128							$.0^{5}36$	$.0^{4}97$	$.0^{3}51$
134								$.0^{4}54$	$.0^{3}37$
146								$.0^{4}11$	$.0^{3}18$
150								$.0^{4}11$	$.0^{3}11$
152								$.0^{4}11$	$.0^{4}85$
158								$.0^{4}11$	$.0^{4}44$
162								$.0^{6}60$	$.0^{4}20$
168									$.0^{4}11$
182									$.0^{5}21$
200									$.0^{7}99$

TABLE 2

Probability that a given value of S will be attained or exceeded for n = 4 and m = 3 and 5

S	$m = 3$	$m = 5$	S	$m = 5$
1	1.000	1.000	61	.055
3	.958	.975	65	.044
5	.910	.944	67	.034
9	.727	.857	69	.031
11	.608	.771	73	.023
13	.524	.709	75	.020
17	.446	.652	77	.017
19	.342	.561	81	.012
21	.300	.521	83	.0087
25	.207	.445	85	.0067
27	.175	.408	89	.0055
29	.148	.372	91	.0031
33	.075	.298	93	.0023
35	.054	.260	97	.0018
37	.033	.226	99	.0016
41	.017	.210	101	.0014
43	.0017	.162	105	$.0^364$
45	.0017	.141	107	$.0^333$
49		.123	109	$.0^321$
51		.107	113	$.0^314$
53		.093	117	$.0^448$
57		.075	125	$.0^530$
59		.067		

example, with $m = 2$ and $n = 3$ we have the following values of the sums of ranks, measured about their mean:

Type			Frequency
−2	0	2	1
−2	1	1	2
−1	0	1	2
0	0	0	1

Here −2, 1, 1, and 2, −1, −1 are taken to be identical types, for they give the same value of S and will also give similar types when we proceed to $m = 3$ as follows.

In the case $m = 3$ each of the above type will appear added to the six permutations of −1, 0, 1; e.g. the type −2, 0, 2 will give one each of −3, 0, 3; −3, 1, 2;

TABLE 3

Probability that a given value of S will be attained or exceeded for $n = 4$ and $m = 2, 4$ and 6

S	$m = 2$	$m = 4$	$m = 6$	S	$m = 6$
0	1.000	1.000	1.000	82	.035
2	.958	.992	.996	84	.032
4	.833	.928	.957	86	.029
6	.792	.900	.940	88	.023
8	.625	.800	.874	90	.022
10	.542	.754	.844	94	.017
12	.458	.677	.789	96	.014
14	.375	.649	.772	98	.013
16	.208	.524	.679	100	.010
18	.167	.508	.668	102	.0096
20	.042	.432	.609	104	.0085
22		.389	.574	106	.0073
24		.355	.541	108	.0061
26		.324	.512	110	.0057
30		.242	.431	114	.0040
32		.200	.386	116	.0033
34		.190	.375	118	.0028
36		.158	.338	120	.0023
38		.141	.317	122	.0020
40		.105	.270	126	.0015
42		.094	.256	128	$.0^{3}90$
44		.077	.230	130	$.0^{3}87$
46		.068	.218	132	$.0^{3}73$
48		.054	.197	134	$.0^{3}65$
50		.052	.194	136	$.0^{3}40$
52		.036	.163	138	$.0^{3}36$
54		.033	.155	140	$.0^{3}28$
56		.019	.127	144	$.0^{3}24$
58		.014	.114	146	$.0^{3}22$
62		.012	.108	148	$.0^{3}12$
64		.0069	.089	150	$.0^{4}95$
66		.0062	.088	152	$.0^{4}62$
68		.0027	.073	154	$.0^{4}46$
70		.0027	.066	158	$.0^{4}24$
72		.0016	.060	160	$.0^{4}16$
74		$.0^{3}94$	.056	162	$.0^{4}12$
76		$.0^{3}94$	.043	164	$.0^{5}80$
78		$.0^{3}94$	.041	170	$.0^{5}24$
80		$.0^{4}72$	.037	180	$.0^{6}13$

TABLE 4

Probability that a given value of S will be attained or exceeded, for n = 5 and m = 3

S	$m = 3$	S	$m = 3$
0	1.000	44	.236
2	1.000	46	.213
4	.988	48	.172
6	.972	50	.163
8	.941	52	.127
10	.914	54	.117
12	.845	56	.096
14	.831	58	.080
16	.768	60	.063
18	.720	62	.056
20	.682	64	.045
22	.649	66	.038
24	.595	68	.028
26	.559	70	.026
28	.493	72	.017
30	.475	74	.015
32	.432	76	.0078
34	.406	78	.0053
36	.347	80	.0040
38	.326	82	.0028
40	.291	86	$.0^390$
42	.253	90	$.0^469$

$-2, -1, 3; -2, 1, 1; -1, -1, 2$; and $-1, 0, 1$. These types are then counted for each of the four basic types of $m = 2$ and we get:

Type			Frequency
−3	0	3	1
−3	1	2	6
−2	0	2	6
−2	1	1	6
−1	0	1	15
0	0	0	2

The case $m = 4$ is treated by considering the numbers of types obtained by adding the six permutations of $-1, 0, 1$ to the types for $m = 3$; and so on.

This method is quite convenient for $n = 2$ and $n = 3$. For $n = 4$ it becomes difficult owing to the labour of considering 24 permutations at each stage and to the increase in the number of types. For $n = 5$ there are 120 permutations and the labour becomes excessive.

The second method employed is a generalisation of a procedure we used for the Spearman coefficient. Taking first of all the case $m = 2$, consider the array

$$\begin{array}{ccccc} a^2 & a^3 & a^4 & \cdots & a^{(n+1)} \\ a^3 & a^4 & a^5 & \cdots & a^{(n+2)} \\ \cdots & \cdots & \cdots & \cdots & \cdots \\ a^{(n+1)} & a^{(n+2)} & a^{(n+3)} & \cdots & a^{2n} \end{array}$$

Any permissible set of values of the sums of ranks is obtained by selecting n entries from this array so that no entry appears more than once in the same row or column. If then, subtracting from each index the mean $(n + 1)$ and squaring, we write

$$E = \left\{ \begin{array}{ccccc} a^{(n-1)^2} & a^{(n-2)^2} & \cdots & a^1 & a^0 \\ a^{(n-2)^2} & a^{(n-3)^2} & \cdots & a^0 & a^1 \\ \cdots & \cdots & \cdots & \cdots & \cdots \\ a^0 & a^1 & \cdots & a^{(n-2)^2} & a^{(n-1)^2} \end{array} \right\} \tag{15}$$

the values of S are the powers of a in E when it is expanded as a sum of $n!$ terms each of which is obtained by multiplying n factors which do not appear in the same row or column. The distribution of S is arrayed by the expansion of E, the number of values of any S being the coefficient of a^S in the expansion.

Similarly, for m rankings, the distribution of S is given by the expansion of an m- dimensional E-function. For example, with $m = 3$ there would be a three-dimensional E-function the bottom plane of which would be

$$\begin{array}{cccc} a^{\left\{3-\frac{3(n+1)}{2}\right\}^2} & a^{\left\{4-\frac{3(n+1)}{2}\right\}^2} & \cdots & a^{\left\{n+2-\frac{3(n+1)}{2}\right\}^2} \\ a^{\left\{4-\frac{3(n+1)}{2}\right\}^2} & a^{\left\{5-\frac{3(n+1)}{2}\right\}^2} & \cdots & a^{\left\{n+3-\frac{3(n+1)}{2}\right\}^2} \\ \cdots & \cdots & \cdots & \cdots \\ a^{\left\{n+2-\frac{3(n+1)}{2}\right\}^2} & a^{\left\{n+3-\frac{3(n+1)}{2}\right\}^2} & \cdots & a^{\left\{2n+2-\frac{3(n+1)}{2}\right\}^2} \end{array}$$

The plane above this would be

$$\begin{array}{ccc} a^{\left\{4-\frac{3(n+1)}{2}\right\}^2} & \cdots & a^{\left\{n+3-\frac{3(n+1)}{2}\right\}^2} \\ \cdots & \cdots & \cdots \\ a^{\left\{n+3-\frac{3(n+1)}{2}\right\}^2} & \cdots & a^{\left\{2n+3-\frac{3(n+1)}{2}\right\}^2} \end{array}$$

and so on.

The E-function is difficult to handle in more than three dimensions, but for the two and three dimensional case it is manageable and we used it to obtain the distribution of S for $n = 5$ and $m = 3$.

4. Adequacy of the z-test. Tables 1–4 provide exact tests for the values of m and n there given. It remains to be seen how good the ordinary z-test applied to W would be for higher values. It may be presumed that if the test is satisfactory for any particular value of m and n for which exact results are available, it will be so for higher values of m and n. Since, for ordinary purposes the significance points of z as tabled by Fisher and Yates [2] would be employed, the most useful comparison would seem to be between those tables and the extreme values of tables 1–4.

For $n = 3$, $m = 9$, the 1% level is given approximately by $S = 78$. We have, testing for such a value, $W = 0.4814$, $z = 1.002$, $n_1 = \frac{16}{9}$, $n_2 = \frac{128}{9}$. By linear interpolation of reciprocals in the tables of z we find for the 1% point and these degrees of freedom $z = 0.954$. The correspondence is hardly satisfactory, and the z test might lead to incorrect inferences in practice. Matters improve a good deal, however, if we make continuity corrections, by subtracting unity from S before calculating W, and increasing by two the divisor $m^2(n^3 - n)/12$, so as to allow for the finite range. In this case $z = 0.979$.

For $n = 4$, $m = 6$ the 1% point is approximately $S = 100$. We have $W = 0.5556$, $z = 0.916$, $n_1 = 8/3$, $n_2 = 40/3$. By linear interpolation as before we find $z = 0.888$.

Continuity corrections again materially improve the agreement, giving a value of $z = 0.893$.

For $n = 5$ $m = 3$ there is no very convenient value of S close to the 1% point. For $P = 0.015$ $S = 74$ and for $P = 0.0078$ $S = 76$.

For $S = 74$ (with continuity corrections) $z = 1.020$
$S = 76$ (" " ") $z = 1.089$

By interpolation from the tables $z = 1.075$. The use of the z test would lead to the correct conclusion that a value of S equal to 74 falls below, and that of 76 above, the 1% point.

For values of m and n not included in Tables 1–4 it thus appears that the z-test with continuity corrections will give sufficiently accurate results, if n is greater than 3, at the 1% points. It may be presumed that the results at the 5% points are equally good and probably better. But for finer values of significance, such as 0.1%, it is doubtful whether the test is sound. The tails of the distribution of S for moderate values of m and n are very irregular.

For instance, the following is the tail of the distribution of S for $n = 3, m = 10$ (the total distribution being 10,077,696):

S	Frequency	S	Frequency
96	11,340	146	740
98	30,090	150	252
104	13,830	152	420
114	7,380	158	240
122	4,200	162	90
126	3,240	168	90
128	1,450	182	20
134	1,860	200	1

and the following is the tail for $n = 4$ $m = 6$ (the total being 7,962,624):

S	Frequency	S	Frequency	S	Frequency
100	5536	122	4100	146	810
102	8160	126	4480	148	225
104	10260	128	240	150	264
106	8850	130	1152	152	120
108	3920	132	660	154	180
110	13344	134	1980	158	60
114	5460	136	300	160	36
116	3870	138	600	162	30
118	3900	140	312	164	45
120	2472	144	100	170	18
				180	1

Irregularities of this kind run all through the distributions we have obtained, and frequency diagrams present the same sort of features we have noticed in the case $m = 2$ (Kendall and others, [4]). The representation of such distributions by continuous functions, no matter how close their lower moments, is obviously to be used with some care. Although the B-distribution or the associated z-distribution will give reasonable significance tests at levels of 1% or greater, they will probably be inadequate to represent frequencies occurring in narrow ranges.

5. **Some Experimental Distributions.** In some previous work we obtained a number of random permutations of the numbers 1–10 and 1–20. These were used to derive some experimental distributions of S which may be worth recording. Table 5 gives the distribution for 200 sets of pentads of 10 and Table 6 that for 100 triads of 20. In the distribution of Table 5, the mean of the grouped distribution is 404. The theoretical mean is 412.5 with a standard error of 12.3. In Table 6 the mean is 1936, the theoretical mean being 1995 with s.e. 53. The distributions accord quite well with expectation.

In conclusion we give two examples to illustrate some points of importance in ranking work. The first is a case in which ranks appear as the primary variate and in which the assumption of normality is clearly illegitimate.

6. **Example 1.** In some experiments in random series a pack of ordinary playing cards was shuffled and the order of the 13 cards of each suit from the top of the pack was noted. The pack was then reshuffled and again the orders noted. This was done 28 times. The question we wished to discuss was whether the shuffling was good, in the sense that the cards were thoroughly mixed at each shuffle.

Here, for each suit, say diamonds, we have 28 rankings of 13. The sums of ranks were 183, 137, 171, 207, 188, 160, 225, 174, 216, 192, 236, 239, 220. The mean is 196, and $S = 11522$, W (without continuity corrections, which are not

TABLE 5

Experimental Distribution of S in 200 sets ($m = 5$, $n = 10$)

S	Frequency
0–	1
50–	2
100–	7
150–	9
200–	21
250–	22
300–	24
350–	26
400–	20
450–	17
500–	12
550–	11
600–	10
650–	4
700–	5
750–	3
800–	3
. . .	. .
1000–	2
. . .	. .
1250–	1
Total	200

TABLE 6

Experimental Distribution of S in 100 sets ($m = 3$, $n = 20$)

S	Frequency
800–	4
1000–	8
1200–	8
1400–	6
1600–	12
1800–	15
2000–	20
2200–	12
2400–	6
2600–	5
2800–	0
3000–	3
3200–	0
3400–	1
Total	100

worth making for these values of m and n) = 0.08075, z (equation (14)) = 0.432. This falls just beyond the 1% point.

Similarly for the clubs W was found to be 0.0535; for the hearts, 0.0245; and for spades, 0.0342. None of these values is significant and we conclude that the randomisation introduced by the shuffling was good, at all events, so far as this test was concerned. It may be added that the shuffling was done with much more care than would be taken in an ordinary game of cards.

In psychological work there has sometimes been a confusion between the determination of a measure of agreement between subjects and that of an objective order based on experimental rankings. It may therefore be as well to point out that in its psychological applications the test of W is one of concordance between judgments. There may be quite a high measure of agreement about something which is incorrect.

7. **Example 2.** A number of students were given 12 photographs of persons unknown to them, and asked to rank them in what they judged from the photographs to be their intelligence. For 16 students the sums of ranks were

112, 94, 101, 84, 97, 75, 104, 84, 102, 146, 125, 124

The mean is 104. $S = 4472$, $W = 0.1222$. $z = 0.368$, and is barely significant, being between the 1% and the 5% points.

For 111 students the sums were

818, 670, 908, 410, 706, 526, 780, 485, 596, 1044, 959, 756

$$W = 0.2378, z = 1.768$$

This is highly significant and it is to be inferred that community of judgment exists between students or groups of students. But there was little relationship between the judgments and the intelligence of the photographed subjects as given by the Binet Intelligence Quotient.

Note added in proof:

While this paper was passing through the press Professor W. Allen Wallis, of Stanford University, kindly drew our attention to some unpublished work of his own on this subject. Professor Wallis had also arrived at the coefficient W which, he points out, is the ranking analogue of the correlation ratio. His paper is, we understand, on the point of publication.

REFERENCES

[1] M. Friedman, "The Use of Ranks to Avoid the Assumption of Normality Implicit in the Analysis of Variance." *Jour. Am. Stat. Assn.*, Vol. 32(1937), p. 675.

[2] R. A. Fisher, and F. Yates, *Statistical Tables for Biological, Agricultural and Medical Research*, 1938, Oliver & Boyd, Edinburgh.

[3] T. L. Kelley, *Statistical Methods*, 1927.

[4] M. G. Kendall, Sheila F. H. Kendall, and Bernard Babington Smith, "The Distribution of Spearman's Coefficient of Rank Correlation in a Universe in which all Rankings Occur an Equal Number of Times," *Biometrika*, Vol. 30(1938), p. 251.

[5] E. J. G. Pitman, "Significance Tests which may be applied to Samples from any Populations: III. The analysis of variance." *Biometrika*, Vol. 29(1938), p. 322.

[6] B. L. Welch, "On the z- test in Randomised Blocks and Latin Squares," *Biometrika*, Vol. 29(1937), p. 21.

LONDON, ENGLAND
AND
UNIVERSITY OF ST. ANDREWS,
SCOTLAND.

From *Biometrika*, Vol. **31**, pp 324–345 (1940)

ON THE METHOD OF PAIRED COMPARISONS

By M. G. KENDALL and B. BABINGTON SMITH

Introduction

1. Suppose we have a number of objects A, B, C, etc. which are to be considered according to the different degrees in which they exhibit some common quality. If the quality is measurable in some objective way the objects will yield a number of variate values, in which case the problem is amenable to treatment by well-understood methods. It may, however, happen either for theoretical or for practical reasons that the quality is not measurable. We then have to rely for a discussion of the variation of the quality on judgments of a more or less subjective kind carried out after a comparison of the objects among themselves.

One of the methods of comparison which has been widely used in this connexion is that of ranking. An observer examines the objects and arranges them in the order in which he judges them to possess the quality under consideration. This arrangement is called a ranking, and when two or more observers provide rankings of the same set of objects there arise the familiar questions of the type: is there any significant resemblance between the judgments of observers? or, do the data furnish any evidence that the objects have a "real" objective ranking?

2. The ranking method suffers from a serious drawback when the quality considered is not known with certainty to be representable by a linear variable. We may, for instance, ask an observer to rank a number of individuals in order of intelligence, and he may comply with the request in the full belief that he is doing something within his powers; but if intelligence is not measurable on a linear scale this ranking may fail to give a real picture either of the observer's preference or of the variation of intelligence among the individuals. It is not impossible that the observer should judge A more intelligent than B, B than C, and C than A, if the individuals are presented for his consideration one pair at a time. The likelihood of this happening is obviously increased when we are dealing with tastes in music, eatables or film stars; and in practice the event is not uncommon. Such "inconsistent" preferences can never appear in ranking, for if A is preferred to B and B to C, then A must automatically be shown as preferred to C. The use of ranking thus destroys what may be valuable information about preferences.

3. In this paper we consider a more general method of investigating preferences. With n objects, we shall suppose that each of the $\binom{n}{2}$ possible pairs is presented to an observer and his preference of one member of the pair noted.

We assume that a choice between two objects can always be made.* With m observers the data then comprise $m\binom{n}{2}$ preferences. The questions to be discussed include:

(*a*) Is there any evidence that a particular observer is capable of forming a reliable judgment of the quality under investigation; and if not, is the fault his, or is it due to the fact that he has been asked to perform an impossible task?

(*b*) Is there any significant concordance of preferences between observers?

(*c*) Can the quality under discussion be represented by a linear variable?

4. The method of offering for judgment objects two at a time is known as the method of paired comparisons. Hitherto it has been used mainly in human psychology, but it has some interesting applications in animal experimentation. For instance, in feeding experiments it is impossible to get an animal to rank a number of foods in order of preference but it is not difficult to offer pairs of foods and to note which is taken first. Experiments of this kind have, of course, to be conducted with great care to ensure that conditions operating when the different pairs are offered are as constant as possible; but the difficulties are far from being insuperable and the method of paired comparisons offers a useful technique in cases where the more usual procedures cannot be applied. From the point of view of theoretical statistics perhaps the most interesting part of the present work is that it offers some lines of approach to the difficult question whether a given quality can be legitimately regarded as based on a linear variable, i.e. whether ranking or scoring methods are justifiable or not.

CONSISTENCE IN PREFERENCES

5. If the object A is preferred to B we write $A \rightarrow B$ or $B \leftarrow A$. The $\binom{n}{2}$ preferences of a single observer may be represented in tabular form as shown in Table I.

In this table, which is shown for the six objects A to F, an entry of unity in column Y and row X means $X \rightarrow Y$, and is thus accompanied by a complementary zero in row Y and column X. The diagonals are blocked out. For example, in Table I, $A \rightarrow B$, $A \rightarrow C$, $D \rightarrow A$, etc.

The arrangement of the objects A to F in the row and column headings is quite arbitrary. There are $(n!)^2$ ways of representing the same configuration of preferences in such a table according to the permutations of objects in row and

* That is, we exclude cases in which an observer cannot make up his mind which object he prefers, just as in the ranking case one excludes the possibility of split ranks. In practice it sometimes happens that an observer is genuinely unable to reach a decision. To allow for this fact in the theoretical discussions would introduce complications of a most intractable kind. When the effect becomes important in practice it can be allowed for by selecting the set of preferences which are most unfavourable to the hypothesis under test.

TABLE I

	A	*B*	*C*	*D*	*E*	*F*
A	—	1	1	0	1	1
B	0	—	0	1	1	0
C	0	1	—	1	1	1
D	1	0	0	—	0	0
E	0	0	0	1	—	1
F	0	1	0	1	0	—

column; but in practice it is generally desirable to have the order in row and column the same, and even among the $n!$ possible arrangements so given there are often practical considerations which determine one order as more convenient than others.

6. Paired comparisons may also be represented geometrically by a method which can be illustrated for the case of the six objects as follows:

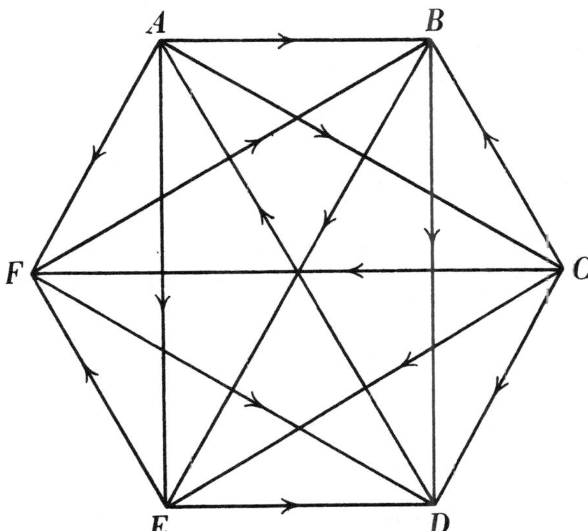

Fig. 1. Geometrical representations of the scheme of preferences of Table I.

We represent the six objects *A* to *F* by the six vertices of a regular hexagon and join the vertices in all possible ways by straight lines. If $A \to B$ we draw an arrow on the line AB pointing from A to B. The arrows shown on Fig. 1 correspond to the preferences shown in Table I.*

7. If an observer makes preferences of type $A \rightarrow B \rightarrow C \rightarrow A$ we say that the triad ABC is inconsistent. In the geometrical representation an inconsistent triad is shown by a triangle in which all the arrows go round in the same direction. We may thus speak of a "circular" triad of preferences. In Fig. 1 the triads ACD, BEF and three others are circular.

It is also possible to have inconsistent triads of greater extent; but any such circuit must contain at least two circular triads. Suppose, for instance, that $ABCD$ is circular, e.g. that $A \rightarrow B \rightarrow C \rightarrow D \rightarrow A$. Then either $A \rightarrow C$ or $C \rightarrow A$. In the first case ACD is circular, in the second ABC. Similarly either ABD or BCD is circular. Thus the circular tetrad must contain just two circular triads. On the other hand it is possible for a tetrad to contain circular triads without being itself circular.

Similarly, if $ABCDE$ is circular either ABC or $ACDE$ is circular and either BCD or $BDEA$ is circular. If the two tetrads are circular there must be at least three circular triads (not necessarily four, because ADE may be common to both). It is easy to see by an actual example based on this configuration that there need not be more than three circular triads; and it is clear that there must be at least three. For if the tetrads are not circular then ABC and BCD must be so and then either CDE is circular or $ABCE$ is so, adding at least one more.

Generally, it appears that a circular n-ad must contain at least $(n-2)$ circular triads; but it may contain more, and the fact that an n-ad contains $(n-2)$ circular triads does not mean that it is itself circular. In discussing inconsistences, therefore, it seems best to confine attention to circular triads, which, so to speak, constitute the inconsistent elements of the configuration, and to ignore the more ambiguous criteria associated with circular polyads of greater extent.

8. We now prove the following theorems:

(1) The maximum possible number of circular triads is $(n^3-n)/24$ if n is odd and $(n^3-4n)/24$ if n is even; and the minimum number is zero.

(2) These limits can always be attained by some configuration of preferences.

(3) For any integral number between the maximum and the minimum there exists at least one preference-configuration with that number of circular triads; and in general there will be more than one.

Consider a polygon of the type shown in Fig. 1 with n vertices. There will

* These preferences were obtained in an experiment on a dog, which was offered the following foods in pairs: meat, biscuit, chocolate, apple, pear and cheese. The members of a pair were cut to the same size and placed equidistantly from the dog, which was then released and allowed to choose. All the pieces of food were eaten avidly, it being that sort of dog, but there were considerable inconsistences in choice. We do not offer these data as more than an illustration of the method.

be $(n-1)$ lines emanating from each vertex. Let $\alpha_1, \alpha_2, \ldots, \alpha_n$ be the number of lines at the respective vertices on which the arrows *leave* the vertex.

Then
$$\mathop{S}_{r=1}^{n}(\alpha_r) = \binom{n}{2}$$

and the mean value of α_r is $(n-1)/2$.

Define
$$T = \mathop{S}_{r=1}^{n}\left(\alpha_r - \frac{n-1}{2}\right)^2$$
$$= S(\alpha_r^2) - \frac{n(n-1)^2}{4}. \qquad \ldots\ldots(1)$$

We now show that if the direction of a preference is altered and the effect is to increase the number of circular triads by d, T is reduced by $2d$; and conversely. Consider the preference $A \to B$. The only triads affected by altering this to $B \to A$ are those containing the line AB. Suppose there are α preferences of type $A \to X$ (including $A \to B$) and β preferences of type $B \to X$. Then four possible types of triad arise:

$A \to X \leftarrow B$, say p in number
$A \leftarrow X \to B$,
$A \to X \to B$, which must number $\alpha - p - 1$
$A \leftarrow X \leftarrow B$, ,, ,, $\beta - p$.

When the preference $A \to B$ is reversed the first two remain non-circular. The third becomes circular, the fourth ceases to be so. The reduction in the value of T is

$$\alpha^2 - (\alpha-1)^2 + \beta^2 - (\beta+1)^2$$
$$= 2(\alpha - \beta - 1)$$
$$= 2d, \text{ say.}$$

The increase in the number of circular triads is

$$(\alpha - p - 1) - (\beta - p) = \alpha - \beta - 1$$
$$= d.$$

More generally, if as the result of reversing any number of preferences T is decreased by $2d$, then d must be an integer and the number of circular triads must be increased by d. This clearly follows from the previous results for the reversal of preferences can take place one at a time and the effect on T and the number of circular triads is cumulative.

We now investigate the maximum and minimum values of T. It is clear from the definition that T is greatest when the α's are the natural numbers $1, 2, \ldots, n$; and this is a possible case because it corresponds to ordinary ranking. Hence max. $(T) = (n^3 - n)/12$.

For the minimum value, consider the polygon $A_1, A_2, \ldots, A_n$. Set up the preferences $A_1 \to A_2 \to \ldots A_n \to A_1$. Clearly at any vertex this results in one arrow entering and one leaving the vertex, i.e. the contribution to α is unity at each vertex. Next set up the preferences $A_1 \to A_3 \to A_5 \to \ldots$. This circuit may either visit each vertex once, or not. In the latter case we proceed to an unvisited vertex and set up the preferences $A_r \to A_{r+2} \to A_{r+4} \to \ldots$ and so on. Again there will be a unit contribution to all the α's.

We then set up the preferences $A_1 \to A_4 \to A_7 \to$ etc. and so on; and in this way we shall ultimately complete the preference scheme.

If n is odd all the preferences described will consist of circular tours of the polygon, and thus the value of α for each vertex will be $(n-1)/2$. If n is even the last preference $A_1 \to A_{\frac{1}{2}n+1}$ will not be a tour but will consist of the single line joining one vertex with the symmetrically opposite vertex. Thus there will be $n/2$ vertices for which $\alpha = n/2$ and $n/2$ vertices for which $\alpha = (n-2)/2$. In this case $T = n/4$.

Now it is clear from the definition of T that it cannot be less than zero, or if n is even, be less than $n/4$. The configuration just given shows that these minima are, in fact, attainable.

Thus T can vary from a maximum of $(n^3-n)/12$ to a minimum of zero or $n/4$. Hence the maximum number of circular triads, being half the variation from maximum to minimum of T (the maximum of T corresponding to the ranking case in which there are no inconsistences) is $(n^3-4n)/24$ if n is even and $(n^3-n)/24$ if n is odd.

This establishes the first two results enunciated at the beginning of this section. To prove the third it is sufficient to give a systematic method of proceeding from the configuration of minimum to that of maximum inconsistence by steps decreasing T two at a time. Consider, for example, the case $n = 8$. For the minimum inconsistence the α's will have the values 0 to 7, which we set out thus:

A	B	C	D	E	F	G	H
0	1	2	3	4	5	6	7

We proceed by reversing the preferences between vertices whose α-values differ by two. This clearly reduces T by two.

Reversing the preferences between C and E we get

A	B	C	D	E	F	G	H
0	1	3	3	3	5	6	7

and between D and F we get

A	B	C	D	E	F	G	H
0	1	3	4	3	4	6	7

which we may rearrange as

A	B	C	E	D	F	G	H
0	1	3	3	4	4	6	7

Now reversing the preferences between B and E and between D and G and rearranging we have

A	B	E	C	F	D	G	H
0	2	2	3	4	5	5	7

and now interchanging A and B, G and H,

A	B	E	C	F	D	G	H
1	1	2	3	4	5	6	6

At this stage we have preserved the α-numbers 2, 3, 4 and 5 in the middle but reduced the extremes A and H. We can now carry out the process again, arriving at the α-numbers.

1 2 2 3 4 5 5 6

and twice again, giving

2 2 3 3 4 4 5 5

whence a final interchange gives

3 3 3 3 4 4 4 4

and this is the position of maximum inconsistence. It is readily verified by following the interchanges on a polygon diagram that the reversals are, in fact, legitimate.

Coefficient of consistence in paired comparisons

9. If d is the number of circular triads in an observed configuration of preferences we define

$$\left.\begin{aligned}\zeta &= 1-\frac{24d}{n^3-n}, \quad n \text{ odd}\\ &= 1-\frac{24d}{n^3-4n}, \quad n \text{ even}\end{aligned}\right\} \quad \ldots\ldots(2)$$

and call ζ the coefficient of consistence. If and only if it is unity there are no inconsistences in the configuration, which may therefore be represented by a ranking. As ζ decreases to zero the inconsistence, as measured by the number of circular triads, increases.

For example, in the configuration of Fig. 1 there are five circular triads, ABD, ACD, AFD, AED and BEF. The maximum possible number is 8. Thus $\zeta = 0{\cdot}375$.

10. ζ can also be interpreted in the light of Table I. Suppose, in that table, we sum the rows. (The column sums are determined by the row sums and add no fresh information.) The sum of any row will be the α-number for that vertex in the polygon which corresponds to the object defining the row. T will then be the value of the sum of squares of deviations of row totals from the mean value $(n-1)/2$, that is to say, will be the variance of the row sums multiplied

by n. ζ is thus a linear function of this variance; but it cannot be tested in the χ^2 distribution as if Table I were a contingency table, for the border cells are not independent or linearly dependent.

11. If an individual observer produces a configuration of preferences which show inconsistence there are usually several explanations; he may be an incompetent judge, the objects may be so alike that consistent differentiation is not possible, or his attention may wander during the course of the experiment. We discuss these questions later. They are mentioned here to explain the motive for the next stage of the mathematics. With what probability can a value of ζ arise by chance if the observer allots his preferences at random with respect to the quality under consideration?

With n objects there are $2^{\binom{n}{2}}$ possible configurations of preferences. We proceed to investigate the distribution of d in this universe of $2^{\binom{n}{2}}$ different members. The method consists of proceeding from the distribution for n to that for $(n+1)$.

For $n = 3$ there are eight configurations, of which two give one circular triad and six no circular triads. Consider the effect of adding a new vertex D to the vertices ABC. Four cases arise:

(1) $D\rightarrow$ all A, B, C.
(2) $D\rightarrow$ two of A, B, C.
(3) $D\rightarrow$ one of A, B, C.
(4) $D\rightarrow$ none of A, B, C.

The last two are symmetrical with the first two and need not be separately considered.

Situation (1) arises in one way and clearly does not add any new circular triads other than those already existing in the configuration ABC. It therefore contributes six values $d = 0$ and two values $d = 1$. So does situation (4).

Situation (2) arises in three ways, according as $D\leftarrow A$, B, or C. The configurations so reached are similar and we may take any one, say $D\leftarrow C$, as the single preference. If $A\leftarrow C$ then DAC is not circular and if $B\leftarrow C$ the DBC is not circular. On the other hand $A\rightarrow C$ and $B\rightarrow C$ will each produce a circular triad. We then have the cases

	No. of circular triplets added
$A\leftarrow C\rightarrow B$	0
$A\rightarrow C\rightarrow B$	1
$A\leftarrow C\leftarrow B$	1
$A\rightarrow C\leftarrow B$	2

We now consider AB. In the first two cases just enumerated the direction of AB does not matter and no circular triads are added. With the third $A \to B$ gives no circular triad but $A \leftarrow B$ adds one. With the fourth $A \to B$ adds one and $A \leftarrow B$ adds none.

Thus the number of circular triads occurring for these four cases is found to be

No. of circular triplets	Frequency
0	2
1	2
2	4

We must multiply the frequency by three and by two to allow for similar symmetrical arrangements, and the final results are

No. of circular triplets	Frequency
0	24
1	16
2	24
Total	64

The principles of this method are clear enough and the work may be formalized by a number of conventions which we omit to save space. In common with many similar combinatorial problems, however, troubles arise from the sheer number of possibilities and the difficulty of ensuring that nothing is overlooked. Up to the present we have found the distribution of d for n up to and including 7. The frequencies and probabilities are given in Table II.

12. For the values already obtained the moments are given by the following formulae:

$$\mu_1' \text{ (about 0)} = \frac{1}{4}\binom{n}{3}, \qquad \text{......(3)}$$

$$\mu_2 = \frac{3}{16}\binom{n}{3}, \qquad \text{......(4)}$$

$$\mu_3 = -\frac{3}{32}\binom{n}{3}(n-4), \qquad \text{......(5)}$$

$$\mu_4 = \frac{3}{256}\binom{n}{3}\left\{9\binom{n-3}{3}+39\binom{n-3}{2}+9\binom{n-3}{1}+7\right\}. \qquad \text{......(6)}$$

We have very little doubt that these results are true in general but can offer no rigorous proof. In so far, however, as the moments are in a sense symmetric sums it appears highly probable that they are given by polynomials

TABLE II

Frequency (f) of values of d and probability (P) that values will be attained or exceeded

Value of d	$n=2$		$n=3$		$n=4$		$n=5$		$n=6$		$n=7$	
	f	P	f	P	f	P	f	P	f	P	f	P
0	2	1·000	6	1·000	24	1·000	120	1·000	720	1·000	5,040	1·000
1			2	0·250	16	·625	120	·883	960	·978	8,400	·998
2					24	·375	240	·766	2,240	·949	21,840	·994
3							240	·531	2,880	·880	33,600	·983
4							280	·297	6,240	·792	75,600	·967
5							24	·023	3,648	·602	90,384	·931
6									8,640	·491	179,760	·888
7									4,800	·227	188,160	·802
8									2,640	·081	277,200	·713
9											280,560	·580
10											384,048	·447
11											244,160	·263
12											233,520	·147
13											72,240	·036
14											2,640	·001
Total	2	—	8	—	64	—	1,024	—	32,768	—	2,097,152	—

in n; and if this is so the values obtained are sufficient to establish polynomials of degree six or less.

It is also to be noted that from the above values of the moments

$$\beta_1 = \mu_3^2/\mu_2^3 \sim 8/n, \qquad \beta_2 = \mu_4/\mu_2^2 \sim 3 + 12/n,$$

from which it appears that a Type III distribution would fit the d-distribution fairly closely for moderate or large values of n. But as the distribution of d is of interest mainly for low values of n, which are all that occur in practice, it hardly seems worth while attempting to fit a curve.

AGREEMENT AMONG SEVERAL OBSERVERS

13. We now consider the investigation of similarities of judgments for m observers. Suppose that in a table of the form of Table I we enter a unit in the cell in row X and column Y whenever $X \rightarrow Y$ and count the units in each cell. A cell may then contain any number from 0 to m. If the observers are in complete agreement there will be $\binom{n}{2}$ cells containing the number m, the remaining $\binom{n}{2}$ cells being zero. The agreement may be complete even if there are inconsistences present.

Suppose that the cell in row X and column Y contains the number γ. Let

$$\Sigma = S\binom{\gamma}{2}, \qquad \ldots\ldots(7)$$

the summation extending over the $n(n-1)$ cells of the table (the diagonal cells being ignored). Σ is then the sum of the number of agreements between pairs of judges. Put

$$u = \frac{2\Sigma}{\binom{m}{2}\binom{n}{2}} - 1. \qquad \ldots\ldots(8)$$

The maximum number of agreements, occurring if $\binom{n}{2}$ cells each contain m, is $\binom{n}{2}\binom{m}{2}$ and thus in the case of complete agreement, and only in this case, $u = 1$. The further we go from this case, as measured by agreements between pairs of observers, the smaller u becomes. The minimum number of agreements occurs when each cell contains $m/2$ if m is even or $(m \pm 1)/2$ if m is odd. That is, if m is even, the minimum number of agreements is

$$2\binom{\frac{m}{2}}{2}\binom{n}{2} = \tfrac{1}{4}m(m-2)\binom{n}{2},$$

and in this case

$$u = -\frac{1}{m-1}. \qquad \ldots\ldots(9)$$

When m is odd the minimum value of u is found to be

$$u = -\frac{1}{m}. \qquad \ldots\ldots(10)$$

14. We propose to call u the coefficient of agreement. It is unity if and only if there is complete agreement in the comparisons. Its minimum value is not -1 except when $m = 2$. This, however, is to be expected in a measure of agreement for there can be no such thing as complete disagreement among three or more observers in paired comparisons. If observer P differs in certain comparisons from observers Q and R, the two latter must agree on those comparisons.

When $m = 2$, u reduces to

$$u = \frac{2\Sigma}{\binom{n}{2}} - 1 \qquad \ldots\ldots(11)$$

and Σ becomes twice the number of cases in which the two observers agree about a comparison. u is thus a generalization of a coefficient τ proposed by Kendall (1938) to measure the correlation between two rankings. For general m, if the entries in the table were constrained to the ranking type, u would be the average intercorrelation τ between observers taken two at a time.

15. In discussing the significance of u it is desirable to know whether the set of preferences which give rise to it could have arisen by chance if the preferences had been assigned at random with respect to the quality under consideration. The procedure which first suggests itself is a generalization of the method used for the case of m rankings (Kendall & Babington Smith, 1939). That is to say, we sum the entries in the rows of the table and consider the variance of these entries. If the preferences are allotted at random we expect to find about equal numbers given to each object, and the variance will be low; in other cases it will be higher.

The difficulty about this suggestion is that it has not been found possible to ascertain the distribution of the variance in the $2^{m\binom{n}{2}}$ possible sets of preferences. The case $m = 1$, corresponding to the distribution of d for inconsistences, is difficult enough to solve. For higher values of m we have failed to find any distributions except in trivial cases.

We can, however, offer a test based on the distribution of u (or Σ). The comparative simplicity of the distributions in this case is in accordance with the remark made by Kendall in the paper under reference that the distribution of τ is much simpler and much more regular than the distribution of the Spearman correlation coefficient ρ.

16. Consider one cell in the table in row X and column Y and let it contain the number γ. Then the corresponding cell in row Y and column X will contain $m-\gamma$. Thus these two contribute to Σ the amount $\binom{\gamma}{2}+\binom{m-\gamma}{2}$.

Now, of the total ways in which the units can be distributed in the first cell there will be $\binom{m}{\gamma}$ in which γ units occur. Consequently the distribution of Σ in the cell and the corresponding cell is given by the expression

$$f = t^{\binom{m}{2}} + \binom{m}{1}t^{\binom{m-1}{2}} + \binom{m}{2}t^{\binom{m-2}{2}+\binom{2}{2}} + \ldots + \binom{m}{\gamma}t^{\binom{m-\gamma}{2}+\binom{\gamma}{2}} + \ldots + t^{\binom{m}{2}}, \quad \ldots\ldots(12)$$

and since the distribution in other pairs of cells is independent if the preferences are allotted at random the distribution of Σ for the whole table is given by

$$D(\Sigma) = f^N, \quad \ldots\ldots(13)$$

where $N = \binom{n}{2}$.

17. The distributions have been worked out for the following values of m and n: $m = 3$, $n = 2$ to 8; $m = 4$, $n = 2$ to 6; $m = 5$, $n = 2$ to 5; $m = 6$, $n = 2$ to 4. Tables III to VI give the probabilities based on these distributions, i.e. the probabilities that a given value of Σ will be attained or exceeded.

For constant n the distribution tends to the Type III form as m tends to infinity. In fact, for a single pair of related cells the variate value corresponding

22-2

TABLE III

The probability P that a value of Σ will be attained or exceeded, for m = 3, n = 2 to 8

n=2		n=3		n=4		n=5		n=6		n=7		n=8	
Σ	P	Σ	P	Σ	P	Σ	P	Σ	P	Σ	P	Σ	P
1	1·000	3	1·000	6	1·000	10	1·000	15	1·000	21	1·000	28	1·000
3	·250	5	·578	8	·822	12	·944	17	·987	23	·998	30	1·000
		7	·156	10	·466	14	·756	19	·920	25	·981	32	·997
		9	·016	12	·169	16	·474	21	·764	27	·925	34	·983
				14	·038	18	·224	23	·539	29	·808	36	·945
				16	·0046	20	·078	25	·314	31	·633	38	·865
				18	$\cdot 0^{3}24$	22	·020	27	·148	33	·433	40	·736
						24	·0035	29	·057	35	·256	42	·572
						26	$\cdot 0^{3}42$	31	·017	37	·130	44	·400
						28	$\cdot 0^{4}30$	33	·0042	39	·056	46	·250
						30	$\cdot 0^{6}95$	35	$\cdot 0^{3}79$	41	·021	48	·138
								37	$\cdot 0^{3}12$	43	·0064	50	·068
								39	$\cdot 0^{4}12$	45	·0017	52	·029
								41	$\cdot 0^{6}92$	47	$\cdot 0^{3}37$	54	·011
								43	$\cdot 0^{7}43$	49	$\cdot 0^{4}68$	56	·0038
								45	$\cdot 0^{9}93$	51	$\cdot 0^{4}10$	58	·0011
										53	$\cdot 0^{5}12$	60	$\cdot 0^{3}29$
										55	$\cdot 0^{6}12$	62	$\cdot 0^{4}66$
										57	$\cdot 0^{8}86$	64	$\cdot 0^{4}13$
										59	$\cdot 0^{9}44$	66	$\cdot 0^{5}22$
										61	$\cdot 0^{10}15$	68	$\cdot 0^{6}32$
										63	$\cdot 0^{12}23$	70	$\cdot 0^{7}40$
												72	$\cdot 0^{8}42$
												74	$\cdot 0^{9}36$
												76	$\cdot 0^{10}24$
												78	$\cdot 0^{11}13$
												80	$\cdot 0^{13}48$
												82	$\cdot 0^{14}12$
												84	$\cdot 0^{16}14$

TABLE IV

The probability P that a value of Σ will be attained or exceeded, for m = 4 and n = 2 to 6
(for n = 6 only values beyond the 1 % point are given)

n=2		n=3		n=4		n=5		n=5		n=6		n=6	
Σ	P	Σ	P	Σ	P	Σ	P	Σ	P	Σ	P	Σ	P
2	1·000	6	1·000	12	1·000	20	1·000	42	·0048	57	·014	79	$\cdot 0^{8}42$
3	·625	7	·947	13	·997	21	1·000	43	·0030	58	·0092	80	$\cdot 0^{8}28$
6	·125	8	·736	14	·975	22	·999	44	·0017	59	·0058	81	$\cdot 0^{9}98$
		9	·455	15	·901	23	·995	45	$\cdot 0^{3}73$	60	·0037	82	$\cdot 0^{9}15$
		10	·330	16	·769	24	·979	46	$\cdot 0^{3}41$	61	·0022	83	$\cdot 0^{9}12$
		11	·277	17	·632	25	·942	47	$\cdot 0^{3}24$	62	·0013	84	$\cdot 0^{10}51$
		12	·137	18	·524	26	·882	48	$\cdot 0^{4}90$	63	$\cdot 0^{3}76$	86	$\cdot 0^{11}30$
		14	·043	19	·410	27	·805	49	$\cdot 0^{4}37$	64	$\cdot 0^{3}44$	87	$\cdot 0^{11}17$
		15	·025	20	·278	28	·719	50	$\cdot 0^{4}25$	65	$\cdot 0^{3}23$	90	$\cdot 0^{13}28$
		18	·0020	21	·185	29	·621	51	$\cdot 0^{5}93$	66	$\cdot 0^{3}13$		
				22	·137	30	·514	52	$\cdot 0^{5}21$	67	$\cdot 0^{4}72$		
				23	·088	31	·413	53	$\cdot 0^{5}17$	68	$\cdot 0^{4}36$		
				24	·044	32	·327	54	$\cdot 0^{6}74$	69	$\cdot 0^{4}18$		
				25	·027	33	·249	56	$\cdot 0^{7}66$	70	$\cdot 0^{5}97$		
				26	·019	34	·179	57	$\cdot 0^{7}38$	71	$\cdot 0^{5}47$		
				27	·0079	35	·127	60	$\cdot 0^{9}93$	72	$\cdot 0^{5}20$		
				28	·0030	36	·090			73	$\cdot 0^{5}10$		
				29	·0025	37	·060			74	$\cdot 0^{6}51$		
				30	·0011	38	·038			75	$\cdot 0^{6}18$		
				32	$\cdot 0^{3}16$	39	·024			76	$\cdot 0^{7}78$		
				33	$\cdot 0^{4}95$	40	·016			77	$\cdot 0^{7}44$		
				36	$\cdot 0^{5}38$	41	·0088			78	$\cdot 0^{7}15$		

TABLE V

The probability P that a value of Σ will be attained or exceeded, for m = 5 and n = 2 to 5

$n=2$		$n=3$		$n=4$		$n=5$		$n=5$	
Σ	P	Σ	P	Σ	P	Σ	P	Σ	P
4	1·000	12	1·000	24	1·000	40	1·000	76	$\cdot 0^{4}50$
6	·375	14	·756	26	·940	42	·991	78	$\cdot 0^{4}16$
10	·063	16	·390	28	·762	44	·945	80	$\cdot 0^{5}50$
		18	·207	30	·538	46	·843	82	$\cdot 0^{5}15$
		20	·103	32	·353	48	·698	84	$\cdot 0^{6}39$
		22	·030	34	·208	50	·537	86	$\cdot 0^{6}10$
		24	·011	36	·107	52	·384	88	$\cdot 0^{7}23$
		26	·0039	38	·053	54	·254	90	$\cdot 0^{8}53$
		30	$\cdot 0^{3}24$	40	·024	56	·158	92	$\cdot 0^{8}12$
				42	·0093	58	·092	94	$\cdot 0^{9}14$
				44	·0036	60	·050	96	$\cdot 0^{10}46$
				46	·0012	62	·026	100	$\cdot 0^{12}91$
				48	$\cdot 0^{3}36$	64	·012		
				50	$\cdot 0^{3}12$	66	·0057		
				52	$\cdot 0^{4}28$	68	·0025		
				54	$\cdot 0^{5}54$	70	·0010		
				56	$\cdot 0^{5}18$	72	$\cdot 0^{3}39$		
				60	$\cdot 0^{7}60$	74	$\cdot 0^{3}14$		

TABLE VI

The probability P that a value of Σ will be attained or exceeded, for m = 6 and n = 2 to 4

$n=2$		$n=3$		$n=4$		$n=4$		$n=4$	
Σ	P	Σ	P	Σ	P	Σ	P	Σ	P
6	1·000	18	1·000	36	1·000	55	·043	74	$\cdot 0^{4}12$
7	·688	19	·969	37	·999	56	·029	75	$\cdot 0^{5}89$
10	·219	20	·832	38	·991	57	·020	76	$\cdot 0^{5}49$
15	·031	21	·626	39	·959	58	·016	77	$\cdot 0^{5}32$
		22	·523	40	·896	59	·011	80	$\cdot 0^{6}68$
		23	·468	41	·822	60	·0072	81	$\cdot 0^{6}17$
		24	·303	42	·755	61	·0049	82	$\cdot 0^{6}12$
		26	·180	43	·669	62	·0034	85	$\cdot 0^{7}34$
		27	·147	44	·556	63	·0025	90	$\cdot 0^{8}93$
		28	·088	45	·466	64	·0016		
		29	·061	46	·409	65	$\cdot 0^{3}83$		
		30	·040	47	·337	66	$\cdot 0^{3}66$		
		31	·034	48	·257	67	$\cdot 0^{3}48$		
		32	·023	49	·209	68	$\cdot 0^{3}26$		
		35	·0062	50	·175	69	$\cdot 0^{3}16$		
		36	·0029	51	·133	70	$\cdot 0^{4}86$		
		37	·0020	52	·097	71	$\cdot 0^{4}68$		
		40	$\cdot 0^{3}58$	53	·073	72	$\cdot 0^{4}48$		
		45	$\cdot 0^{4}31$	54	·057	73	$\cdot 0^{4}16$		

to a frequency $\binom{m}{\gamma}$ is $\binom{m-\gamma}{2}+\binom{\gamma}{2}$, which is a quadratic in γ. Were the variate value a linear function of γ the distribution for the single cell would tend to normality in accordance with the well-known property of the binomial. The case of the quadratic value corresponds to a transformation of the variate of the type $x^2 = y$ and the transform of the normal form $\exp(-x^2)\,dx$ becomes the Type III form $\exp(-y)\,y^{-\frac{1}{2}}dy$. Since the N cells are independent and the sum of variates in the same Type III form is also distributed in that form, it follows that Σ is in the limit distributed as $\exp(-\Sigma)\,\Sigma^{\frac{N}{2}-1}d\Sigma$ except perhaps for some constants. Thus Σ or some multiple of it is distributed as χ^2.

For constant m the distribution tends to normality with increasing n.

18. The first of these results suggests that the Type III distribution will provide an approximation to the distribution (13) when m is moderately large. We proceed to find the first four moments of (13).

It is sufficient to find the first four moments of (12), those of (13) being obtainable therefrom in virtue of the relationships which connect seminvariants of independent distributions.

The rth moment of (12) about the origin is given by

$$2^m\mu_r' = \left[\left(t\frac{\partial}{\partial t}\right)^r f\right]_{t=1}, \qquad \text{......(14)}$$

since 2^m is the total frequency. Thus we have

$$2^m\mu_1' = \mathop{S}_{r=0}^{m}\binom{m}{r}\left(r^2-mr+\frac{m^2-m}{2}\right) = 2^m\binom{m}{2}+S\binom{m}{r}(r^2-mr). \qquad \text{......(15)}$$

Sums such as $S\binom{m}{r}r^p$ can be obtained by operating on the binomial $(1+x)^m$ p times by $x\frac{\partial}{\partial x}$, e.g. we find

$$S\left\{\binom{m}{r}r\right\} = 2^m\frac{m}{2},$$

$$S\binom{m}{r}r^2 = 2^m\left\{\frac{m}{2}+\frac{1}{2}\binom{m}{2}\right\}$$

and hence, substituting in (15),

$$\mu_1' = \frac{1}{2}\binom{m}{2}.$$

Thus the mean of the distribution (13) is given by

$$\mu_1' = \tfrac{1}{2}N\binom{m}{2}. \qquad \text{......(16)}$$

In a similar way we find

$$\mu_2 = \tfrac{1}{4}N\binom{m}{2}, \qquad \ldots\ldots(17)$$

$$\mu_3 = \tfrac{3}{4}N\binom{m}{3}, \qquad \ldots\ldots(18)$$

$$\mu_4 = N\binom{m}{2}\left\{\frac{3m^2-15m+17}{8} + \tfrac{3}{32}N(m^2-m)\right\}. \qquad \ldots\ldots(19)$$

These are the moments of Σ. Those of u are obtained by dividing by an appropriate power of $N\binom{m}{2}$ and it may be noted in particular that the mean of u is zero.

We have directly from (17), (18) and (19)

$$\beta_1 = \frac{8}{N}\frac{(m-2)^2}{m(m-1)},$$

$$\beta_2 = \frac{4}{Nm(m-1)}\left\{3m^2-15m+17+\frac{3N}{4}m(m-1)\right\}.$$

For constant m, as $N \to \infty$,

$$\beta_1 \to 0, \quad \beta_2 \to 3$$

and for constant N, as $m \to \infty$,

$$\beta_1 \to \frac{8}{N}, \quad \beta_2 \to \frac{12}{N} + 3,$$

confirming the tendency towards the Type III distribution.

19. The first four moments of the Type III distribution

$$dF = ke^{-px}\,x^{q-1}\,dx$$

are

$$\frac{q}{p}, \frac{q}{p^2}, \frac{2q}{p^3}, \frac{3q(q+2)}{p^4}.$$

Equating the second and third moments to those given by (17) and (18) we find

$$q = \frac{Nm(m-1)}{2(m-2)^2}, \qquad \ldots\ldots(20)$$

$$p = \frac{2}{m-2}. \qquad \ldots\ldots(21)$$

To make the first moments correspond we move the origin of the Σ distribution a distance $\frac{1}{2}N\binom{m}{2}\frac{m-3}{m-2}$ to the right. We thus reach the approximation to the Σ distribution, coinciding in the first three moments

$$dF = ke^{-\frac{2x}{m-2}}x^{\frac{Nm(m-1)}{2(m-1)^2}-1}dx,$$

where $$x = \Sigma - \tfrac{1}{2}N\binom{m}{2}\frac{m-3}{m-2}$$

or, transforming to the more usual χ^2 form by putting $\chi^2 = 4x/(m-2)$, we find that

$$\left\{\Sigma - \tfrac{1}{2}N\binom{m}{2}\frac{m-3}{m-2}\right\}\frac{4}{m-2} \qquad \text{......(22)}$$

is distributed as χ^2 with

$$\nu = \frac{Nm(m-1)}{(m-2)^2} \qquad \text{......(23)}$$

degrees of freedom.

The fourth moments of Σ and the χ^2 approximation differ by terms of order N^{-1} and m^{-1} compared with their absolute values.

20. It only remains to be seen how large m and n must be for this to provide a satisfactory approximation.

Consider first the distributions for $m = 3$. When $n = 8$, $N = 28$, we have, for the approximation, 4Σ distributed with 168 degrees of freedom. From Table III we see that for $\Sigma = 54$, $P = 0{\cdot}011$ and for $\Sigma = 58$, $P = 0{\cdot}0011$. Applying a continuity correction by deducting unity from Σ we find for the χ^2 approximation with $\chi^2 = 4 \times 53$, $\nu = 168$, $P = 0{\cdot}011$, and with $\chi^2 = 4 \times 57$, $P = 0{\cdot}00114$. The correspondence is very close, in spite of the low value of m.

For $m = 4$, $n = 5$, $N = 10$, the approximation gives $2\Sigma - 30$ distributed with 30 degrees of freedom. For $\Sigma = 40$ and 41, this gives, with continuity corrections of 0·5, half the variate-interval, $\chi^2 = 49$ and 51, $\nu = 30$. From the diagram given in Yule & Kendall's "Introduction to the Theory of Statistics" (1937) it is seen that these values lie one on either side of the 1 % value; and this is in accordance with the exact values of P, which are seen from Table IV to be 0·016 and 0·0088. Similarly we find that the values of Σ, 37 and 38, lie on either side of the 5 % level, which is again in accordance with the exact values, $P = 0{\cdot}060$ and 0·038.

For $m = 6$, $n = 4$, $N = 6$, the approximation gives $\Sigma - 33{\cdot}75$ distributed with 11·25 degrees of freedom. For $\Sigma = 59$ and 60 the corresponding χ^2 values are seen to lie on either side of the 1 % point, which accords with the exact value of Table VI.

We conclude that the χ^2 approximation provides an adequate test of significance for the values of m and n outside the range for which Tables III–VI give exact values.

21. As a matter of theoretical interest we may record the results for the distribution of u when the data are ranked. It appears that in this case

$$\frac{12}{m-2}\frac{2n+5}{2n^2+6n+7}\left[\Sigma-\frac{1}{2}\binom{m}{2}\binom{n}{2}\left\{1-\frac{1}{3(m-2)}\frac{(2n+5)^2}{2n^2+6n+7}\right\}\right]$$

is distributed approximately as χ^2 with

$$\binom{m}{2}\frac{2}{(m-2)^2}\binom{n}{2}\frac{(2n+5)^3}{(2n^2+6n+7)} \text{ degrees of freedom.}$$

This result is not of much practical value. The case of m rankings can be more simply treated by other methods.

INTERPRETATION OF RESULTS OF PAIRED COMPARISONS

22. In the light of the foregoing theory we may discuss the interpretation of the results of a paired comparison experiment.

If for each observer the coefficient of consistence is unity the comparisons reduce to rankings and may be discussed by known methods. But if some or all the coefficients are not unity we have to consider the following possibilities:

(*a*) Some of the observers may be bad judges and the inconsistences reflect their shortcomings in making comparisons.

(*b*) Some of the objects may differ by amounts which fall below the threshold of distinguishability for some observers.

(*c*) The property under judgment may not be a linear variate at all and we may be getting the sort of confusion which would result if observers were asked to compare English towns according to the bivariate concept "geographical position".

(*d*) Several of the effects may be operating simultaneously.

23. If we have only one observer and have no prior knowledge of his capabilities it is not in general possible to apportion his inconsistences among these causes. Exceptions may occur when the inconsistences are of a marked and peculiar kind; for instance if they involve only four objects out of 15, we may suspect that the four are practically indistinguishable rather than that the observer is unable to make distinctions at all and avoided inconsistences among the others by sheer chance. But even here conclusions drawn *a posteriori* after inspection of the data are dubious. Table II gives a test of the hypothesis that an observer is incapable of making judgments. For example, with $n = 7$, the chances are 983 in 1000 that if the preferences are made at random there will be more than two inconsistent triads, so that if we find two or less, it is improbable that the observer is completely incapable of judgment. We might then be led to suppose that his small deviation from internal consistence is due to fluctuation of attention, very close resemblance to the objects giving rise to the inconsistences, or both.

24. With m observers the investigation can be taken a good deal further. If all the observers show inconsistences we suspect that the objects are at fault or that the observers are being asked to perform an impossible task. On the other hand, if most of the observers show a small or zero inconsistence we suspect that the others are just bad judges and may reject their data accordingly.

As between indistinguishability of objects and non-linearity of variate, a choice of explanations would depend largely on the extent to which inconsistences were concerned with the same set of objects. If there is a high value of u, indicating concordance of judgment, we expect to find most of the inconsistences confined to certain objects, and common to observers. In this case we suspect that the objects are close together in the degree to which they exhibit the quality under consideration. But if the observers scatter their inconsistences over the whole field u will be moderate or low and we suspect that the observers are being asked to do something beyond their capacity; and this brings us to question the validity of regarding the quality as a linear variable.

25. When a quality such as "bravery" or "intelligence" is insusceptible to measurement there is frequently doubt of this kind. But this has not deterred investigators from assuming that such statistical variables exist, or from requiring observers to rank objects according to them, or in some cases from replacing such rankings by quantiles of the normal curve. We are never tired of criticizing this Principle of the Hypostasis of Plausible Terminology. Hitherto it has flourished largely because of the difficulty of adducing evidence against it; and we hope that the inconsistence of paired comparisons will provide a criterion, however rough, of the legitimacy of the methods to which it leads.

But we would emphasize that our approach to the method of paired comparisons has a somewhat different object from that elaborated by Thurstone (1927 and many subsequent papers). As we understand it, his method is appropriate where one is entitled to assume *a priori* or by reason of precautions taken in the selection of material that a linear variable is involved and that there exist perceptible differences between the items presented for comparison. Our object is to make it possible to dispense with such assumptions and precautions.

26. A few words may be added about the case in which an objective order is known to exist (as, for instance, in judging individuals according to age or weight). In such circumstances the appearance of inconsistences will indicate unreliability of the part of the observer or subliminal differences between objects. A measure of the observer's reliability may be obtained by calculating u between known and observed comparisons. If ζ is high enough to enable us to accept his judgments as internally consistent on the whole, u may still be low enough to reject his judgments as accurate.

27. We conclude with an example of the application of the foregoing theory to some experimental material.

Classes of children (ages 11 to 13 inclusive) were asked to state their preferences with respect to certain school subjects. Each child was given a sheet on which were written the possible pairs of subjects and asked to underline the one preferred in each case. Two classes gave the following results:

(*a*) 21 boys, 13 school subjects. The preferences are shown in Table VII, which is in the form described in section 13; e.g. there were 18 boys who preferred Art to Religion.

TABLE VII

Preferences of 21 *boys in* 13 *subjects*

	1	2	3	4	5	6	7	8	9	10	11	12	13	Total
1. Woodwork	—	14	20	15	15	16	16	18	18	18	20	21	20	**211**
2. Gymnastics	7	—	14	12	13	18	14	16	16	20	16	18	19	**183**
3. Art	1	7	—	10	14	10	16	18	16	16	17	16	19	**160**
4. Science	6	9	11	—	11	12	15	14	13	13	17	17	16	**154**
5. History	6	8	7	10	—	14	11	12	14	15	13	14	16	**140**
6. Geography	5	3	11	9	7	—	14	14	13	13	16	15	17	**137**
7. Arithmetic	5	7	5	6	10	7	—	9	11	13	15	13	15	**116**
8. Religion	3	5	3	7	9	7	12	—	12	14	14	16	14	**116**
9. English Literature	3	5	5	8	7	8	10	9	—	10	13	13	15	**106**
10. Commercial subjects	3	1	5	8	6	8	8	7	11	—	10	10	14	**91**
11. Algebra	1	5	4	4	8	5	6	7	8	11	—	10	13	**82**
12. English Grammar	0	3	5	4	7	6	8	5	8	11	11	—	13	**81**
13. Geometry	1	2	2	5	5	4	6	7	6	7	8	8	—	**61**
													Total	**1638**

The calculation of Σ for this table, in which the objects are arranged in order of total number of preferences, may be shortened by noting that Σ as given by equation (7) may be transformed into the form

$$\Sigma = S(\gamma^2) - mS(\gamma) + \binom{m}{2}\binom{n}{2},$$

where the summation now takes place over the half of the table below the diagonal. Since the numbers in this half are smaller than those in the other half there is a considerable saving in arithmetic.

We find $$\Sigma = 9718$$

and hence $$u = \frac{2 \times 9718}{\binom{21}{2}\binom{13}{2}} - 1 = 0{\cdot}186.$$

There is thus a certain amount of agreement among the children, indicated by the positive value of u. Is this significant?

We note first of all that this distribution of preferences could not have arisen by chance to any acceptable degree of probability. In fact, $\chi^2 = 412{\cdot}4$ (equation (22)) and $\nu = 90{\cdot}7$. The large value of ν justifies the use of the normal approximation to the χ^2 distribution and we find $\sqrt{(2\chi^2)} - \sqrt{(2\nu - 1)} = 15{\cdot}3$ a very improbable result on the hypothesis of a random allocation of preferences.

The distribution of circular triads was as follows:

No. of triads	Frequency	No. of triads	Frequency
0	1	12	1
1	1	17	3
4	5	21	1
6	2	25	1
7	2	29	1
8	1	39	1
10	1	Total	21

The total number of circular triads was 242 with a mean of 11·5. Only one boy was entirely consistent. On the other hand, for $n = 13$ the maximum number of circular triads is 91, with a mathematical expectation of 71·5. It is thus clear that, except perhaps for one boy, we cannot suppose that any boy allotted preferences at random. We are again led to conclude that the boys are genuinely capable of making distinctions, and that consistently on the whole. Half the boys have coefficients of consistence ζ greater than 0·92.

We conclude that the boys can make preferences and that in their view the subjects are sufficiently different to enable a reasonably consistent set of preferences to be made. So far as these data are concerned we would see no objection to the assumption that a *scale* of preferences can be set up. With this in mind we can say that the value of u indicates a certain amount of agreement, though not a strong one, between the boys as to which subjects they prefer.

(*b*) 25 girls, 11 school subjects. Table VIII shows the data.

TABLE VIII

Preferences of 25 girls in 11 subjects

	1	2	3	4	5	6	7	8	9	10	11	Total
1. Gymnastics	—	10	19	17	20	17	21	21	21	18	22	186
2. Science	15	—	12	15	17	15	21	19	18	16	17	165
3. Art	6	13	—	16	16	18	10	17	16	19	16	147
4. Domestic Science	8	10	9	—	16	11	13	15	14	11	14	121
5. History	5	8	9	9	—	14	18	12	13	15	18	121
6. Arithmetic	8	10	7	14	11	—	12	13	12	16	18	121
7. Geography	4	4	15	12	7	13	—	14	15	14	14	112
8. English Literature	4	6	8	10	13	12	11	—	14	13	14	105
9. Religion	4	7	9	11	12	13	10	11	—	11	17	105
10. Algebra	7	9	6	14	10	9	11	12	14	—	12	104
11. English Grammar	3	8	9	11	7	7	11	11	8	13	—	88
											Total	1375

We find $\Sigma = 8928$, $u = 0{\cdot}082$.

For the χ^2 significance test, $\chi^2 = 180{\cdot}3$, $\nu = 62{\cdot}4$ and $\sqrt{(2\chi^2)} - \sqrt{(2\nu - 1)} = 7{\cdot}9$, as before a very significant result.

The distribution of circular triads was

No. of triads	Frequency	No. of triads	Frequency
1	2	17	1
2	2	19	1
3	1	22	1
4	1	23	2
6	1	27	1
8	1	32	1
9	1	35	1
11	2	37	1
12	2	38	1
13	1		—
14	1	Total	25

The total number of circular triads is 382 with a mean 15·28. For $n = 11$ the maximum number of circular triads is 55 with an expectation of 41·25. Several of the girls come very close to this the worst having a coefficient of consistence equal to 0·31.

We are, however, again led to conclude that the preferences were not allotted at random and that most of the girls are capable of exercising a judgment which is on the whole consistent. There is only a very slight agreement in preferences.

Thus the girls are less consistent and less alike in preferences than the boys.

REFERENCES

KENDALL, M. G. (1938). "A New Measure of Rank Correlation." *Biometrika*, **30**, 81.

KENDALL, M. G. & BABINGTON SMITH B. (1939). "On the Problem of *m* Rankings." *Ann. Math. Statist.* **10**, 273.

THURSTONE, L. L. (1927). "A Law of Comparative Judgment." *Psychol. Rev.* **34**, 275.

From *Biometrics*, Vol. 11, pp. 43–62 (1955)

FURTHER CONTRIBUTIONS TO THE THEORY OF PAIRED COMPARISONS[1]

M. G. KENDALL

Visiting Professor, Institute of Statistics, North Carolina State College

1. When a pair of objects is presented for comparison and the two are placed in the relationship preferred: not-preferred, we have what is known as a *paired comparison.* A set of n objects can be compared, a pair at a time, in some or all of the possible $n(n - 1)/2$ ways of choosing a pair, and the set of paired comparisons so derived gives us a picture of the interrelationships of the objects under preference. A paired-comparison scheme is more general than a ranking; for with the latter A-preferred-to-B and B-preferred-to-C automatically ensures A-preferred-to-C, whereas with paired comparisons it might happen that C was preferred to A. The existence of these departures from the ranking situation may be due to various reasons, such as the fact that 'preference' is a complicated comparison being made with reference to several factors simultaneously; and one reason for using paired comparisons is to give such effects a chance to show themselves.

2. Situations often occur in which a set of m observers express preferences among n objects and we have to select that object, or perhaps that sub-set of objects, which are, in some sense, "most preferred." The simplest case is the one where there are only two objects, A and B, and every observer votes for either A or B as president of an institution. If 51 per cent of the votes are cast for A and 49 per cent for B we declare A elected. In doing so we have satisfied 51 per cent of the preferences but have had to proceed contrary to 49 per cent; we may say that 49 per cent of the preferences were *violated.* More generally, when we have to select a subset of the n objects as "elected" we shall in general, in the absence of complete unanimity, violate a number of preferences. Circumstances force us to do so to some extent. The problem is to do so to the least possible extent.

3. Consider the case in which 8 members of a body have to elect a committee of three from among themselves. We will suppose that no member votes for himself (though this makes no essential difference) and that there are no abstentions (though this too makes no essential

[1]This research was supported by the United States Air Force, through the Office of Scientific Research of the Air Research and Development Command.

difference). If the 8 members are represented by the letters *A* to *G* they might vote as follows:

Member	*Members Preferred*
A	*BDE*
B	*DAF*
C	*DGA*
D	*CBE*
E	*ABC*
F	*ACD*
G	*BAC*

(1)

Here, for the moment, we suppose that there is no preference expressed among the triplets of members preferred; that is to say, *A* prefers *B*, *D*, *E* but does not say whether *B* is preferred to *D* or *E*, or *D* to *E*. He might then have written down his nominees in any order.

Under this system each elector expresses 9 preferences. *A*, for example, says, in effect, that he prefers *B* to *C*, *F*, and *G*, prefers *D* to *C*, *F*, and *G*, and prefers *E* to *C*, *F*, and *G*. There are thus 63 preferences altogether. We will represent this scheme in a two-way array of the following kind:

	A	*B*	*C*	*D*	*E*	*F*	*G*	No. of preferences
A	—	11		111	1111	111	111	15
B	1	—	1	11	1	1111	111	12
C	1	1	—	111	111	111	1111	15
D		11	1	—	11	11	11	9
E	1		1		—	11	11	6
F				1	1	—	1	3
G		1			1	1	—	3
Totals	3	6	3	9	12	15	15	63

(2)

Here, if *A* is preferred to *B* (a relationship we shall henceforward write as *A* pref. *B* or $A \rightarrow B$) we write a unit in the row *A*, column *B*. For example *C* prefers *D*, *G*, *A* to each of *B*, *E*, *F*. We therefore have units in row *D*, Col. *B*; row *D*, Col. *E*; row *D*, Col. *F*; row *G*, Col. *B*;

row G, Col. E; row G, Col. F; row A, Col. B; row A, Col. E; row A, Col. F. The totality of preferences expressed in (1) is given in the array (2), together with row and column totals.

Notice that: (a) the sum of row and column totals for each letter is 18. This provides a check. The reason is that each of the letters is compared with three others by each of six observers, so that each letter has 18 preferences (one way or the other).

(b) each column or row total is a multiple of three; for if any letter is preferred at all by an observer it is preferred to three others.

4. From the array (2) we see that A and C had 15 preferences each. If all preferences expressed by all observers have equal weight there is nothing to choose between them. B comes next with 12 preferences. All the others have fewer. Thus, if we have to elect three out of the seven to form a committee, we elect A, B and C.

5. The procedure we have followed exhibits the structure of the preference scheme most clearly, but for the purposes of electing a committee of three we can proceed much more expeditiously. In fact, from array (1) we see that the voting is as follows:

Member	*Number of votes*	
A	5	
B	4	
C	5	
D	3	
E	2	
F	1	
G	1	
	21	(3)

A comparison of this with (2) shows that in the latter the row totals are thrice the number of votes. The reason is easy to see, for if any letter gets a vote it is thereby preferred to three others.

6. Now let us suppose that the rules of election are altered slightly and that each elector writes down the three members he prefers *in order of preference.* Such an order might be that of array (1) where, for example, A gives B his first preference, D his second and E his third. Each elector now expresses 12 preferences, three among the set he names and 9 by implication between those three and the three he omits. If we now form an array of preferences we get, instead of (2)

The antisymmetry of the table has now been lost and row or column totals are no longer divisible by three. But we could still pick out the three members with the greatest number of preferences (A, C, B as before) without constructing a full table. In fact from (1) we score for A the following preferences allotted by the electors B to G:

$$4 + 3 + 0 + 5 + 5 + 4 = 21$$

and so for the other letters. The scores are the preference totals in the final column of (4).

	A	B	C	D	E	F	G	Totals
A	—	3	3	4	4	4	3	21
B	2	—	3	3	3	4	3	18
C	2	2	—	4	4	4	4	20
D	1	2	1	—	3	2	3	12
E	1	0	1	0	—	2	2	6
F	0	0	0	1	1	—	1	3
G	1	1	0	0	1	1	—	4
Totals	7	8	8	12	16	17	16	84

(4)

7. The same method can obviously be applied to any number of voters and any size of committee. Under the condition that there are no abstentions and that nobody votes for himself, the total number of preferences expressed by m voters for a committee of n (no preferences between committee nominees) is $mn(m - n - 1)$; or if preferences are expressed by ranking nominees, is $mn(m - n/2 - \frac{3}{2})$. We may now, if we wish, relax some of the conditions without affecting essentials.

(a) If every man is allowed to vote for himself nothing new is introduced so long as we adhere to the principle of giving each preference the same weight;

(b) The same principles apply when a number of electors express preferences concerning a group of individuals who are not members of themselves. If m judges express preferences for k out of n objects (without ordering them) the number of preferences is $mk(n - k)$.

(c) If there are any abstentions we can continue as before to count those preferences which are expressed. Suppose, for example, that instead of (1) we had the following preferences expressed (second column):

Member	Preferences	Corrected Preferences
A	*BDE*	*BDE*
B	*CA*	*CA*
C	*DGAB*	*DGA*
D	*CBE*	*CBE*
E	*AB*	*AB*
F	*ACD*	*ACD*
G	*BBB*	*B*

(5)

We suppose that these are in order. Member *C* has overstepped the mark. Unless we reject his ballot as spoiled we delete *B* from his ordering. Member *B* prefers *C* to *A* and both to the other four, but cannot express a preference between those other four and hence submits only two names. Member *G* tries to "plump" but we disallow this and count his expression as a preference for *B* only. We now have the preferences in the third column of (5) giving the following:

Prefer-ences for								
A		4	+3		+5	+5		= 17
B	5			+4	+4		+5	= 18
C		5		+5		+4		= 14
D	4		+5			+3		= 12
E	3			+3				= 6
F								= 0
G			4					= 4
								71

(6)

A, *B* and *C* are still elected but *B* now gains more preferences than *A*.

We notice that election on this principle maximizes the number of satisfied preferences as before.

(d) If any voter "ties" certain nominees, this is equivalent to expressing no preference between them and everything proceeds as before. For example, if in (5) member *D* tied *C*, *B*, *E* there would be two fewer preferences for *C* and one fewer preference for *B* in (6).

(e) In particular this method covers the case when each of a set of judges ranks all the objects, and not merely a preferred sub-set of them. The whole method, in fact, is very flexible in this respect. So long as any preferences are expressed we can pursue the same technique. The only thing to take particular care about is that one judge has the same *opportunities* as another for expressing the same number of preferences, even though he may not avail himself of them. We clearly introduce bias if we give one judge a chance to express two preferences and another only one. The system proposed is in accordance with the best democratic principles in that each judge has the same number of votes, and all votes have the same weight.

(f) It is possible to order the members, according to the number of preferences allotted to them, in a ranking (which may itself contain tied members). Thus we constrain a paired-comparison system into a ranking at the expense of violating a number of preferences. The fewer the violations the nearer the scheme to an actual ranking. In tables of the type of (2) or (4) a perfect and unanimous ranking would correspond to a situation in which all the non-zero cells were above the main diagonal.

(g) In those cases where we choose to regard any object as compared with itself, as for example if we wish to complete the diagonals in (2), we may allot $\frac{1}{2}$ to the cell in the same row and column. This will clearly not affect the order of the objects according to numbers of preferences received, for each object then receives an extra $\frac{1}{2}$ for each observer.

(h) Likewise, if an observer cannot express a preference between a given pair A, B we may allot $\frac{1}{2}$ to each of the cells in row A, column B and row B, column A in arrays of type (2).

(i) We can, if we wish, give effect to differences in reliability between judges. For example, if in array (2) we regard D as twice as important in his preferences as the others, we enter 2 for each preference instead of unity in the table.

8. Finally, let us note that the number of preferences can be used to calculate a coefficient of agreement among judges. This is another aspect of the coefficient of agreement in paired comparisons proposed by Babington Smith and myself some years ago. (See my *Advanced Theory of Statistics*, vol. 1, chapter 16). In fact if the total possible number of agreements is N and the actual number of agreements is M, the coefficient of agreement would be simply $2M/N - 1$ which varies from $-1/m$ or $-1/(m - 1)$ to 1. In table (2) for example the cells (A, B) and (B, A) have respectively 2 and 1 members. The pair A, B are compared three times and of these comparisons two are in agreement; there is thus one agreement out of a possible 3; likewise for AG, there are three agreements, each in the all AG, out of a possible 3. For the whole table it will be found that there are 47 agreements out of a possible 74 and the coefficient of agreement is 0.270.

9. We may also use the table to calculate a coefficient of departure from the ranking situation. Suppose we arrange the table so that rows and columns follow the order of the number of preferences expressed; in the case of table (2) this merely amounts to interchanging the rows and columns corresponding to B and C. The number of units below the diagonal is then 13 and that above the diagonal is 50. No other arrangement of rows and columns can divide the 63 preferences so unequally. If all were above the diagonal the preferences would be consistent with a ranking. We might then take as our measurement of departure from the ranking situation the coefficient $(13/63) \times 2 = 0.413$. We have multiplied the factor 13/63 by two because the furthest situation from ranking occurs when *one half* of the total preferences are allotted to the cells below the diagonal.

10. So much for the elements of the subject. I now proceed to consider sundry developments which are necessary to enable a more penetrating study of a paired-comparison situation to be made. The first arises from the nature of paired comparisons in themselves and may best be introduced by an example.

Let us suppose that six players A to F are engaged in a chess tournament in which each plays the other once. The set of scores (1 for a win, $\frac{1}{2}$ for a draw and 0 for a loss) then represents a set of paired comparisons made in all possible ways between them. We assume that all games reach a decision so that there are no missing values. A possible set of results is as follows:

	A	B	C	D	E	F	Total score	
A	$\frac{1}{2}$	1	1	0	1	1	$4\frac{1}{2}$	
B	0	$\frac{1}{2}$	0	1	1	0	$2\frac{1}{2}$	
C	0	1	$\frac{1}{2}$	1	1	1	$4\frac{1}{2}$	
D	1	0	0	$\frac{1}{2}$	0	0	$1\frac{1}{2}$	
E	0	0	0	1	$\frac{1}{2}$	1	$2\frac{1}{2}$	
F	0	1	0	1	0	$\frac{1}{2}$	$2\frac{1}{2}$	(7)

The simple way of arranging the competitors in order of success is to add up their scores, as is done in the final column. If we had three prizes we should divide the first and second between A and C and divide the third among B, E and F. Only D does not qualify for a share of the prize money. Such a procedure would be adopted in most tournaments of the kind.

11. But we now notice one rather anomalous effect. D, the only

player to receive nothing, has in fact beaten one of the winners, A. We are not allowed to dismiss this as a mere fluke, because all preferences are equally valid. Furthermore A has beaten C but is nevertheless ranked with him. Vague but genuine feelings for general equity lead us to inquire whether something should not and cannot be done to restore the balance. Such a method was suggested by Dr. T. H. Wei (1952) in an unpublished thesis successfully submitted to the University of Cambridge for the Ph.D. degree. In effect Wei's procedure amounts to this:

We recalculate a score for each player by giving him the score of every player he has beaten and half the score of every player with whom he has drawn. This leads to the following new scores:

$$
\begin{aligned}
A &= \tfrac{1}{2}(4\tfrac{1}{2}) + 2\tfrac{1}{2} + 4\tfrac{1}{2} + 0 + 2\tfrac{1}{2} + 2\tfrac{1}{2} = 14\tfrac{1}{4}\\
B &= 0 + \tfrac{1}{2}(2\tfrac{1}{2}) + 0 + 1\tfrac{1}{2} + 2\tfrac{1}{2} + 0 = 5\tfrac{1}{4}\\
C &= 0 + 2\tfrac{1}{2} + \tfrac{1}{2}(4\tfrac{1}{2}) + 1\tfrac{1}{2} + 2\tfrac{1}{2} + 2\tfrac{1}{2} = 11\tfrac{1}{4}\\
D &= 4\tfrac{1}{2} + 0 + 0 + \tfrac{1}{2}(1\tfrac{1}{2}) + 0 + 0 = 5\tfrac{1}{4}\\
E &= 0 + 0 + 0 + 1\tfrac{1}{2} + \tfrac{1}{2}(2\tfrac{1}{2}) + 2\tfrac{1}{2} = 5\tfrac{1}{4}\\
F &= 0 + 2\tfrac{1}{2} + 0 + 1\tfrac{1}{2} + 0 + \tfrac{1}{2}(2\tfrac{1}{2}) = 5\tfrac{1}{4}
\end{aligned}
\qquad (8)
$$

We now arrange the players in order of new scores; and we now notice that A and C have separated, A being first and C second, while D has moved up to equality with B, E, and F.

This is as far as one would wish to go on practical grounds, perhaps, but now a further point raises itself. We have re-allocated the scores once. Why not do so again? If we re-allocate the scores of (8) by the same method we find

$$
\begin{aligned}
&A \quad \tfrac{1}{2}(14\tfrac{1}{4}) + 5\tfrac{1}{4} + 11\tfrac{1}{4} + 0 + 5\tfrac{1}{4} + 5\tfrac{1}{4} = 34\tfrac{1}{8}\\
&B \quad 0 + \tfrac{1}{2}(5\tfrac{1}{4}) + 0 + 5\tfrac{1}{4} + 5\tfrac{1}{4} + 0 = 13\tfrac{1}{8}\\
&C \quad = 26\tfrac{5}{8}\\
&D \quad = 16\tfrac{7}{8}\\
&E \quad = 13\tfrac{1}{8}\\
&F \quad = 13\tfrac{1}{8}
\end{aligned}
\qquad (9)
$$

A and C are still first and second but D takes third place and B, E, F share the fourth position.

If we re-allocate the scores once more we find scores

A	824.375	
B	365.625	
C	695.625	
D	425.625	
E	365.625	
F	365.625	(10)

The order is now the same as we derived from (9); and if we ascertain new scores on the same principle we shall find that no new ordering has appeared. Later I shall prove that after a time the situation always "settles down" in this way.

13. There are two interesting features of this procedure. Let us revert to the preference scheme of (7) and regard the scores as a matrix. If we square this matrix we obtain

$$\begin{bmatrix} \frac{1}{4} & 3 & 1 & 4 & 2 & 3 \\ 1 & \frac{1}{4} & 0 & 2 & 1 & 1 \\ 1 & 2 & \frac{1}{4} & 4 & 2 & 2 \\ 1 & 1 & 1 & \frac{1}{4} & 1 & 1 \\ 1 & 1 & 0 & 2 & \frac{1}{4} & 1 \\ 1 & 1 & 0 & 2 & 1 & \frac{1}{4} \end{bmatrix} \qquad \begin{matrix} \textit{Row totals} \\ 14\frac{1}{4} \\ 5\frac{1}{4} \\ 11\frac{1}{4} \\ 5\frac{1}{4} \\ 5\frac{1}{4} \\ 5\frac{1}{4} \end{matrix} \qquad (11)$$

and the row totals are those previously obtained in (8) by the first re-allocation of scores. The reason for this will be obvious to anyone familiar with the rules of matrix multiplication and the result is generally true for all preference matrices. Furthermore, if we multiply (11) again by the matrix (7) and add row totals we shall get the scores of (9); and so on. The continual re-allocation of scores is equivalent to taking successive powers of the matrix.

14. Let us now consider what interpretation can be given to the process in terms of comparisons. The following diagram shows the scheme of (7) in geometrical form. The six players are represented by the six vertices of a regular hexagon, which are joined by straight lines in all possible ways. If *A* pref. *B* we draw an arrow from *A* towards *B*. If no preference was expressed (or the game was drawn) we do not draw an arrow.

15. It will be seen that the score of any player in (7) is the number of arrows *leaving* his vertex, together with $\frac{1}{2}$ (as the conventional score in the diagonal, when he is compared with himself) and $\frac{1}{2}$ for any line passing through his vertex on which no arrow is drawn. When we proceed to the next stage we count the number of paths leaving the

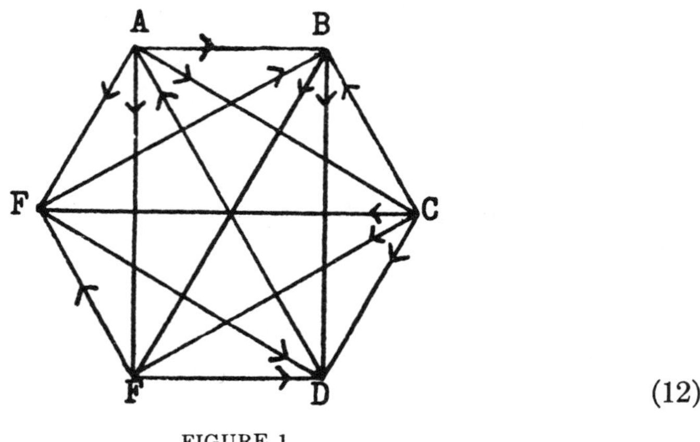

(12)

FIGURE 1

vertex and taking two steps. For example, for A we have the following paths leaving A and also leaving the vertex next visited:

$$ABD, ABE; \quad ACB, ACD, ACE, ACF; \quad AED, AEF; \quad AFB, AFD.$$

There are ten of these "transitive" preferences. We also count the preference of B with itself, C with itself, etc., as $\frac{1}{2}$ each, making a further score of 2; and finally we score $\frac{1}{2}$ of $\frac{1}{2}$ for the double preference of A with itself. The total score is $14\frac{1}{4}$, which is the score for A in (11). It may be verified that the same procedure gives the other scores in that array.

Similarly the scores obtained by the next re-allocation, as given in (9), are the numbers of paths of three lines leaving the respective vertices, all arrows going the same way, with similar conventions about vertices taken with themselves; and so on. Our re-allocation is equivalent to powering the matrix or to counting paths of transitive preferences of increasing extent.

16. From the geometrical viewpoint it is seen that in proceeding by re-allocation we are extending our concept of comparison. We began by considering comparisons of pairs by themselves. When we proceed to the next stage we compare pairs which form part of triads; but we do not compare the triads by considering them as three pairs (which would bring us back to the first situation). Thus it is possible to

"compare" A and C by the route $A \to B \to C$ or A and B by $A \to C \to B$. Both of these "comparisons" do not count in our score because they cannot both happen together; but either counts when it occurs.

17. Or we may put it another way by saying that we compare two members AB not directly, but through their comparisons with other members, e.g. by ACB, ADB, AEB and AFB. We choose the leading members in the final order so as to maximize the agreement with transitive preferences. Whether this is the right thing to do depends to some extent on practical circumstances. The process of continual re-allocation has the advantage that it results in an objective final ordering; but whether this is what we want depends on whether we are considering a situation in which direct comparison is the basic generator of the data, or whether we wish to give scope for more reflective judgment in roundabout comparisons involving other members.

18. Let us now consider the case when several judges make paired comparisons, or several tournaments are played between the same set of players. For each observer we shall have a preference matrix of the type of (7). To obtain a composite picture, on the supposition that the judges are equally reliable, we superpose the matrices. Thus if (7) represents the preferences of a judge for 6 varieties of ice cream when offered to him in pairs, two additional judges might have the following preference matrices:

	A	B	C	D	E	F	Totals
A	$\frac{1}{2}$	1	0	1	1	0	$3\frac{1}{2}$
B	0	$\frac{1}{2}$	$\frac{1}{2}$	1	0	1	3
C	1	$\frac{1}{2}$	$\frac{1}{2}$	1	0	1	4
D	0	0	0	$\frac{1}{2}$	1	$\frac{1}{2}$	2
E	0	1	1	0	$\frac{1}{2}$	1	$3\frac{1}{2}$
F	1	0	0	$\frac{1}{2}$	0	$\frac{1}{2}$	2

(13)

	A	B	C	D	E	F	Totals
A	$\frac{1}{2}$	0	1	1	$\frac{1}{2}$	1	4
B	1	$\frac{1}{2}$	1	0	0	1	$3\frac{1}{2}$
C	0	0	$\frac{1}{2}$	1	1	1	$3\frac{1}{2}$
D	0	1	0	$\frac{1}{2}$	$\frac{1}{2}$	$\frac{1}{2}$	$2\frac{1}{2}$
E	$\frac{1}{2}$	1	0	$\frac{1}{2}$	$\frac{1}{2}$	1	$3\frac{1}{2}$
F	0	0	0	$\frac{1}{2}$	0	$\frac{1}{2}$	1

(14)

Adding these and (7) together we get

	A	B	C	D	E	F	Totals
A	$1\frac{1}{2}$	2	2	2	$2\frac{1}{2}$	2	12
B	1	$1\frac{1}{2}$	$1\frac{1}{2}$	2	1	2	9
C	1	$1\frac{1}{2}$	$1\frac{1}{2}$	3	2	3	12
D	1	1	0	$1\frac{1}{2}$	$1\frac{1}{2}$	1	6
E	$\frac{1}{2}$	2	1	$1\frac{1}{2}$	$1\frac{1}{2}$	3	$9\frac{1}{2}$
F	1	1	0	2	0	$1\frac{1}{2}$	$5\frac{1}{2}$
Totals	6	9	6	12	$8\frac{1}{2}$	$12\frac{1}{2}$	54

(15)

On the basis of simple paired comparisons we should place A and C as bracketed equal, E as third, B as fourth, D as fifth and F as last.

19. The question now arises whether we should re-allocate the scores by powering the matrix (15); or whether it would be preferable to power each matrix and then amalgamate the rankings at the end. The two processes will not always lead to identical results, although in practice they should not differ very much. Arithmetically it is simpler to power just the one matrix (15), and in cases where there are many judges this would be almost decisive. This is the procedure I would recommend myself, but if there were any serious doubts I would perform the analysis both ways and compare the results. A wide disparity would, in my view, suggest that neither was very reliable. It would arise mostly in cases where there were substantial disagreements among judges.

20. I now prove that the process of repeated powering does in fact converge to a limiting ranking. Dr. Wei offered a proof of the result for one observer and a complete set of preferences in his thesis.

First of all we define a matrix A of non-negative elements to be *indivisible* if it cannot be expressed in the form (by rearrangement of rows and columns)

$$A = \begin{bmatrix} A_{11} & A_{12} \\ 0 & A_{22} \end{bmatrix} \qquad (16)$$

If a preference matrix of type (15) is divisible in this sense the members of one block of objects are always preferred to every member of another. In such a case we divide the data into the two blocks and operate on each, finally ranking the members of the first group and then the members of the second. Similarly, if one of these blocks is itself divisible we divide it up; and so on. We clearly lose no generality by doing this, and divisibility is not a handicap in our preference situations.

21. I now require a theorem of Frobenius (cf. Wielandt, 1950[2]) which says that for indivisible matrices A with non-negative elements and positive elements in the diagonal there exists a unique simple positive root of the equation $| A - \lambda I | = 0$ which is greater than all other roots in absolute value; and that the corresponding characteristic vector has all its elements of the same sign (which we may take to be positive).

Let λ_1 be this largest root and Y_1 the corresponding vector. Then if $\lambda_2, \cdots \lambda_p$ are the other roots and $Y_2 \cdots Y_p$ the corresponding vectors, and if P be the preference matrix, we have

$$PY = \Lambda Y \tag{17}$$

where Λ is the diagonal matrix

$$\Lambda = \begin{bmatrix} \lambda_1 & & & 0 \\ & \lambda_2 & & \\ & & \ddots & \\ 0 & & & \lambda_p \end{bmatrix} \tag{18}$$

It is now easy to show that for any positive integer k

$$P^k Y = \Lambda^k Y \tag{19}$$

As the powering proceeds the major root λ_1 becomes dominant and (19) tends to the equation

$$P^k Y_1 = \lambda_1^k Y_1 \tag{20}$$

Thus from some k onwards the final ordering will be determined by the vector Y_1, which has non-negative elements.

22. We notice that the proof remains applicable to preference matrices in which some preferences may be missing, or when ties are present, provided that the matrix is not divisible. If any cell in a combined preference matrix contains no entries we insert a zero.

23. It is also of some interest to note that we may prove that the preference matrix is never singular. In fact, we can always express it (apart from positive numerical factors) in the form

$$(Q + U) \tag{21}$$

[2]I am indebted to Professor A. C. Aitken and Dr. F. G. Foster for some references on this subject. The preference matrices are similar to, but not identical with, the matrices of transition probabilities studied in the theory of stationary stochastic processes.

where Q is an anti-symmetric matrix and U is the matrix all of whose elements are unity. For example (15), after division of rows by $1\frac{1}{2}$, can be expressed as U plus the matrix

$$Q = \begin{bmatrix} 0 & \frac{1}{3} & \frac{1}{3} & \frac{1}{3} & \frac{2}{3} & \frac{1}{3} \\ -\frac{1}{3} & 0 & 0 & \frac{1}{3} & -\frac{1}{3} & \frac{1}{3} \\ -\frac{1}{3} & 0 & 0 & 1 & \frac{1}{3} & 1 \\ -\frac{1}{3} & -\frac{1}{3} & -1 & 0 & 0 & -\frac{1}{3} \\ -\frac{2}{3} & \frac{1}{3} & -\frac{1}{3} & 0 & 0 & 1 \\ -\frac{1}{3} & -\frac{1}{3} & -1 & \frac{1}{3} & -1 & 0 \end{bmatrix} \qquad (22)$$

We reduce $Q + U$ systematically by subtracting the first column from the second column, then the first row from the second row; then the first column from the third column, then the first row from the third row; and so on. The effect on Q is to reduce it to another anti-symmetric matrix, say Q', and the effect on U is to reduce it to a unit in the top left-hand corner and zero elsewhere. Thus the determinant of $Q + U$ is the determinant of Q' plus the determinant of the principal minor obtained by omitting the first row and column, which is also antisymmetric.

Now the determinant of $p \times p$ antisymmetric matrix is zero if p is odd and positive if p is even. Hence the determinant of $Q + U$ is the sum of two components, one zero and the other positive; and hence it does not vanish.

22. In practice the number of paired comparisons arising from n objects may be inconveniently large and the question arises whether it is possible to economize in the number of comparisons made. In the example of the chess tournament which has been mentioned above (paragraph 10) if each player is to play every other, 15 games must be played. But only three can be conducted at once, so at best 5 sessions are necessary. If this is too long, and, say, three sessions are all that can be allowed, only nine games can be played and six have to be sacrificed. The question is, which six? Or again, if an individual is comparing items by taste testing, his patience or his palate may not endure the presentation of all the possible pairs, and a problem arises as to how best to cut down the number of pairs and which pairs to present.

23. Problems like this arise in many fields of experimentation and are usually dealt with by incomplete balanced blocks. Some new points, however, arise in paired-comparison work. Durbin (1951) has considered

the use of Youden designs in ranking experiments. More recently Benard and van Elteren (1953) have discussed tests of significance where incomplete rankings are concerned. Without trying to exhaust the subject I proceed to consider the use of incomplete balanced blocks in preference schemes.

24. Consider first of all the case of a single observer. Of the $n(n - 1)/2$ preferences which he could make we require to pick out a sub-set. Certain elementary principles of choice at once suggest themselves:

(a) every object should appear equally often. In this sense the design should be balanced;

(b) the preferences should not be divisible in the sense that we can split the objects into two sets and no comparison is made between any object in one and any object in the other.

In terms of preference matrices (a) means that there should be the same number of non-empty items in each row and column; (b) means that the matrix does not divide into two blocks and become of the form $\begin{pmatrix} X & 0 \\ 0 & Y \end{pmatrix}$ when the zeros represent empty cells. In terms of the preference diagram (a) means that there are the same number of paths direct between points leaving or entering each vertex and (b) means that the figure does not separate into two distinct polygons.

25. When possible I add a further condition of symmetry to the situation, that is to say

(c) In the preference diagram the number of paths of length l proceeding from any point to any other point shall be the same for all pairs of points.

The length l here means the number of lines traversed in the path, e.g. the path (in Figure 1, section 14) ABC from A to C is of length 2 and $AEBDC$ from A to C is of length 4. Where no pair of objects is compared in these "partial" situations we omit the line between them. If they are joined by a line without an arrow this means that they have been compared but that no preference has been expressed.

In terms of preference matrices this condition implies a kind of symmetry of interlocking. A path ABC implies entries in row A, column B and column C (and the reflections column A, row B and row C); and analogous entries must occur in other rows in such a way that all the objects are symmetrically involved.

26. Under these conditions we can meet a requirement suggested to me in conversation by Dr. R. C. Bose: if all the preferences are exerted at random (e.g. if we toss up for it which of a pair shall be preferred) all possible final orderings of the objects produced by powering the matrix should be equally probable. This follows from the symmetry

of the situation, for we can interchange two objects in the designs without altering the preference matrix, so far as concerns the underlying probabilities, and all final orders are therefore equally probable.

27. In a sense, it seems to me, condition (c) is necessary as well as sufficient for a proper design. If it is not obeyed certain objects become subject to different schemes of preference from others and their final positions are not determined on an unbiased basis. In terms of powered preference matrices, the sums of rows are not based on the same number of transitive comparisons of length l.

28. The conditions laid down above impose certain restrictions on the scope of a paired-comparison experiment. For instance, if there are six objects and the numbers of entries in the rows of the preference matrix are equal, the number of comparisons necessary to obtain a balanced experiment must be a multiple of three. Anything else destroys the balance. The connectivity condition (b) further limits the freedom of choice; for example, with six objects at least six comparisons are required.

29. The setting up of incomplete designs is most easily thought of in terms of tours round the preference polygon. Consider the case $n = 7$. (Prime numbers are easier to deal with in most experimental designs.) There are 21 comparisons altogether. To obtain a balanced design we must have either 7 or 14 comparisons (or, of course, the full 21). The first 7 may, without loss of generality, be taken as the tour $ABC \cdots G$ round the preference heptagon. (No generality is lost

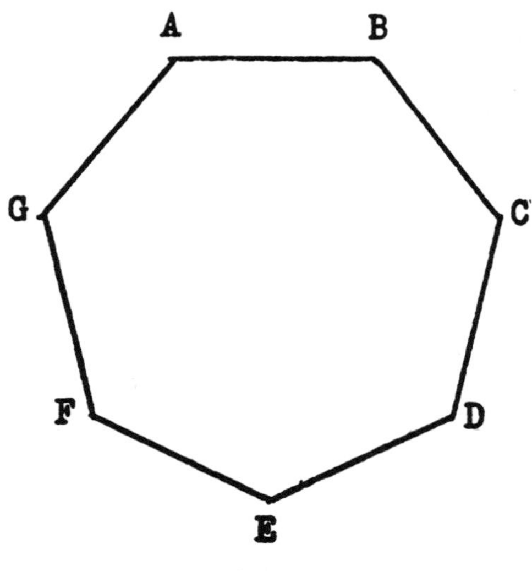

FIGURE 2

because each member must be connected to two others and hence they are on a chain which may be taken to be the order A to G.) For the next 7 we have two possibilities: (a) start from A, miss a vertex and go to C, miss a vertex and go to E and so on; (b) start from A, miss two vertices and go to E, then two vertices and go to G and so on. We do not obtain new designs by tours missing three or more vertices because they are equivalent to (a) or (b). The two schemes are shown in Figure 3.

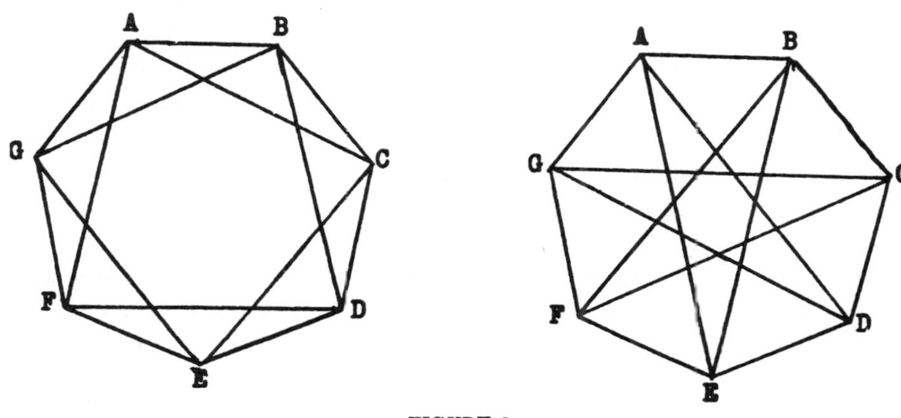

FIGURE 3

These schemes are not identical. In the former there are two triangular tours connecting any pair, e.g. ACB and AGB, whereas in the second there is only one, e.g. AEB. In terms of time taken in performance there is nothing to choose between them. For example if they represented a chess tournament, each round requires three games, one player having a bye, and for 14 games 5 rounds are required. Such might be

Scheme 1			Bye	Scheme 2			Bye	
AB,	*CD*,	*EF*	*G*	*AB*,	*CD*,	*EF*,	*G*	
AC,	*BD*,	*EG*	*F*	*AD*,	*BC*,	*FG*,	*E*	
BC,	*DE*,	*FG*	*A*	*BE*,	*CF*,	*DG*,	*A*	
AF,	*CE*,	*BG*	*D*	*AE*,	*BF*,	*CG*,	*D*	
AG,	*DF*		*B, C, E*	*AG*,	*DE*		*B, C, F*	(23)

30. It remains to be considered whether one scheme is preferable to the other by some other criterion. There is nothing to choose between them in relation to balance or the application of the powered-matrix method. We note, however, that the patterns of transitive preferences are different. In the first any pair is connected by two triangles, three

quadrilaterals, etc., in the second by one triangle, four quadrilaterals, etc. On the whole, I should be inclined to select the second design from a feeling that it has higher connectivity, but an exact criterion awaits further investigation.

31. When we have several judges, an obvious extension of symmetry requirements necessitates that each participates to an equivalent extent: in some sense the design should be balanced by judges as well as by comparisons. Something depends on whether we require to compare judges in addition to objects. If so, each pair of judges must have certain comparisons in common. With two judges and seven objects, for example, one simple way would be to allot to each 14 comparisons, one judging according to each of the designs of Figure 3. They would then have 7 comparisons in common and all possible comparisons could be made.

32. I do not propose on this occasion to attempt a systematic exposition of the design problems involved in paired comparisons. Designs of an optimum kind which balance by numbers of comparisons, objects compared, numbers of observers on given comparisons and so forth are probably rather rare; and if something has to be sacrificed it depends on what is the point of major interest whether we sacrifice symmetry in comparisons or in judges. A final example will make clear a few of the principles involved.

Consider again the case of seven objects, *ABCDEFG*. There are three distinct tours round the preference polygon,

$$\begin{matrix} A & B & C & D & E & F & G \\ A & C & E & G & B & D & F \\ A & D & G & C & F & B & E \end{matrix} \tag{24}$$

Each tour involves seven comparisons and each object is compared with two others in a tour.

For a complete set of comparisons each observer would have to make 21. If this is felt to be too much we may allocate 14, consisting of two tours each. And if the tours are represented by a, b, c, we may allocate to the observers 1, 2, 3

$$\begin{matrix} 1: & a, & b \\ 2: & b, & c \\ 3: & c, & a \end{matrix} \tag{25}$$

With these schemes every comparison is made equally often (twice); every tour is made equally often (twice); every observer makes the same number of comparisons (14); every observer has a tour in common with every other observer; and thus every observer can be compared with every other observer in respect of two comparisons involving any specified object.

If we have more than three observers, we take a number equal to a multiple of three and replicate the design.

Now suppose we had eleven objects, A to K. The full set of comparisons numbers 55. There are five distinct tours round the preference polygon

$$\begin{array}{llllllllllll} a: & A & B & C & D & E & F & G & H & I & J & K \\ b: & A & C & E & G & I & K & B & D & F & H & J \\ c: & A & D & G & J & B & E & H & K & C & F & I \\ d: & A & E & I & B & F & J & C & G & K & D & H \\ e: & A & F & K & E & J & D & I & C & H & B & G \end{array} \tag{26}$$

Now if we try to allot two tours to each of five observers we lose symmetry; for there are 10 pairs of tours choosable from these five. We have, to preserve complete balance, to allot four tours to each observer 1, 2, 3, 4, 5

$$\begin{array}{lllll} 1: & b, & c, & d, & e \\ 1: & c, & d, & e, & a \\ 3: & d, & e, & a, & b \\ 4: & e, & a, & b, & c \\ 5: & a, & b, & c, & d \end{array} \tag{27}$$

Again the tours are balanced, but we have not achieved very much. Each observer now makes 44 comparisons, against the full set of 55.

We can sacrifice symmetry in several ways. We may, for instance, allot two tours to each observer, e.g.

$$\begin{array}{lll} 1: & a, & b \\ 2: & b, & c \\ 3: & c, & d \\ 4: & d, & e \\ 5: & e, & a \end{array} \tag{28}$$

Here every observer can be compared with two other observers but not every pair can be compared. Or if we have, say, 10 observers we may allot all the 10 possible pairs of tours one to each. Each observer then makes 22 comparisons and can be compared with four other observers. If 22 comparisons are still felt to be too many for one observer we may allocate the 55 preferences according to a linked design, e.g. (numbering the preferences 1 to 55) with 11 observers, 10 preferences each

1 :	1,	2,	3,	4,	5,	6,	7,	8,	9,	10	
2 :	1,	11,	12,	13,	14,	15.	16,	17,	18,	19	
3 :	2,	11,	20,	21,	22,	23,	24,	25,	26,	27	
4 :	3,	12,	20,	28,	29,	30,	31,	32,	33,	34	
5 :	4,	13,	21,	28,	35,	36,	37,	38,	39,	40	
6 :	5,	14,	22,	29,	35,	41,	42,	43,	44,	45	
7 :	6,	15,	23,	30,	36,	41,	46,	47,	48,	49	
8 :	7,	16,	24,	31,	37,	42,	46,	50,	51,	52	
9 :	8,	17,	25,	32,	38,	43,	47,	50,	53,	54	
10 :	9,	18,	26,	33,	39,	44,	48,	51,	53,	55	
11 :	10,	19,	27,	34,	40,	45,	49,	52,	54,	55	(29)

Here we have cut down the comparisons for each observer to 10 and each comparison is made twice. But we have lost a good deal of the comparison between judges; every judge can be compared with every other judge but only on one comparison of objects.

REFERENCES

Benard, A. and Van Elteren, Ph. (1953), A generalization of the method of m rankings. *Kon. Neder. Ak. van Weterschappen.* A, 56, 358.

Durbin, J. (1951), Incomplete blocks in ranking experiments. *Brit. Jour. Psych.* 4, 85.

Frobenius, G. (1912), Über Matrizen aus nicht negativen Elementen. *Sitz. Preuss. Akad. Wiss.*, 456.

Wei, T. H. (1952), The algebraic foundations of ranking theory. Unpublished thesis, Cambridge University, England.

Wielandt, H. (1950), Unzerlegbare, nicht negative Matrizen. *Math. Zeit.*, 52, 642.

From *Journal of the Royal Statistical Society*, Vol. **104**, pp 43–52 (1941)

MISCELLANEA

CONTENTS

The Effect of the Elimination of Trend on Oscillations in Time-Series

By M. G. Kendall

1. A time-series usually exhibits to a greater or less degree three main features: (1) a trend, or "secular" movement; (2) oscillations * of some regularity about this trend; and (3) casual or random fluctuations. One of the principal problems in the analysis of such a series is the isolation of these three constituents for separate study; and the first factor which falls for isolation is the trend.

The concept of "trend," like that of time itself, is one of those ideas which are generally understood but difficult to define with exactitude. A movement which has the evolutionary appearance of a trend over a period of thirty or forty years may in reality be one phase of an oscillatory movement of greater extent. A good deal depends on the length of the series under consideration whether we regard any particular tendency in the series as a trend, or long-term movement, or an oscillation, or short-term movement. But in any case we require of a trend curve that it shall exhibit only the general direction of the time-series, and in practice this amounts to saying that it must be representable, at least locally, by a smooth non-periodic function such as a polynomial or a logistic curve.

2. In recent years the mathematics of trend fitting have been brought to an advanced stage of development. There are various methods advocated by different writers for different purposes; but, putting aside graphical methods, which involve personal judgment and cannot claim serious attention for theoretical work, they all depend on one of two processes: either (*a*) a polynomial of chosen degree is fitted to the whole series, almost invariably by least squares; or (*b*) a moving average of chosen extent and with chosen weights is used to determine trend values at different points of the series. The

* I follow Udny Yule (1921) in using the word "oscillation" instead of the more usual word "cycle." The latter, strictly speaking, implies a regularity of period which is rarely found in time-series. A good deal of the argument whether the trade "cycle" exists or not seems to me to depend on disagreement about the meaning of the word.

weights are usually selected to ensure that a polynomial of predetermined degree shall represent the series *locally*.

In practice the method of moving averages has decisive advantages over the fitting of a curve to the whole series. For one thing, a curve of sufficient flexibility to give an adequate fit to many time-series must be of high degree and will involve laborious arithmetic and considerable uncertainty in the stability of the resulting formula. What seems fatal to the extended practical use of the method (*a*), however, is the fact that if extra members are added to the series the calculations for a curve including these new data have to begin almost *ab initio*; so that if, as very frequently happens, one wishes to bring one's work up to date the arithmetic has nearly all to be done afresh.

3. It seems, therefore, that until a better method has been found, there is no choice but to use some variant involving moving averages. A thorough account of the different possibilities has been given by Sasuly (1934). The principal interest in the work which has led up to our present knowledge of the subject seems to have been the study of trend itself, not the elimination of trend in order that attention may be concentrated on the residual oscillations. In consequence some effects of the moving-average method on these residuals are in danger of escaping notice; and the object of this paper is to bring them into greater prominence.

4. Suppose we have a time-series $\phi(t)$ defined for integral values of the variable t, and let

$$\phi(t) = \phi_1(t) + \phi_2(t) + \phi_3(t) \quad . \quad . \quad . \quad . \quad (1)$$

where ϕ_1 represents the trend, ϕ_2 the oscillatory part and ϕ_3 the residual attributable to casual fluctuations. The method of moving averages consists of taking the weighted average of each set of p consecutive terms of ϕ and assigning this average to be the trend value at some point on the range covered by the p values, usually the middle point, so that it is generally convenient to take p odd. The value of p and the weights employed are determined by the accuracy desired. The ideal would be to find weights which, when applied to a range of p consecutive values of ϕ_1, reproduce the value of ϕ_1 at the middle point of the range wherever that range was situated on the curve ϕ_1; in other words, the trend of ϕ_1 ought to be ϕ_1 itself. Without prior knowledge of ϕ_1 it is impossible to determine a moving average which can be guaranteed to do this; but it is possible to find weights which will reproduce a parabola of any required order and a moving average based on them will give satisfactory results if ϕ_1 can be approximated locally by such a parabola. We may even take the trend to be defined by this property.

Consider then the optimum case in which the moving average does reproduce the trend curve exactly. Denote by T the operation of taking a moving average. Then from (1) we have

$$T\phi = T\phi_1 + T\phi_2 + T\phi_3 \quad . \quad . \quad . \quad . \quad (2)$$

and on subtracting this expression from (1) "to eliminate trend," we get

$$\phi - T\phi = \phi_2 + \phi_3 - T\phi_2 - T\phi_3 \quad . \quad . \quad . \quad (3)$$

ϕ_1 and $T\phi_1$ being equal by hypothesis.

It is $\phi - T\phi$ which we study when considering oscillatory movements; and the point to be emphasized is that the existence of the terms $T\phi_2$ and $T\phi_3$ in (3) may introduce oscillatory terms which were not, or annihilate oscillatory terms which were, in the original ϕ. That is to say, the method of moving averages may induce into the data oscillations which are entirely spurious or may reduce or remove oscillations which are entirely genuine.

5. Consider first of all the effect of moving averages on the residuals ϕ_3. To simplify the discussion I will consider the simplest kind of moving average, namely the arithmetic mean of p values centred at the point concerned, and will suppose that the residuals are the components of a random variable with variance V. Writing $\xi_j^{(1)}$ for the component at t_j we have for the moving average at that point, say $\xi_j^{(2)}$

$$\xi_j^{(2)} = \frac{1}{p} \sum_{k=-[p/2]}^{[p/2]} \xi_{j+k}^{(1)} \quad . \quad . \quad . \quad . \quad . \quad (4)$$

where $[p/2]$ is the integral part of $p/2$. Consecutive values of $\xi^{(1)}$ are independent but consecutive values of $\xi^{(2)}$ are not. In fact $\xi_j^{(2)}$ and $\xi_{i+k}^{(2)}$ have $p - k$ values of $\xi^{(1)}$ in common and will be correlated if $p > k$. Whereas $\xi^{(1)}$ was a random series $\xi^{(2)}$ will be much smoother; and if we proceed to take moving averages of $\xi^{(2)}$, resulting say in $\xi^{(3)}$ we shall get a smoother series still.

This effect has been considered at length by Yule (1921) and Slutzky (1937). One does not have to repeat the averaging many times before the resulting series becomes very smooth and presents to the eye just those features which are characteristic of oscillatory time-series—fluctuations with varying amplitude and phase, the periods (peak to peak or trough to trough) being distributed with central tendency about some modal value.* Slutzky's main concern in dealing with this subject was to show that effects resembling trade fluctuations can be generated by the additive action of purely random causes. For the present purpose it is considered from a different

* Slutzky establishes this last result empirically. It has not yet been demonstrated as a mathematical theorem; but it has been confirmed by Dodd (1939) and by experiments of my own on random numbers. For the parallel phenomenon in trade cycles see Mitchell (1927).

point of view—the generation of spurious oscillatory movements in the residual $\phi - T\phi$ owing to the term $T\phi_3$.

6. Dodd (1939) has recently made a useful contribution to this subject by considering the periods of oscillations generated by moving averages. If we have a set of variables ξ which are independent and normally distributed about zero with variance v the probability that

$$y_p = \sum_{j=1}^{p} a_j\xi_j < 0 \quad . \quad . \quad . \quad . \qquad (5)$$

and

$$y_{p+1} = \sum_{j=1}^{p} a_j\xi_{j+1} > 0 \quad . \quad . \quad . \quad . \qquad (6)$$

i.e., that the generated series changes sign from negative to positive, is the proportional frequency of

$$dF = \frac{1}{(2\pi)^{\frac{p+1}{2}}} exp - \frac{1}{2v}\left(\sum_{j=1}^{p+1} \xi_j^2\right) d\xi_1 \ . \ . \ . \ d\xi_{p+1}$$

in $(p + 1)$ dimensions between the planes (5) and (6). This is evidently equal to θ, the angle between the two planes, which is given by

$$\cos\theta = \frac{\Sigma(a_j a_j + 1)}{\Sigma(a_j)^2} \quad . \quad . \quad . \quad . \quad . \qquad (7)$$

Thus the mean distance from one " upcross " to the next is $2\pi/\theta$.

Similarly, the probability that

$$\Delta y_p = y_{p+1} - y_p < 0 \quad . \quad . \quad . \quad . \quad . \qquad (8)$$

and

$$\Delta y_{p-1} > 0 \quad . \quad . \quad . \quad . \quad . \quad . \quad . \qquad (9)$$

is, the angle between these two planes, say θ_1, and thus the mean distance from peak to peak is $2\pi/\theta_1$.

Dodd also considers the elimination of minor oscillations or " ripples." If in addition to the conditions (8) and (9) we also require that

$$y_p > y_{p+k} \quad . \quad . \quad . \quad . \quad . \quad . \quad . \qquad (10)$$

where k is some arbitrarily chosen number, the probability of (8), (9), and (10) being satisfied simultaneously is the area on the unit three dimensional sphere cut off by these three planes. If the angles between pairs of planes are A, B, C, this is equal to $A + B + C - 2\pi = \theta_2$, say. Hence the probability of the conditioned maximum is $\theta_2/4\pi$ and the reciprocal is the mean distance between maxima, ripples excepted.

Dodd supports his conclusions with some experimental material derived by the application of some well-known graduation formulæ to *rectangular* material. The agreement is good, and suggests that the results of the theory are applicable to other than normal random variation.

7. What we define as the " length " of an oscillation is to some extent arbitrary, but Dodd's results allow a certain freedom of choice. A minimum would seem to be a mean distance from peak to peak, *i.e.*, $2\pi/\theta_1$. A maximum would be the mean distance from upcross to upcross, *i.e.*, $2\pi/\theta$. To obtain greater precision within these limits we need to decide what is to be regarded as a ripple, and it looks as if every case would have to be settled on its merits.

8. Consider now the amplitude of the induced oscillations in $T\phi_3$. Since $\xi^{(2)}$ is the sum of p independent random variables it will have a variance V/p. $\xi^{(3)}$ is the sum of p correlated variables and the expressions for its variance are more complicated.

It will be evident on a little reflection that if a simple moving average of extent p is iterated q times, the result is equivalent to an average of the series with extent $q(p - 1) + 1$ with weights which are the successive coefficients in $(1 + x + x^2 + \ldots x^{p-1})^q$. The variance of $\xi^{(q+1)}$ is therefore the sum of the squares of these coefficients multiplied by V/p^q. Thus

$$\text{var.}\ (\xi^{(q+1)}) = \frac{V}{p^q} \cdot \text{coeff. of } x^{q(p-1)} \text{ in } (1 + x + x^2 + \ldots x^{p-1})^{2q}$$

$$= \frac{V}{p^q} \text{ coeff. of } x^{q(p-1)} \text{ in } \frac{(1 - x^p)^{2q}}{(1 - x)^{2q}} \quad \ldots \quad (11)$$

The following table shows the ratio of the variance of $\xi^{(q+1)}$ to V for a few values of p and q.

$p =$ \ $q =$	1	2	3	4	5
2	0·33	0·23	0·19	0·17	0·15
4	0·25	0·17	0·14	0·12	0·11
5	0·20	0·14	0·11	0·10	0·09
6	0·17	0·11	0·09	0·08	0·07
7	0·14	0·10	0·08	0·07	0·06

The variance of the series derived from $\xi^{(1)}$ is reduced very considerably by the first averaging but less so by subsequent averagings, and this is what we might expect from the correlations between members of the series. For example, when $p = 7$, the first averaging reduced the variance by $\frac{1}{7}$, whereas the next four averagings reduce it only by little more than a further $\frac{1}{2}$.

9. The general inference is that the oscillatory movements in $T\phi_3$ will be small compared with the random fluctuations of ϕ_3 itself if p is large. But it is important to realize that they are not on that account negligible. For example a periodogram analysis of ϕ_3 will reveal no periodicities whereas one of $T\phi_3$ may and probably will.

To reduce the effect of $T\phi_3$ as much as possible it is desirable to make p large, and if both p and q are at our disposal, it will be more use extending p than iterating to increase q. In other words, for a given total weight the individual weights should be as small and as equal as possible, as is otherwise evident from the consideration that in this way their total sums of squares will be minimized.

But unfortunately this requirement runs contrary to the necessities of the weighting to eliminate trend. To fit a moving quartic or sextic, for instance, we require a system of weights which are very far from equal (cf. Sasuly (1934)), and in fact some of the best-known formulæ for trend elimination such as Spencer's 21-point formula (1904) or Macaulay's 43-point formula (1931) use several iterated averages. The weights of Macaulay's formula are

$$\frac{1}{9600}\begin{array}{l}[7, 18, 30, 40, 45, 28, -8, -60, -122, -178, -205, -190, \\ \quad -127, -6, 163, 360, 562, 760, 928, 1050, 1127, 1156, \\ \quad 1127 \ldots 7]\end{array}$$

and if applied to a random series would reduce the variance to about 0·11 of its original value, *i.e.*, about as much as a simple average of nines.

10. Consider now the effect of the moving average method on ϕ_2 the genuinely oscillatory part of the original series. Yule (1921) has already discussed a similar problem from a different point of view, namely the effect of variate differencing on harmonic series. Consider the simple case when the oscillatory part is a sine term $\sin(\alpha + \lambda t)$, t being in our convention an integer. Since

$$\sum_{j=1}^{p} \sin(\alpha + j\lambda) = \frac{\sin \frac{1}{2}p\lambda}{\sin \frac{1}{2}\lambda} \sin\{\alpha + \tfrac{1}{2}(p-1)\lambda\}$$

a simple moving average of p consecutive terms centred at the middle term will result in a sine series of the same period and phase as the original, but with amplitude reduced by the factor

$$\frac{1}{p}\frac{\sin \frac{1}{2}p\lambda}{\sin \frac{1}{2}\lambda} \quad . \quad . \quad . \quad . \quad . \quad . \quad . \quad (12)$$

Iteration q times will reduce the amplitude by the qth power of this factor.

Thus the term $T\phi_2$ will be small if p is large, q is large or if $\frac{1}{2}p\lambda = 0 \pmod{\pi}$, *i.e.*, if the extent of the moving average is a period of the oscillation. But if λ is small and $p\lambda$ is small the amplitude is barely reduced at all, and $\phi_2 - T\phi_2$ will largely disappear, *i.e.*, the moving average will partially obliterate the harmonic term in ϕ_2. In this case, $p\lambda$ being small, the extent of the moving average is small compared with the period of the harmonic, *i.e.*, the oscillation is a very slow one. This result is what we should expect. A slow oscillation is really treated as a trend by the moving average and eliminated accordingly. The moving average will emphasize the shorter oscillations at the expense of the longer ones. If the moving average is taken over rather more than the time period the moving averages may have the original oscillation with the sign reversed and consequently the difference from the trend may somewhat exaggerate the true oscillations.

11. To sum up: in the study of oscillations obtained from a time-series by eliminating trend with moving averages it is desirable to safeguard against the introduction of spurious effects and the distortion of genuine effects due respectively to the random and oscillatory terms in the original series. This can best be done by extending the moving average so far as possible and by making it approximate to a multiple of any cycles which are suspected to exist. Iteration rapidly reduces the distortion of genuine oscillatory movements, but does not exert such a great effect on the spurious cycles due to random fluctuations.

These considerations support the desirability of extending the moving average as far as possible; but other considerations will work in the reverse direction. The saving of arithmetic; the avoidance of sacrificing terms at the beginning and end of the series; and the nature of the weighting dictated by trend elimination itself are factors of this kind.

12. Some interesting material illustrating the foregoing theory may be drawn from agricultural statistics. I took the sheep population of England and Wales for each of the years 1867 to 1939 and, after some trials of alternative methods, eliminated trend by a simple nine-year moving average. The residuals were as follows:

Year	Residual (000)	Year	Residual (000)	Year	Residual (000)
1871	−176	1893	+ 34	1915	+ 19
72	−112	94	−103	16	+128
73	+ 50	95	−104	17	+ 97
74	+141	96	− 15	18	+ 69
75	+ 60	97	− 23	19	− 29
76	− 20	98	+ 17	20	−174
77	+ 12	99	+ 71	21	−107
78	+ 82	1900	+ 35	22	−142
79	+130	01	+ 16	23	−109
80	− 14	02	− 27	24	− 23
81	−166	03	− 32	25	+ 60
82	−179	04	− 49	26	+121
83	− 84	05	− 61	27	+ 94
84	+ 38	06	− 52	28	− 25
85	+ 97	07	− 24	29	− 90
86	+ 8	08	+ 68	30	− 75
87	− 5	09	+141	31	+ 72
88	−105	10	+119	32	+152
89	− 99	11	+ 66	33	+112
90	+ 35	12	− 52	34	− 64
91	+159	13	−117	35	− 87
92	+167	14	− 61		

This is a fairly typical specimen of the oscillatory movements in English agriculture, though the regularity of the fluctuations is rather better marked than usual for livestock numbers and crop acreages. The alternation of groups of positive and negative signs is alone sufficient to suggest the existence of a cycle of about eight years. The question is, can this apparent oscillatory movement have been induced by the moving average employed? The question may be answered decisively in the negative by two quite distinct approaches.

13. Consider in the first place the period of oscillations induced in random series by a nine-point average. In equation (7) the a's are all equal, so that $\cos\theta = 8/9$, $\theta = 27° 40'$, and the mean distance between upcrosses is about 13. In equation (8) it will be found that $\Delta y_r = x_1 + x_9$, $\Delta y_{r-1} = x_0 + x_8$ so that the angle between the planes so defined is $\pi/2$ and the mean distance between maxima is 4. The mean period of induced oscillations would then be between 4 and 13, probably about 10. This is not different enough from the observed value to dispose of the suggestion that the observed oscillation is spurious.

14. We have then to consider the variance of the series. The variance of the residuals given in the foregoing table in 8474. If these are generated by averaging a random series that series must have a variance nine times as great, *i.e.*, 76,266. We require for comparison with this the true value of the variance of random elements in the original series.

Such a value may be obtained by finite differences. If we difference the original series (before trend is eliminated) k times, the random element ϕ_3 grows in importance with k at the expense of the systematic elements ϕ_1 and ϕ_2 and the variance of the differenced series will tend to $\binom{2k}{k}$ times that of ϕ_3. The following are the sums of squares of deviations from zero up to $k = 10$ divided by $\binom{2k}{k}$.

k	S (deviation)$^2/\binom{2k}{k}$
1	3467
2	1442
3	954
4	629
5	518
6	448
7	401
8	371
9	356
10	347

The ratios are evidently tending to a value in the neighbourhood of 347, and a comparison of this figure with that required by the explanation of generation by a moving average, 76,266, rejects the explanation decisively.

15. An entirely different approach leads to the same conclusions. Yule (1921) has introduced a very fruitful method of dealing with oscillations in time-series by considering the correlations of the members of the series among themselves. If we calculate the product moment correlation between each member of the series and the succeeding member we have the autocorrelation coefficient of order 1, denoted by r_1; generally the correlation between each member and the next but $(k - 1)$ is the autocorrelation of order k, r_k. If a moving average of extent p is taken of a random series the successive terms of the resultant are correlated up to r_p; but within sampling limits $r_k = 0$, $k > p$. If however, there is a harmonic term in the series it will be reproduced in the autocorrelations. Yule's method therefore offers a criterion for distinguishing between the two cases and we can determine whether the oscillations in a series are induced or not by examining whether the autocorrelations decay to zero. The following are the coefficients for the sheep data here considered :

Order of autocorrelation	Coefficient	Order of autocorrelation	Coefficient
1	+0·5952	16	+0·2764
2	−0·1513	17	+0·4394
3	−0·6005	18	+0·2926
4	−0·5372	19	−0·0743
5	−0·1376	20	−0·3589
6	+0·1441	21	−0·3809
7	+0·2028	22	−0·1176
8	+0·1183	23	+0·1728
9	+0·0064	24	+0·3433
10	−0·0779	25	+0·3519
11	−0·1419	26	+0·1536
12	−0·1724	27	−0·2030
13	−0·1857	28	−0·4556
14	−0·1280	29	−0·4150
15	+0·0518	30	−0·1844

The coefficients show no signs of tending to zero and in fact the oscillations are well defined, with a period of between eight and nine years. Whether this evidence is sufficient to establish the existence of a sheep cycle I will not attempt to discuss on the present occasion; but it is clearly sufficient to dispose of the supposition that the oscillations in the series obtained by eliminating trend with a moving average are induced from random fluctuations by the averaging process.

References

Dodd, E. L. (1939), " The Length of the Cycles which Result from the Graduation of Chance Elements," *Ann. Math. Statist*, **10**, 254.
Macaulay, F. R. (1931), *The Smoothing of Time Series*, New York.
Mitchell, W. C. (1927), *Business Cycles*, New York.
Sasuly, M. (1934), *Trend Analysis of Statistics*, Washington, D.C.
Slutzky, E. (1937), " The Summation of Random Causes as the Source of Cyclic Processes," *Econometrika*, **5**, 105.
Spencer, J. (1904), " On Graduation of Rates of Mortality, Manchester Unity of Oddfellows," *J. Inst. Act.*, **38**, 334; and (1907), **41**, 361.
Yule, G. Udny (1921), " On the Time Correlation Problem," *J. Roy. Stat. Soc.*, **84**, 497.

From *Biometrika*, Vol. **33**, pp 105–22 (1944)

ON AUTOREGRESSIVE TIME SERIES

By M. G. KENDALL

1. The product-moment correlation coefficients obtained by correlating the members of a time-series among themselves provide a useful method of investigating the behaviour of the series, especially in regard to oscillatory movements. Consider a series of values $x_1 \ldots, x_n$, measured about their mean, trend-free and defined at equal intervals of time $t = 1, \ldots, n$. The kth serial correlation r_k is defined as the correlation between members of the series k intervals apart,

$$r_k = \frac{\sum_{j=1}^{n-k} x_j x_{j+k}}{\left\{\sum_{j=1}^{n-k} x_j^2 \sum_{j=1}^{n-k} x_{j+k}^2\right\}^{\frac{1}{2}}}. \tag{1}$$

The figure obtained by plotting r_k as ordinate against k as abscissa and joining each point to the next is known as the correlation diagram or correlogram. Since $r_0 = 1$ the correlogram always starts from the point (0, 1). When the series is infinite in extent, the *serial* correlations become *auto*-correlations (denoted by a Greek ρ), and the series for which such correlations do not vanish may be said to be autocorrelated.

2. If the time series is random the correlogram presents no systematic appearance. If it consists of a simple harmonic the correlogram reproduces that harmonic. If it results from a moving average of finite extent d of a random series the correlogram will, within sampling limits, be zero after $k = d$. The correlogram thus provides a criterion for distinguishing between various kinds of oscillatory time series. In this paper I propose to consider the correlogram of a series defined by the difference equation

$$u_{t+2} + a u_{t+1} + b u_t = \epsilon_{t+2}, \tag{2}$$

where a and b are constants and ϵ_{t+2} is a random variable. This equation I shall call the generating equation and the coefficients a and b generating coefficients. The generated series may be said to be autoregressive. The notion of generating series in this way was introduced by Yule (1927) in a classical paper on sunspot periodicities and has been applied to meteorological and economic series with some success. The autoregressive scheme can, in fact, explain a typical phenomenon of such series for which Fourier and periodogram analysis cannot satisfactorily account without great artificiality, namely, the continual shift in phase and variation of amplitude which occur even when the series is smooth. Accounts of the autoregressive scheme have been given in the books by Wold (1938) and Davis (1941).

3. The equation (2) may be solved by the ordinary methods appropriate to difference equations. If the roots of

$$\xi^2 + a\xi + b = 0$$

are $\alpha + i\beta$, $\alpha - i\beta$, the complementary function of (2) is

$$p^t(A \cos \theta t + B \sin \theta t), \tag{3}$$

where $p = +\sqrt{b}$, $\theta = \arctan\frac{\beta}{\alpha} = \arctan\sqrt{\left(\frac{4b}{a^2} - 1\right)}$, and A and B are arbitrary constants.

It is here assumed that b is positive and that $4b > a^2$. It is also to be assumed that $\sqrt{b} = p$ is not greater than unity. The complementary function (3) then represents a damped harmonic. I shall call $2\pi/\theta$ the fundamental period of the generated system.

Let ξ_t be a particular value of (3) such that

$$\xi_0 = 0, \quad \xi_1 = 1, \tag{4}$$

i.e. such that

$$\xi_t = \frac{2}{\sqrt{(4p^2 - a^2)}} p^t \sin \theta t. \tag{5}$$

Then a particular integral of (2) will be found to be

$$\sum_{j=0}^{\infty} \xi_j \epsilon_{t-j+1}. \tag{6}$$

The complete solution is

$$u_t = p^t(A \cos \theta t + B \sin \theta t) + \sum_{j=0}^{\infty} \xi_j \epsilon_{t-j+1}. \tag{7}$$

In practical cases we may assume that the series was 'started up' some time ago, so that the complementary function has been damped out of existence. The series is then given by

$$u_t = \sum_{j=0}^{\infty} \xi_j \epsilon_{t-j+1}. \tag{8}$$

This is a moving sum of a random series with damped harmonic weights. It has been generally assumed, apparently on the basis of experimental evidence, that the mean period of the generated series will be $2\pi/\theta$, the same as the fundamental period present in the term ξ. This, however, is not necessarily so. Something depends on what we call a period in the generated series, whether, for example, we decide to exclude small ripples on the main wave. One possibility is to define the period as the distance between successive 'upcrosses', i.e. points where the series changes sign from negative to positive. The mean distance between major peaks or major troughs will not, probably, be very different from this in the majority of practical cases.*

Some idea of the mean length of the period as so defined can be obtained for particular distributions of ϵ, though a general discussion presents great difficulties. Consider the sum, which will be required again later,

$$\sum_{j=0}^{\infty} \xi_j \xi_{j+k} = \frac{4}{4p^2 - a^2} \Sigma\{p^{2j+k} \sin \theta j \sin \theta(j+k)\} = \frac{2p^k}{4p^2 - a^2} \Sigma[p^{2j}\{\cos \theta k - \cos \theta(2j+k)\}]$$

$$= \frac{2p^k}{4p^2 - a^2} \left\{ \frac{\cos \theta k}{1 - p^2} - \frac{\cos \theta k - p^2 \cos \theta(k-2)}{1 - 2p^2 \cos 2\theta + p^4} \right\}. \tag{9}$$

We have

$$\Sigma \xi_j \xi_{j+1} / \Sigma \xi_j^2 = \frac{2p \cos \theta}{1 + p^2} = \frac{-a}{1+b} = \cos \phi, \text{ say}. \tag{10}$$

Now if ϵ is normally distributed, the mean period (from one upcross to the next) is $2\pi/\phi$ (Dodd, 1939). Also it is easily seen that $\cos \theta = \dfrac{-a}{2p} = \dfrac{-a}{2\sqrt{b}}$. Thus the period given by the

* Dodd (1939) points out that in series generated by moving averages of random series there sometimes occur oscillations above or below the x-axis which would not be taken into account by counting upcrosses. The difficulty is to decide which of these is not to be ignored as a 'ripple'. I think that for harmonic weights such as are given by the autoregressive scheme the method of upcrosses is satisfactory, as indicated in paragraph 4.

generating equation is not in general the same as the period in the generated series, defined as the mean difference between upcrosses. The ratio of the first to the second is

$$\frac{\arccos \dfrac{a}{1+b}}{\arccos \dfrac{a}{2\sqrt{b}}}. \tag{11}$$

4. One would expect differences of a similar kind when the variation of ϵ is not normal. Experiments indicate, for example, that rectangular variation of ϵ gives much the same periods as normal variation. It does not seem to have been previously remarked that the mean period of the generated series is not that of the fundamental, but part of the explanation is no doubt due to the fact that for ranges of b encountered in practice (say $0{\cdot}5 \leqslant b \leqslant 1{\cdot}0$) the value of the ratio (11) is not very different from unity. In the extreme case $b = 1$ (when the series becomes undamped) the ratio is exactly unity. If $b = 0{\cdot}5$, $a = -1$, the ratio is 1·07. Notwithstanding the theoretical difference exhibited by equation (11), I think we may take it that for most practical purposes a good estimate of the observed period (upcross to upcross) is that given by the generated equation, *if it is known*. Below I give examples of two artificial series of the type of equation (2) which support this conclusion.

5. Equation (9) provides one further interesting item of information, namely, the relationship between the variance of ϵ and the variance of the generated series. We have

$$\operatorname{var} u = \lim_{n\to\infty} \frac{1}{n} \sum_{t=1}^{n} \left(\sum_{j=0}^{\infty} \xi_j \epsilon_{t-j+1} \right)^2 = \sum_{j=0}^{\infty} \xi_j^2 \operatorname{var} \epsilon,$$

cross-product terms in ϵ vanishing since it is a random variable:

$$\frac{\operatorname{var} u}{\operatorname{var} \epsilon} = \Sigma \xi_j^2 = \frac{2}{4p^2-a^2}\left\{\frac{1}{1-p^2} - \frac{1-p^2\cos 2\theta}{1-2p^2\cos 2\theta + p^4}\right\} = \frac{1+b}{(1-b)\{(1+b)^2-a^2\}}.$$

Thus
$$\frac{\operatorname{var} \epsilon}{\operatorname{var} u} = \frac{1-b}{1+b}\{(1+b)^2-a^2\}. \tag{12}$$

In an actual series of 65 terms (referred to in detail below) with $a = -1{\cdot}1$, $b = 0{\cdot}5$, the observed ratio of variances was 0·43. The value given by (12) is 0·35. In another series of the same length for which $a = -1{\cdot}5$, $b = 0{\cdot}9$, the observed ratio was 0·04, the value given by (12) was 0·07. It is difficult to judge how good the agreement is, but to me it appears satisfactory for such short series. It is noticeable that the generated series may have a very much larger variance than the random series on which it is based.

6. Consider now the autocorrelations of the generated series. We have

$$\operatorname{cov}(u_j u_{j+k}) = \lim \frac{1}{n} \sum_t \{ \sum_j (\xi_j \epsilon_{t-j+1}) \sum_j (\xi_j \epsilon_{t+k-j+1}) \} = \operatorname{var} \epsilon \sum_{j=0}^{\infty} \xi_j \xi_{j+k}$$

$$= \operatorname{var} \epsilon \frac{2p^k}{4p^2-a^2}\left\{\frac{\cos\theta k}{1-p^2} - \frac{\cos\theta k - p^2\cos\theta(k-2)}{1-2p^2\cos 2\theta + p^4}\right\}.$$

Hence
$$\rho_k = \frac{p^k\{\cos\theta k + \cos\theta(k-2) - 2\cos 2\theta \cos k\theta + p^2\cos\theta k - p^2\cos\theta(k-2)\}}{(1+p^2)(1-\cos 2\theta)}$$

$$= \frac{p^k\{\sin(k+1)\theta - p^2\sin(k-1)\theta\}}{(1+p^2)\sin\theta}.$$

9-2

Writing
$$\tan\psi = \frac{1+p^2}{1-p^2}\tan\theta,$$

we have
$$\rho_k = \frac{p^k(1-2p^2\cos 2\theta+p^4)^{\frac{1}{2}}}{(1+p^2)\sin\theta}\sin(k\theta+\psi) = p^k\frac{\sin(k\theta+\psi)}{\sin\psi}. \tag{13}$$

Apart from the constant factor, ρ_k is thus the product of the damping factor p^k and a harmonic term which has the fundamental period of the generating equation. It is noteworthy that, although the point $\rho_0 = 1$ is a peak at the beginning of the correlogram the existence of the phase angle ψ implies that the interval from $k = 0$ to the next maximum of the correlogram is not equal to the fundamental period. In judging the length of the period from the correlogram it is therefore better to measure from upcross to upcross or from trough to trough; or, if peaks are preferred, not to count the maximum at $k = 0$ as a peak.

The same result may be obtained in two other ways. Multiplying equation (2) by u_{t-k} and summing for all t we have
$$\rho_{k+2}+a\rho_{k+1}+b\rho_k = \frac{\Sigma(\epsilon_{t+2}u_{t-k})}{\operatorname{var} u}.$$

Since u_{t-k} depends only on ϵ_{t-k} and terms with lower subscripts, the expression on the right vanishes if k is not less than -1. We then have
$$\rho_{k+2}+a\rho_{k+1}+b\rho_k = 0 \quad (k \geqslant -1). \tag{14}$$

This result is due to Walker (1931). It was pointed out by Wold that if we multiply (2) by u_{t+k+2} and sum we get
$$\rho_k+a\rho_{k+1}+b\rho_{k+2} = \frac{\Sigma(\epsilon_{t+2}u_{t+k+2})}{\operatorname{var} u}$$

The expression on the right no longer vanishes. In fact u_{t+k+2} contains the term $\xi_{k+1}\epsilon_{t+2}$ and we have
$$\rho_k+a\rho_{k+1}+b\rho_{k+2} = \frac{\operatorname{var}\epsilon}{\operatorname{var} u}\xi_{k+1} \quad (k \geqslant -1). \tag{15}$$

If $k = -1$ the two equations become identical, for $\rho_{-1} = \rho_1$ and we have
$$\rho_1(1+b)+a = 0, \quad \rho_1 = -\frac{a}{1+b}. \tag{16}$$

If we now solve either of the difference equations (14) and (15), making use of the initial conditions $\rho_0 = 1$ and (16) we arrive back at equation (13).

7. The foregoing result (13) would lead us to suppose that the correlogram of an autoregressive series would be damped according to the factor p^k, and this is true for an infinite series. In a number of practical cases, however, I was puzzled by the fact that correlograms of series which appeared on the face of it to be of type (2) did not damp out in the required way. Figs. 1 and 2 show the correlograms of two series for wheat prices and sheep population. The original series were taken from the agricultural returns for England and Wales, trend eliminated by a nine-years moving average and the first thirty serial correlations computed for the resulting series of 64 or 65 terms. The data and the correlations for what prices are given in Tables 1 and 2; those for sheep have been given in a previous paper (Kendall, 1941). I found a similar effect as regards non-damping in nine other agricultural time series, though the correlograms were not so regular as in these two cases. It was always possible, however, that the failure of the fluctuations to damp out in the correlogram was due wholly or partly

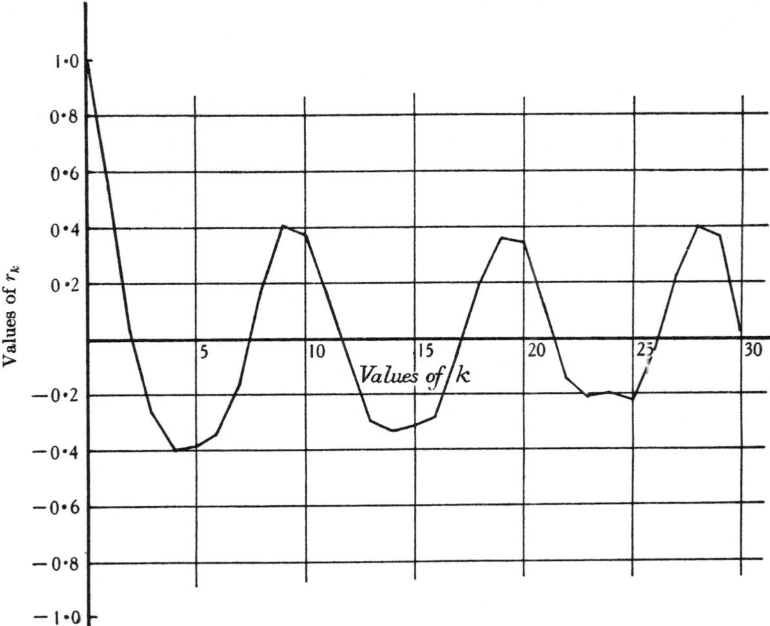

Fig. 1. Correlogram of the wheat price data of Tables 1 and 2.

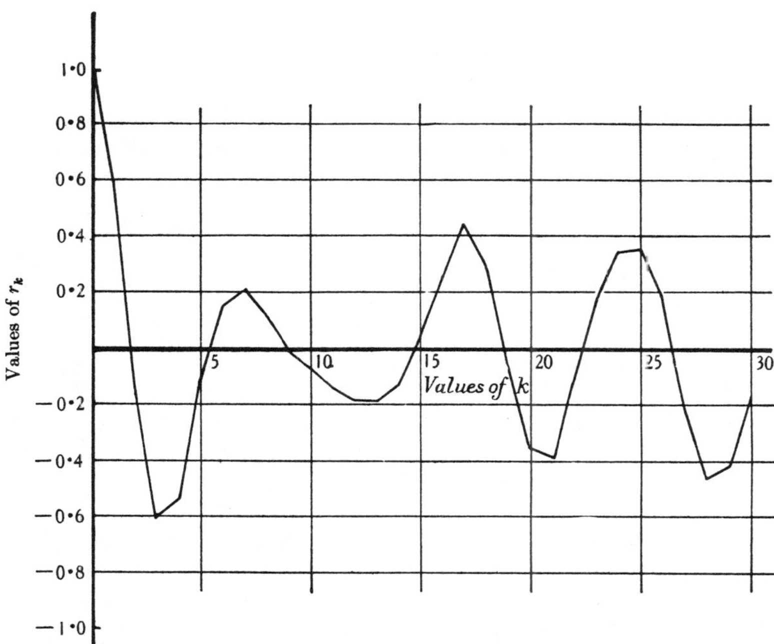

Fig. 2. Correlogram of the sheep population data referred to in paragraph 7.

to failure on the part of the original series to conform to the scheme of equation (2). To avoid such complications I constructed an artificial series with equation

$$u_{t+2} = 1\cdot 1u_{t+1} - 0\cdot 5u_t + \epsilon_{t+2}. \tag{17}$$

The ϵ's were taken from tables of random numbers and consisted of positive or negative numbers ranging by units from $-9\cdot 5$ to $+9\cdot 5$. The series was 'started up' from zero by

Table 1. *Wheat prices*, 1871–1934 *inclusive, deviations from nine-years moving average in pence per hundredweight*

Year	Price	Year	Price	Year	Price
1871	−5	1893	6	1915	−3
2	−13	4	18	6	−7
3	−21	5	15	7	−40
4	−6	6	1	8	−28
5	19	7	−11	9	−27
6	11	8	−17	1920	−51
7	−20	9	5	1	−14
8	3	1900	3	2	37
9	8	1	4	3	42
1880	2	2	2	4	11
1	−6	3	3	5	−9
2	−11	4	−1	6	−17
3	−7	5	2	7	−15
4	6	6	5	8	−5
5	9	7	0	9	−10
6	9	8	−3	1930	−2
7	4	9	−9	1	19
8	3	1910	1	2	14
9	4	1	9	3	21
1890	−3	2	11	4	19
1	−20	3	32		
2	−5	4	33		

Table 2. *Serial correlations of the wheat price series of Table* 1

Order of correlation	r	Order of correlation	r	Order of correlation	r
1	+0·577	11	+0·171	21	+0·115
2	+0·025	12	−0·075	22	−0·144
3	−0·267	13	−0·302	23	−0·204
4	−0·402	14	−0·332	24	−0·195
5	−0·389	15	−0·317	25	−0·221
6	−0·340	16	−0·282	26	−0·052
7	−0·174	17	−0·049	27	+0·217
8	+0·166	18	+0·200	28	+0·404
9	+0·411	19	+0·360	29	+0·364
10	+0·372	20	+0·343	30	+0·024

assuming $u_0 = u_{-1} = 0$. The effect of assuming two consecutive terms equal to zero is not important, for the contribution to a term of the series of terms far back is very small. We are therefore entitled to expect that an artificial series constructed in this way should conform to the foregoing theory. The resultant series and the serial correlations are given in Tables 3 and 4 and the correlogram in Fig. 3. Now this series is heavily damped, p being $\sqrt{(0\cdot 5)} = 0\cdot 7071$.

At the twentieth serial correlation, according to equation (14), r_{20} should be less than 0·002 in absolute magnitude. Actually it is one hundred times as big.

8. The explanation lies, I think, in the shortness of the series. In arriving at equation (13) it was assumed that product sums such as $\Sigma\epsilon_j\epsilon_{j+k}$ were zero; and this is so for a series of infinite length. But when the series is short these sums may differ quite appreciably from zero. For instance, in samples from a normal population with zero correlation, the standard error of the correlation in samples of 65 is about 0·125, so that values of 0·25 are not very improbable and even values as high as 0·375 are not impossible.

Table 3. *Artificial series* $u_{t+2} = 1\cdot1u_{t+1} - 0\cdot5u_t + \epsilon_{t+2}$ *constructed as described in the text*

No. of term	Value of series	No. of term	Value of series	No. of term	Value of series
1	7	23	−4	45	−13
2	6	24	−5	46	1
3	−6	25	−9	47	6
4	−4	26	−4	48	4
5	3	27	−4	49	11
6	−4	28	3	50	15
7	−5	29	9	51	9
8	−1	30	4	52	8
9	10	31	−8	53	4
10	10	32	−6	54	−1
11	6	33	−3	55	4
12	−4	34	−2	56	7
13	−4	35	0	57	11
14	−7	36	−1	58	0
15	−2	37	−3	59	1
16	6	38	3	60	0
17	17	39	−1	61	−5
18	24	40	−8	62	−11
19	17	41	−3	63	−8
20	4	42	−8	64	−3
21	1	43	−10	65	5
22	−5	44	−16		

Table 4. *Serial correlations of the series of Table* 3

Order of correlation	r	Order of correlation	r	Order of correlation	r
1	+0·70	11	−0·05	21	+0·05
2	+0·29	12	−0·17	22	−0·12
3	+0·01	13	−0·27	23	−0·28
4	−0·17	14	−0·31	24	−0·43
5	−0·27	15	−0·30	25	−0·57
6	−0·25	16	−0·18	26	−0·56
7	−0·13	17	+0·12	27	−0·26
8	+0·07	18	+0·29	28	+0·02
9	+0·12	19	+0·33	29	+0·17
10	+0·05	20	+0·22	30	+0·27

Consider then the series of n terms such as

$$u_j = \xi_1\epsilon_j + \xi_2\epsilon_{j-1} + \xi_3\epsilon_{j-2} + \dots.$$

The product moment of u_j and u_{j+k} will be

$$\sum_j u_j u_{j+k} = \sum_j \sum_l \xi_l \epsilon_{j-l+1} \sum_m \xi_m \epsilon_{j+k-m+1} = \sum_j \sum_{l,m} \xi_l \xi_m \epsilon_{j-l+1} \epsilon_{j+k-m+1}.$$

The terms in ϵ^2 give the expression (13). The others will be sums of products of the ξ's multiplied by the serial covariances of the ϵ's themselves. The dominating terms in these latter will be those containing the larger ξ's. These terms are themselves damped harmonics of the fundamental type and when applied to the sums $\Sigma \epsilon_j \epsilon_{j+k}$ may be expected to generate oscillatory movements of about the same period as the original series. We may therefore expect that for short series the correlogram of the autoregressive system may not decay very rapidly, but that the product terms may themselves result in a small fluctuation. This appears to be happening in both the practical and the artificial examples given above.

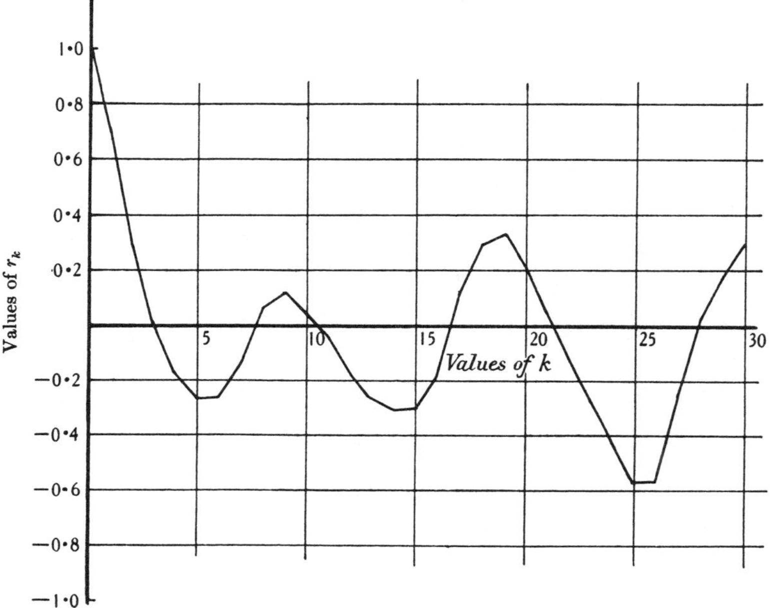

Fig. 3. Correlogram of the artificial series of Tables 3 and 4.

9. Putting aside the mathematics for a moment and looking at the point in a general way, we can, I think, appreciate that something of the kind is to be expected in short series. The variance of a short series should not differ systematically from the variance of the infinite series; but the covariances may systematically exceed in absolute value the value for the infinite series. In fact, as we proceed along the series, the oscillations change in phase, and when we have gone far enough will be quite unrelated in phase to the initial oscillations; but if we only go so far as the second or third oscillation the last oscillation may not, so to speak, have had time to get very much out of phase with the first. The consequence will be that the correlations for such a short space of time will tend to be higher than those for the series as a whole. For the generated system (and indeed for time series in general) we have to be careful not to assume too lightly that values calculated for a part of the series are typical of the corresponding values for the series as a whole; and this notwithstanding that the part of the series was chosen 'at random'.

One would like to know how long a practical series must be for the damping to show itself decisively. An exact discussion of this point presents difficulty, because in a finite series the serial covariances $\Sigma\epsilon_j\epsilon_{j+k}$ are not only non-vanishing but are correlated among themselves. Suppose, however, that we have a series with a damping factor p. Then the kth serial coefficient will not exceed p^k in absolute value, and the difference of the correlogram as a whole from expectation will not at any point exceed the error due to the sampling fluctuation of the serial correlation between u_j and u_{j+k}. This in turn will not exceed the sampling error of the correlation in samples of n from an uncorrelated series, which for large samples has a standard error $n^{-\frac{1}{2}}$. For example, if we took a series of 650 terms instead of 65, correlations up to the 30th would have a standard error not exceeding $1/\sqrt{(650-30)} = 0{\cdot}04$, and hence fluctuations of 0·10 would not be impossible. But by the 30th term in the correlogram the serial correlation has been practically damped out of existence. The fair inference is, I think, that except for the first few serial correlations of the main series, which are not damped very much, the serial correlations are seriously affected even for long series, for the sampling errors are not negligible compared with the small damped 'true' values. In general one would expect the damping effect to show itself for the first five, ten or twenty terms and then to be obliterated by the sampling effect. Whether this happens at the fifth, tenth or twentieth term depends not only on ϵ but on p, the rapidity of damping. In most of the natural series I have examined the damping is fairly rapid, so that the damping effect in the correlogram disappeared after the first few terms. The economic examples given by Wold (1938) and the meteorological examples of Walker (1931) appear to me to support this conclusion.

10. I turn now to consider another effect which may obscure the presence of the generated system of type (2) and may exert an important influence on the correlogram. The random element ϵ considered up to this point is what Yule calls a disturbance, and is integrated into the course of the series by the autoregressive scheme. There may also be a component η *superposed* on the system. This superposed element, if random, is like an error of observation in that its value at any point is unrelated to its value at other points.

If a random element with variance $\operatorname{var}\eta$ is superposed on an infinite series with variance $\operatorname{var} u$ the variance of the whole will be $\operatorname{var}\eta + \operatorname{var} u$. The autocovariances, however, will not be affected (except by sampling effects for short series). Consequently all the autocorrelations except ρ_0 will be reduced in the ratio

$$c = \frac{\operatorname{var} u}{\operatorname{var} u + \operatorname{var}\eta}. \tag{18}$$

For short series there will still be a reduction of the same type, but the value of c may differ from its theoretical value for sampling reasons.

11. An autoregressive series of 65 terms was constructed according to the formula

$$u_{t+2} = 1{\cdot}5u_{t+1} - 0{\cdot}9u_t + \epsilon_{t+2}, \tag{19}$$

where the ϵ_{t+2} were the values of random numbers proceeding by units from $-49{\cdot}5$ to $49{\cdot}5$ with a theoretical variance of 833. On to the series so derived there were superposed (a) a rectangular random element $-49{\cdot}5(1)\ 49{\cdot}5$ and (b) a further rectangular random element $-199{\cdot}5(1)\ 199{\cdot}5$, additional to the first, the latter series then being divided by ten and rounded up to the nearest integer. The resultant series and serial correlations are shown in Tables 5 and 6, and the correlograms in Fig. 4.

According to (12) the variance of u for an infinite series generated by (19) should be $13{\cdot}97 \operatorname{var}\epsilon$. The values of c for the two series here considered are then, respectively,

$13{\cdot}97/14{\cdot}963 = 0{\cdot}93$ and $13{\cdot}97/30{\cdot}963 = 0{\cdot}45$. In the second case the effect of the superposed variation is to halve the observed correlations. The effect on the actual series of 65 terms is somewhat irregular (see para. 14 below).

Table 5. *Artificial series* $u_{t+2} = 1{\cdot}5u_{t+1} - 0{\cdot}9u_t + \epsilon_{t+2}$, *(a) with small superposed element* η, *(b) with large element* η, *constructed as described in the text*

No. of term	Value of series		No. of term	Value of series		No. of term	Value of series	
	(a)	(b)		(a)	(b)		(a)	(b)
1	5	16	23	−215	−34	45	88	3
2	36	7	24	−219	−15	46	76	−5
3	8	10	25	−76	−17	47	93	21
4	−81	−5	26	95	−4	48	34	8
5	−89	7	27	239	15	49	60	15
6	−15	−7	28	316	23	50	−69	−13
7	35	10	29	289	21	51	−120	−10
8	112	13	30	169	13	52	−54	12
9	146	17	31	49	−7	53	−56	−13
10	100	19	32	−114	−26	54	12	7
11	1	7	33	−259	−26	55	5	−11
12	−131	1	34	−290	−26	56	13	−18
13	−195	−6	35	−208	−29	57	3	−1
14	−259	−27	36	−31	−4	58	13	11
15	−258	−37	37	21	12	59	−4	9
16	−118	−28	38	109	18	60	−71	−13
17	32	6	39	26	19	61	−94	−7
18	110	−3	40	33	−5	62	15	19
19	245	6	41	22	0	63	45	−3
20	166	24	42	−30	−10	64	138	19
21	79	15	43	51	−15	65	115	0
22	−177	−34	44	49	17			

Table 6. *Serial correlations of series (a) and (b) of Table 5*

Order of correlation	*r*		Order of correlation	*r*		Order of correlation	*r*	
	(a)	(b)		(a)	(b)		(a)	(b)
1	+0·78	+0·49	11	+0·33	+0·16	21	+0·21	+0·05
2	+0·33	+0·13	12	−0·04	−0·16	22	−0·07	−0·28
3	−0·22	−0·13	13	−0·38	−0·37	23	−0·28	−0·33
4	−0·63	−0·42	14	−0·58	−0·50	24	−0·32	−0·29
5	−0·75	−0·46	15	−0·52	−0·36	25	−0·22	−0·20
6	−0·59	−0·39	16	−0·23	−0·14	26	−0·01	−0·07
7	−0·22	−0·01	17	+0·14	+0·19	27	+0·16	+0·11
8	+0·19	+0·28	18	+0·43	+0·42	28	+0·23	+0·20
9	+0·48	+0·38	19	+0·55	+0·41	29	+0·13	+0·01
10	+0·53	+0·52	20	+0·45	+0·37	30	−0·11	−0·11

The correlograms run according to expectation. The effect of the bigger random element is to reduce the amplitude at the beginning of the series and to introduce some minor irregularities in the data, but not to affect substantially the lengths of the correlogram oscillations.

12. But here arises one important difficulty. Suppose we are given such series as these and require to estimate a and b, the constants of the generating equation. The procedure adopted by Yule was as follows: we take the observed serial correlations r_1 and r_2 as estimates of the autocorrelations. We then find the regression equation of u_{t+2} on u_{t+1} and u_t by the usual methods, assuming that the variance of the series is the same as the variance for the series less its first term and that the serial correlation r_1 for the whole series is the same as that for the series less its first term. The regression equation is

$$u_{t+2} = \frac{r_1(1-r_2)}{1-r_1^2}u_{t+1} + \frac{r_2 - r_1^2}{1-r_1^2}u_t. \tag{20}$$

The observed regression equation is then taken as an estimate of the generating equation so that we have as estimates of a and b

$$-a = \frac{r_1(1-r_2)}{1-r_1^2}, \quad -b = \frac{r_2-r_1^2}{1-r_1^2} = \frac{r_2-1}{1-r_1^2}+1. \tag{21}$$

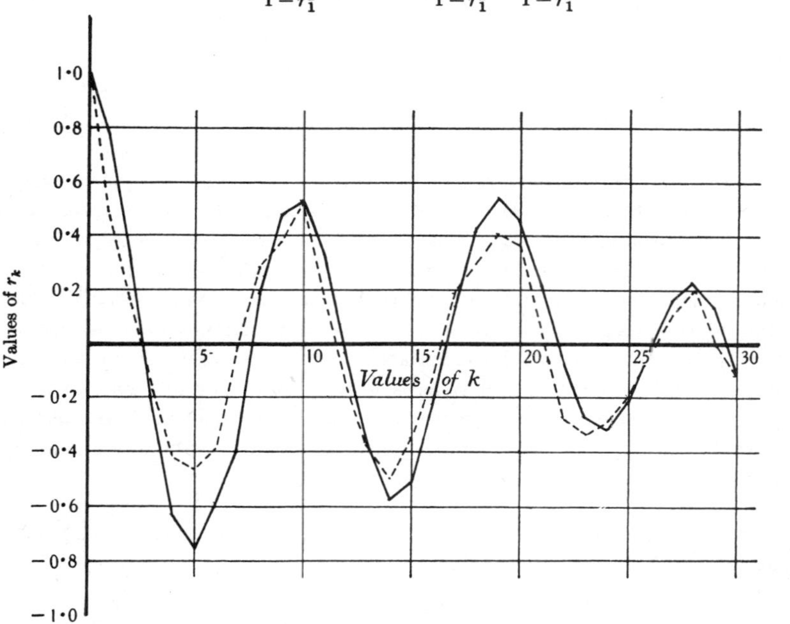

Fig. 4. Correlogram of the two artificial series of Tables 5 and 6, the full line representing series (a) with slight superposed variation, the broken line series (b) with large superposed variation.

From the above it will be evident that there are two sources of possible error in the use of these equations: (1) if the series is short r_1 and r_2 may not be reliable estimates of the autocorrelations for the infinite series, (2) if there is any superposed variation the observed r's will be lower than the true r's for the autoregressive system, being in fact cr_1 and cr_2 where c is given by (18). Consider the second of these effects.

13. In the case of superposed variation the use of (21) will lead to the equations

$$-a' = \frac{cr_1(1-cr_2)}{1-c^2r_1^2}, \quad -b' = \frac{cr_2-c^2r_1^2}{1-c^2r_1^2}.$$

The estimated fundamental period of the generating equation is then given by

$$4\cos^2\theta' = \frac{a'^2}{b'} = \frac{cr_1^2(1-cr_2)^2}{(1-c^2r_1^2)(r_2-cr_1^2)}. \tag{22}$$

If we expand (22) in powers of $\gamma = 1-c$ we find, to first order in γ,

$$\frac{a'^2}{b'} = \frac{a^2}{b}\left\{1-\gamma\frac{(1+b)(3b^2-b-a^2)}{b\{(1+b)^2-a^2\}}\right\}.$$

Hence, if $3b^2-b-a^2>0$ the effect of a superposed variation (equivalent to positive γ) is to give

$$\frac{a'^2}{b'} < \frac{a^2}{b},$$

or in other words to result in a shortening of the observed period. The condition that $3b^2-b-a^2>0$ is equivalent to

$$b > \tfrac{1}{6}\{-1+\sqrt{(12a^2+1)}\}$$

and is not very restrictive since in any case $a^2 \leqslant 4$ and $4b \geqslant a^2$. The inequality is obeyed by all the examples I have met in practice.

We therefore reach the interesting conclusion that if there is any superposed random variation present, the period calculated from the observed regression equation according to formulae (21) will probably be too short even for long series. Yule himself found too short a period for his sunspot material and, suspecting that it was due to superposed variation, attempted to reduce that variation by graduation. The result was to give a longer period more in accordance with observation. It does not appear, however, that the superposed variation in his case was very big. In a number of agricultural time series which I have examined it is sometimes about half the variation of the series and the effect on the period as calculated from the serial correlations is very serious. For instance, in the cases of wheat prices and sheep population referred to above, formulae (21) give periods of 7·0 and 6·8 years, whereas the correlograms indicate periods of about 9·5 and 8·5 years respectively.

14. As an example, consider the second of the artificial series referred to in paragraph 11. For the observed serial correlations I find

$$r_1' = 0{\cdot}486, \quad r_2' = 0{\cdot}133,$$

giving, according to (21),

$$-a' = 0{\cdot}552, \quad b' = 0{\cdot}135, \quad \cos\theta' = \frac{-a'}{2\sqrt{b'}} = 0{\cdot}751, \quad \theta' = 41{\cdot}3^\circ,$$

which corresponds to a period of about 8·7 years, whereas we know from the construction of the series that

$$a = 1{\cdot}5, \quad b = 0{\cdot}9,$$

giving $\theta = 37{\cdot}7^\circ$ and a period of 9·5 years.

Considering the profound effect which the superposed variation has had on the first two serial coefficients, reducing r_1 from 0·78 to 0·49 and r_2 from 0·33 to 0·13, one might have expected the period to have been affected even more than appears from this result. But the example serves to bring out once again the difficulties associated with short series and the unreliability of coefficients calculated from the first two serial correlations in such cases.

If, for instance, we had found $r_2' = 0{\cdot}18$ instead of $0{\cdot}13$ we should have obtained a period of about 12 years, and if $r_2' = 0{\cdot}20$ the solution becomes impossible, for a' and b' then assume values such that $a'^2 > 4b'$ and $\cos\theta' > 1$. Furthermore, such a value as $0{\cdot}18$ for r_2' increases the period instead of decreasing it. We note, in fact, that r_2 has been reduced by $0{\cdot}13/0{\cdot}33 = 0{\cdot}40$ so that it is not legitimate to assume that r_1 and r_2 are reduced by a constant c. The errors introduced by neglect of the autocovariances of the superposed random element may be so serious as to destroy the value of calculations based on the observed regression coefficients.

15. Even in long series, when it is legitimate to suppose that r_1 and r_2 are reduced in the same proportion, the length of the period is very sensitive to superposed variation. Consider, for example, the effect of small variations in c near $c = 1$. We have from (22), differentiating logarithmically and putting $c = 1$,

$$-2\tan\theta\frac{d\theta}{dc} = 1 - \frac{2r_2}{1-r_2} + \frac{2r_1^2}{1-r_1^2} + \frac{r_1^2}{r_2 - r_1^2},$$

which on substituting $$r_1 = -\frac{a}{1+b}, \quad r_2 = -b + \frac{a^2}{1+b}$$

reduces to $$-\tan\theta\frac{d\theta}{dc} = \frac{(1+b)(3b^2+b-a^2)}{2b\{(1+b)^2-a^2\}}. \tag{23}$$

Now the period $$P = 2\pi/\theta \quad \text{and} \quad \tan\theta = \sqrt{\left(\frac{4b}{a^2}-1\right)}.$$

Hence $$\frac{dP}{dc} = -\frac{P^2a(1+b)(3b^2+b-a^2)}{4\pi\sqrt{(4b-a^2)}\{(1+b)^2-a^2\}}. \tag{24}$$

When $a = -1{\cdot}5$, $b = 0{\cdot}9$, $P = 9{\cdot}5$, this reduces to

$$\frac{dP}{dc} = 14 \text{ intervals.}$$

Thus if $c = 0{\cdot}9$, i.e. the superposed variation is only about 10 % of the total, the period may be shortened by something of the order of one interval.

16. The position may then be summarized as follows:

(*a*) The correlogram of a generated system of the type of equation (2) will be damped according to the damping factor of the equation; but if the series is short the damping may be considerably less than the theoretical value.

(*b*) The correlogram will show a period equal, within limits of error, to that of the fundamental period of the system. The distance from the unit ordinate at $k = 0$ and the first maximum of the correlogram may not, however, be a full period and in estimating periods from the correlogram it is better to reckon from upcross to upcross.

(*c*) It does not appear that in short series the periodic movement is substantially affected, at any rate not to such an extent as the damping.

(*d*) When superposed random variation is present the fundamental period calculated from the observed regressions will be too short and may be very considerably so.

(*e*) The period of the generated series, defined as the mean distance from upcross to upcross, is not quite the same as the fundamental period: but the difference is not likely to be important in practice.

17. Armed with these results we may consider the difficult problem of determining, for a given time series, the autoregressive scheme which may have generated it.

The first step is to calculate the serial correlations and examine the correlogram. If the latter shows fairly regular oscillations there is a presumption that the series is of the autoregressive type, and this conclusion need not be rejected solely because the oscillations do not damp so rapidly in the later portion of the correlogram as at the beginning. To take the wheat price material given above, an examination of the correlogram (Fig. 1) strongly suggests a simple damped harmonic. The sheep series (Fig. 2) is not so clearly defined and there is a suggestion of more than one period in the behaviour of the correlogram. This may, however, be due to the shortness of the series and one would be inclined to consider it in the first place as a single damped harmonic.

The length of the period, as pointed out above, is best determined by the mean distances between upcrosses. In the wheat price data there are upcrosses at about 7·5 years, 17·2 years and 26·1 years, giving periods of 9·7 and 8·9 years with a mean of 9·3 years. Much the same result is arrived at by counting the periods between troughs.

18. The second step is to calculate a' and b' by equations (21) and to find the period of the fundamental harmonic as given by the observed regression equation. For wheat prices we have

$$r'_1 = 0{\cdot}5773, \quad r'_2 = 0{\cdot}0246.$$

Hence

$$a' = -0{\cdot}8446, \quad b' = 0{\cdot}4630,$$

whence

$$\cos\theta' = 0{\cdot}6206, \quad \theta' = 51{\cdot}63^\circ,$$

giving a fundamental period of 6·97 years.

This is too small, and we are immediately led to suspect the existence of superposed variation. The problem then arises of determining the variance of the superposed element. If this is random, and there are no periodic terms of very short period the variate difference method may be used. For the wheat price material, taking differences up to the 10th on the primary series (before trend was eliminated) I find for an estimate of the random variance $\operatorname{var}\eta = 27{\cdot}72$. The total variance of the series is 272·8. The constant c of equation (18) is

$$1-\frac{27{\cdot}72}{272{\cdot}8} = 0{\cdot}90.$$

The putative serial correlations of the autoregressive series will then be

$$r_1 = 0{\cdot}641, \quad r_2 = 0{\cdot}027$$

and

$$a = -1{\cdot}059, \quad b = 0{\cdot}652,$$

whence

$$\cos\theta = 0{\cdot}6551, \quad \theta = 49{\cdot}07^\circ,$$

giving a period of 7·34 years.

This is still too short. It would require a random superposed variance of about 25 % of the total, instead of the observed 10 %, to produce a period of 9·3 years.

19. This illustrates very well a constantly recurring difficulty in the theory of the variate difference method and of time series generally. The superposed variation may not be random. Indeed, we have little ground for expecting that it should be. A positive correlation between the successive values of η will reduce the variance shown as random by the variate difference method and unless we have prior reason to suppose that η is random the values given by the variate difference method are quite likely to be too small. Unfortunately we rarely have any prior knowledge of η, but from general economic considerations one would not be surprised

to find that there do exist positive correlations from one year to the next, owing to the enduring nature of some of the causes which can give rise to superposed variation. I conclude generally that discrepancies of the type here considered support the view that the period is to be determined from the correlogram, not from solution of the regression equation.

20. Two points may be mentioned incidentally. One, a matter of technique, is that the arithmetic of serial correlations and variate differences are closely linked together and the results of the one can be used to derive those of the other. This means an enormous saving in arithmetic and the method is described in the Appendix to this paper.

The second point concerns the removal of superposed random variation by graduation formulae. It will be clear that if the superposed variation is not random graduation may only make matters worse; but even if it is random, graduation formulae may induce spurious cyclical effects into the data. It seems to me that as a general rule graduation is to be undertaken with great caution.

21. Reverting to the main topic, it would seem that if the period shown by the correlogram and the period calculated from the observed regression equation disagree, and cannot be reconciled by the assumption of a superposed random element, there is little further to be done to dissect the superposed element from the autoregressive part of the system. If, however, the variate difference method supplies a variance of η which can satisfactorily explain the difference in periods, we may go forward. The constant c can be calculated and the constants of the autoregressive part of the system determined. Equation (12) then gives an estimate of the ratio between the variance of the basic random element ϵ and the variance of the generated series. If there is a superposed element it will not be possible to find the values of that element at every point and so to determine ϵ at every point; but if η is non-existent, ϵ can be explicitly determined, except for the first two terms of the series. In fact, we merely apply the generating equation to the observed series, the residuals between prediction and observation being the values of ϵ. An examination of these residuals will confirm whether they may be regarded as a random series.

22. The foregoing treatment can be extended to the case of a more general linear regressive system

$$u_{t+m}+a_1 u_{t+m-1}+\ldots+a_m u_t = \epsilon_{t+m},$$

or to cases where the regression is curvilinear; and the necessity for more general schemes in representing observed data can, as shown by Yule, be discussed in terms of partial autocorrelations and scatter diagrams. A more serious problem arises if the series ϵ is itself not random, a state of affairs which one fears might be fairly common in economic series. To take the wheat price data once again, it would not be surprising to find that the wheat price oscillations were regenerated by a series of disturbances, part of which were attributable to variations in acreages, yields, or the prices of other crops. Such disturbances might themselves be oscillatory. For such cases the problem becomes exceedingly complicated. To discuss it at all satisfactorily one would require a long series or collateral evidence in the form of other series of a similar character. If there is a royal road in this subject it has not yet been discovered.

APPENDIX

Relationship between variate differences and serial correlations

If we have a series of values $x_1 \ldots x_n$ the first differences are $x_1 - x_2$, etc., the second differences $x_1 - 2x_2 + x_3$, etc., and so on. Let S_j be the sum of the squares of the jth differences and write for the product-sums

$$P_j = \sum_{k=1}^{n-j} x_k x_{k+j}. \tag{25}$$

The sums S are those appearing in variate difference analysis and the quantities P appear in serial correlation analysis. Either set can be exp·essed in terms of members of the other, as follows:

We have
$$S_0 = P_0. \tag{26}$$

$$S_1 = \sum (x_j - x_{j+1})^2 = \sum_{j=1}^{n-1} x_j^2 - 2\sum_{j=1}^{n-1} x_j x_{j+1} + \sum_{j=2}^{n} x_j^2 = 2P_0 - x_1^2 - x_n^2 - 2P_1, \tag{27}$$

$$\begin{aligned} S_2 = \sum (x_j - 2x_{j+1} + x_{j+2})^2 &= \sum_{j=1}^{n-2} x_j^2 + 4\sum_{j=2}^{n-1} x_j^2 + \sum_{j=3}^{n} x_j^2 - 4\sum_{j=1}^{n-2} x_j x_{j+1} + 2\sum_{j=1}^{n-2} x_j x_{j+2} - 4\sum_{j=2}^{n-1} x_j x_{j+1} \\ &= 6P_0 - 8P_1 + 2P_2 - x_1^2 - x_n^2 - (2x_1 - x_2)^2 - (2x_n - x_{n-1})^2 \end{aligned} \tag{28}$$

and so on. For the purpose of expressing the general formulae of this kind it is convenient to modify the sums S. Suppose we write the series $x_1 \ldots x_n$ preceded and followed by a number of zeros. The difference table will then appear as follows:

0				
	0			
0		0		
	0		$-x_1$	
0		$+x_1$		
	$-x_1$		$3x_1 - x_2$	etc.
x_1		$-2x_1 + x_2$		
	$x_1 - x_2$		$-3x_1 + 3x_2 - x_3$	
x_2		$x_1 - 2x_2 + x_3$		
	$x_2 - x_3$		$x_1 - 3x_2 + 3x_3 - x_4$	
x_3		$x_2 - 2x_3 + x_4$		

with a symmetrical effect at the other end. Writing now T_1, T_2, etc., for the sum of the squares of members in the first, second, etc., column of differences we see that

$$T_j = \sum_k \left\{ x_k - \binom{j}{1} x_{k+1} + \binom{j}{2} x_{k+2} \ldots \right\}^2, \tag{29}$$

where the summation now takes place over all values of x and there are no complications introduced by end effects. In fact, we have thrown the end effects into the sums T which replace the S's. In actually calculating the T's from the S's it is very little trouble to add the extra terms to the tables giving the latter; and when calculating the S's from the T's only the differences at the end of the table need be worked out.

We have then from (29), on expansion

$$T_j = P_0\left\{\binom{j}{0}^2+\binom{j}{1}^2+\ldots\right\}+2(-1)^k\sum_k\left[P_k\left\{\binom{j}{0}\binom{j}{k}+\ldots+\binom{j}{j-k}\binom{j}{j}\right\}\right]. \tag{30}$$

The coefficients of the various P's are easily seen to be equal to corresponding powers of t in

$$\left\{\binom{j}{0}t^0-\binom{j}{1}t+\binom{j}{2}t^2\ldots\right\}\left\{\binom{j}{0}t^j-\binom{j}{1}t^{j-1}+\ldots\right\},$$

i.e. in $(-1)^j(1-t)^{2j}$, and we find, on substitution in (30),

$$T_j = P_0\binom{2j}{j}-2P_1\binom{2j}{j-1}+2P_2\binom{2j}{j-2}+\ldots+2(-1)^jP_j. \tag{31}$$

For example

$$T_0 = P_0,$$
$$T_1 = 2P_0-2P,$$
$$T_2 = 6P_0-8P_1+2P_2,$$
$$T_3 = 20P_0-30P_1+12P_2-2P_3,$$
$$T_4 = 70P_0-112P_1+56P_2-16P_3+2P_4,$$
$$T_5 = 252P_0-420P_1+240P_2-90P_3+20P_4-2P_5,$$
$$T_6 = 924P_0-1584P_1+990P_2-440P_3+132P_4-24P_5+2P_6,$$
$$T_7 = 3432P_0-6006P_1+4004P_2-2002P_3+728P_4-182P_5+28P_6-2P_7,$$
$$T_8 = 12870P_0-22880P_1+16016P_2-8736P_3+3640P_4-1120P_5+240P_6-32P_7+2P_8,$$
$$T_9 = 48620P_0-87516P_1+63648P_2-37128P_3+17136P_4-6120P_5+1632P_6-306P_7+36P_8-2P_9,$$
$$T_{10} = 184756P_0-335920P_1+251940P_2-155040P_3+77520P_4-31008P_5+9690P_6-2280P_7+380P_8-40P_9+2P_{10}. \tag{32}$$

The coefficients check in virtue of the fact that they sum to zero.

Conversely we have

$$2P_0 = 2T_0,$$
$$2P_1 = -T_1+2T_0,$$
$$2P_2 = T_2-4T_1+2T_0,$$
$$2P_3 = -T_3+6T_2-9T_1+2T_0,$$
$$2P_4 = T_4-8T_3+20T_2-16T_1+2T_0,$$
$$2P_5 = -T_5+10T_4-35T_3+50T_2-25T_1+2T_0,$$
$$2P_6 = T_6-12T_5+54T_4-112T_3+105T_2-36T_1+2T_0,$$
$$2P_7 = -T_7+14T_6-77T_5+210T_4-294T_3+196T_2-49T_1+2T_0,$$
$$2P_8 = T_8-16T_7+104T_6-352T_5+660T_4-672T_3+336T_2-64T_1+2T_0,$$
$$2P_9 = -T_9+18T_8-135T_7+546T_6-1287T_5+1782T_4-1386T_3+540T_2-81T_1+2T_0,$$
$$2P_{10} = T_{10}-20T_9+170T_8-800T_7+2275T_6-4004T_5+4290T_4-2640T_3+825T_2-100T_1+2T_0. \tag{33}$$

The coefficients sum in turn to 2, 1, −1, −2, −1, 1, 2, etc. The coefficient of T_k in $2P_j$ is

$$(-1)^{j-k}\frac{2j}{j+k}\binom{j+k}{2k}.$$

REFERENCES

DAVIS, H. T. (1941). *The Analysis of Economic Time Series.* Bloomington, Indiana.

DODD, E. L. (1939). The length of the cycles which result from the graduation of chance elements. *Ann. Math. Statist.* **10**, 254.

KENDALL, M. G. (1941). The effect of the elimination of trend on oscillations in time series. *J. Roy. Statist. Soc.* **104**, 43.

YULE, G. UDNY (1927). On a method of investigating periodicities in disturbed series, with special reference to Wolfer's sunspot numbers. *Philos. Trans.* A, **226**, 267.

WALKER, SIR GILBERT (1931). On periodicity in series of related terms. *Proc. Roy. Soc.* A, **131**, 518.

WOLD, H. (1938). *A Study in the Analysis of Stationary Time-series.* Uppsala.

From *Journal of the Royal Statistical Society*, Vol. **108**, pp 93–141 (1945)

On the Analysis of Oscillatory Time-Series

By M. G. Kendall

[Read before the Royal Statistical Society, January 17th, 1945, Dr. David Heron in the Chair]

CONTENTS

INTRODUCTION

1. The problems discussed in this paper may best be introduced by the case which first directed my attention to them. In 1922 Sir William Beveridge [4] read to this Society a well-known paper on " Crop Yields and Rainfall in Western Europe," in the course of which he gave a periodogram analysis of one of his own series of wheat-price indices extending over about 300 years. This work was at the time, and so far as I know still remains, the most thorough application of the Schuster technique for detecting periods in time series ever made. On the basis of the resulting periodogram, supported by other considerations which I shall discuss later, Sir William was led to the conclusion that his data contained 19 cycles with lengths ranging from 2·735 to 68 years.

In 1942 I subjected the series to a correlogram analysis, the details of which are given below, and could find evidence of only two oscillatory movements with mean periods of about 5 and 15 years. The difference between the results of the two techniques was very striking, and all the more difficult to understand because for infinite series they are theoretically equivalent. Only three explanations seemed possible:

(*a*) the correlogram was insensitive and had failed to reveal existent oscillations, at least to my eye;

(*b*) the periodogram had been wrongly interpreted, and 17 of Sir William's 19 periods were not significant;

(*c*) both (*a*) and (*b*) were true to some extent, and the number of real periods which could account for the data lay somewhere between 2 and 19.

2. The work described below was prompted by a wish to clear up this difficulty. Like all investigations into the behaviour of time-series, it led me rather further than I expected to have to go; and in my opinion the conclusions to be drawn from it undermine not merely Sir William's investigation, but nearly all similar attempts to detect periodic movements in economic, meteorological and geophysical time-series by the periodogram method. I shall show that most of the 19 periods proposed by Sir William can be accounted for as sampling effects, and may not be significant of anything other than the unreliability of the Schuster technique in economic investigations. I shall show that the accepted tests of significance of the periodogram are not valid. I shall explain the great

importance of defining clearly what is meant by a " period " in time-series by showing that a five-year "cycle" between peaks can arise in simple auto-regressive series almost irrespective of their fundamental period. I shall re-emphasize the importance of Yule's contributions to the theory of time-series as reflecting a fundamental departure from the classical methods of analysis. Finally I shall suggest some further lines of enquiry which may repay exploration in this, perhaps the most difficult, and certainly one of the most important, branches of statistical theory.

SKETCH OF THE DEVELOPMENT OF TECHNIQUES FOR THE ANALYSIS OF OSCILLATORY SERIES

3. To make the later argument more intelligible, I begin with a brief account of the development of modern techniques for the detection of oscillatory movements in observed time-series. The history may conveniently start with the work of Buys-Ballot [5] on periodic changes in temperature, since it was he who first made effective use of the tabular form which bears his name and which forms the basis of the modern periodogram.* His idea is simple but fundamental. Suppose we wish to test whether a series contains an element of (integral) period p. We write down the series in rows of p thus:

$$\begin{array}{lllllll}
 & u_1 & u_2 & u_3 & \cdot\;\cdot\;\cdot & u_p \\
 & u_{p+1} & u_{p+2} & u_{p+3} & \cdot\;\cdot\;\cdot & u_{2p} \\
 & \cdot & \cdot & \cdot & \cdot\;\cdot\;\cdot & \cdot \\
 & u_{(m-1)p+1} & u_{(m-1)p+2} & u_{(m-1)p+3} & \cdot\;\cdot\;\cdot & u_{mp} \\
\hline
\text{Totals} & U_1 & U_2 & U_3 & \cdot\;\cdot\;\cdot & U_p
\end{array} \qquad (1)$$

If the number of terms N in the series is not an exact multiple of p, we ignore the few that are left over after completing as many rows as possible.

If there is present a term of period p in the series, the column totals U will cumulate the periodic effect; but if the remaining element is random, the effect of summing m rows will be to reduce the relative contribution of that element to the column totals, and similarly if there are other elements with different periods, they will get out of step in successive rows and tend to cancel out in the totals. Hence, if there are enough rows, we may expect that the totals U will reveal the periodic effect and will reduce any masking effects due to random components or oscillatory components of different period which would have prevented us from discerning it in the primary series.

4. The methods of Buys-Ballot were developed in the latter part of the nineteenth century, but the next important step was taken by Schuster [16, 17]. We construct the sums

$$A(p) = \frac{2}{mp}\sum_{j=1}^{p}\left\{U_j \cos\frac{2\pi j}{p}\right\} \qquad (2)$$

$$B(p) = \frac{2}{mp}\sum_{j=1}^{p}\left\{U_j \sin\frac{2\pi j}{p}\right\} \qquad (3)$$

and form the *intensity*

$$I(p) = A^2 + B^2 \qquad (4)$$

* The Buys-Ballot table is usually referred to as such in continental works but not, so far as I know, in books or papers written by Englishmen. It is quite possible that Buys-Ballot was not the first to employ the table, but I have followed the continental practice without bothering to inquire into questions of priority.

There is a value of I corresponding to each trial period p, and if we calculate a set of values for a certain range of trial periods, we have what is known as a periodogram analysis. The graph of $I(p)$ as ordinate against p as abscissa is called the periodogram.*

There are a few variants of these formulae which need not be noticed here, and it may be remarked that the method may be extended to give the intensity for rational non-integral trial periods. One useful modification of Schuster's original formula is to divide the intensity by twice the variance of the primary series:

$$E(p) = \frac{I(p)}{2 \text{ var } u} \quad . \quad . \quad . \quad . \quad . \quad . \quad . \quad . \quad (5)$$

so as to standardize it and permit of the comparison of different periodograms. The factor 2 is introduced for reasons connected with Fourier analysis, and the letter E has been used by Davis [7] as the first letter of the word " energy." I shall not follow Davis in his analogy between the energy of a dynamical system and the standardized ordinate of the periodogram because it seems to me to be false and misleading, but I preserve his notation.

5. In effect what Schuster's method does is to find the correlation between the primary series and a harmonic term of period p. If this correlation is higher than for neighbouring trial periods, there occurs a peak in the periodogram; and in particular, if the primary series contains a harmonic term of period p, the periodogram should exhibit a peak of width $2p^2/(mp)$. The Schuster technique consists essentially of examining the periodogram for peaks, dismissing minor ones as side-bands of the main peaks or as sampling effects, and attributing to each remaining peak a constituent harmonic term of corresponding period in the primary series. This is the method employed by Sir William Beveridge.

6. There have been several modifications of Schuster's method proposed, mainly with the object of obviating the rather tedious arithmetic which is involved. All those I have seen are no better than indifferent approximations, and some of them are not even that. An attempt of rather a different kind was made by Whittaker [20], who derives from the column totals U in the Buys-Ballot table a function

$$\eta^2 = \frac{\text{var}\left(\frac{U}{m}\right)}{\text{var } u} \quad . \quad . \quad . \quad . \quad . \quad . \quad . \quad . \quad . \quad (6)$$

and constructs a diagram by graphing η against p, a figure which he also calls a periodogram. The reason for employing the symbol η is that Whittaker was under the mistaken impression that he was calculating the correlation ratio of the Buys-Ballot table. He shows that if the primary series contains a harmonic of period p and the remaining constituents of the series are uncorrelated with this term, there is a peak in his version of the periodogram at the trial value p. But it does not follow that if there is a peak there is any periodicity in the data, as is evident from the consideration that η is independent of the order of the sums U in the Buys-Ballot table. For this reason alone I regard the Whittaker periodogram as unsatisfactory.

* More correctly, in Schuster's terminology, a periodograph, but it is now customary to use the one word " periodogram " both for the set of values and for the diagram which exhibits them graphically.

7. Schuster's periodogram was applied to a number of geophysical and astronomical phenomena, but it was not very long before investigators ran into difficulties. Schuster himself, for instance [17], considered a sunspot series of 150 terms, and computed the periodograms for each of the two halves of 75 terms. The periodogram of the whole series showed a very significant peak at about 11 years, but those for the two halves were quite different, one having a single peak at 11 years, the other peaks at about 9·25 and 13·75 years. Schuster himself says that "looking at them, we might think that they illustrate two entirely different phenomena." The same effect has been observed by other writers, and in particular Beveridge found large differences between the intensities of some of his periods in the two halves of his series.

8. The interesting thing about these results is that it was concluded by nearly everyone, not that there was anything suspicious about the way in which the periodograms were being interpreted, but that the series themselves were behaving in an anomalous way. Turner, for instance, developed a theory in which cycles are supposed to happen in groups and then to disappear to make way for a new "chapter" in the history of the phenomenon. Other writers, among whom Bartels [3] is prominent, have hypostasized the effect by drawing a distinction between "true" periodicity and "quasi" periodicity. There has also been a good deal of discussion, which I find most unconvincing, of the reasons why the phase of the oscillations varies during the course of the series.

9. In considering the work of these writers, we have to remember that the ideas of all investigators of the theory of cyclical movements in the nineteenth century, and of most of those in the twentieth, were conditioned by the mathematics of Fourier analysis. Their object was to exhibit the series as a sum of sine or cosine terms, and if the fit between observation and theory was not exact, the differences were regarded as random errors of observation. Their methods have met with success in the domain of the very small, such as the oscillations which we believe to compose light-waves, and also in the domain of the very large, such as tidal effects generated by planetary motion. But I am sorry to say that in most of the intervening domain Nature does not seem to have studied the mathematical theory of harmonic analysis nearly so thoroughly as she ought. If there is one thing more clear than another in economic series, it is that they do *not* behave as if they were composed of harmonics with a superposed random element, for they shift continually in amplitude, distance between peaks and phase. By taking enough harmonics, of course, one can represent almost anything by a series of sine and cosine terms, but I do not think anyone who has tried to obtain a close fit to an economic series of any length would feel that he was really probing into the essential causes of the generating process.

10. Although I am not competent to express an opinion on matters of meteorology and geophysics, I interpret the difficulties which have been encountered in those subjects as evidence of effects similar to those in economic series. It appears that all these phenomena are subject to disturbances of the "casual" or "stochastic" type which, once they have occurred, are integrated into the system and influence its future motion. Any scheme of representation which seeks to exhibit these disturbances as errors of observation superposed on a functional scheme is therefore likely to prove inadequate.

11. We owe the new approach to the problem to Yule and to Frisch, working independently. I should also mention the work of Slutzky [18], who emphasized that oscillatory movements of the economic type could be generated by taking

a moving average of a random series, and either by perseverance or by a fluke found such a generated series which fits surprisingly well to an index of British business activity for 1855–77. Frisch [11], who attributes the fundamental idea to Knut Wicksell, has considered disturbed oscillations mainly from the theoretical viewpoint. It is in the papers of Yule [22, 23, 24], however, that we find both the fundamental idea and a practical technique which makes its application possible.

12. Yule considers a system which is capable of damped oscillations under its own internal forces, but is subjected to a stream of external shocks which continually regenerate the oscillations. The qualities of the system itself are supposed to be expressible by the fact that the value at any point of time is a function of the values at previous points, and the shocks by a disturbance function which may be a random variable. Such a system I call autoregressive. For the case of a series defined at equi-distant intervals of time we may write

$$u_{t+m} = f(u_{t+m-1}, \ldots, u_t) + \varepsilon_{t+m} \quad . \quad . \quad . \quad . \quad . \quad (7)$$

where ε is the disturbance function. In particular, we have the linear autoregressive scheme

$$u_{t+m} + a_1 u_{t+m-1} + \ldots + a_m u_t = \varepsilon_{t+m} \quad . \quad . \quad . \quad . \quad (8)$$

which has been used by Yule himself [24] on sunspot numbers, by Walker [19] on barometric pressure and by myself [14] on agricultural prices, acreages, crop yields and livestock numbers. Yule and I found that our series were sufficiently closely represented by the simple form

$$u_{t+2} + a u_{t+1} + b u_t = \varepsilon_{t+2} \quad . \quad . \quad . \quad . \quad . \quad . \quad (9)$$

which I have used for the experiments described below.

13. The autoregressive scheme (which includes that of hidden periodicities as a very special case) generates a series with the typical variation in amplitude and distance from peak to peak, and the typical shift in phase. I have no space to give an account of its properties here, but it may be useful to state without proof that if the generated series is stationary in the sense that it fluctuates about a fixed mean value with constant variance, the effective solution of (8) is a moving average of the variable ε with infinite extent. The scheme of moving averages is thus very similar to the autoregressive scheme, and for present purposes the two can be considered together.

14. The coefficient of product-moment correlation between members of the series k intervals apart is called the serial correlation of order k. For the infinite series we have the corresponding autocorrelation.

$$\rho_k = \frac{\operatorname{cov}(u_j, u_{j+k})}{\operatorname{var} u} \quad . \quad . \quad . \quad . \quad . \quad . \quad . \quad . \quad (10)$$

and for the finite series

$$r_k = \frac{\sum_{j=1}^{N-k} (u_j\, u_{j+k})}{\left\{ \sum_{j=1}^{N-k} (u_j^2) \sum_{j=1}^{N-k} (u_{j+k}^2) \right\}^{\frac{1}{2}}} \quad . \quad . \quad . \quad . \quad . \quad . \quad (11)$$

where the values of u are measured about their mean. For every $k < N$ we may compute a value of r_k and the diagram obtained by graphing r_k as ordinate against k as abscissa is called the correlogram.* If the primary series is a sum

* Yule called it a correlation diagram, the word correlogram, so far as I know, being first employed by Wold [21].

of harmonics, the correlogram will also consist of a sum of harmonics of the same periods as the original, but with different amplitudes. On the other hand, if the series is autoregressive, the correlogram shows damped oscillations, and in particular if the series is generated by a moving average of finite extent, the serial correlations vanish after a certain order, within sampling limits. The correlogram thus provides a criterion for determining the nature of the oscillations. Or at least, it would do so for an infinite series; the correlogram of a finite series is more difficult to interpret.

15. For an infinite series everywhere defined the correlogram and the periodogram are mathematically related, one being a kind of Fourier transform of the other (cf. Davis [7], p. 112). We should thus hardly expect them to give apparently inconsistent results for practical series of any great length. That they can do so must mean, I think, that their interpretation requires much more care than has been exercised in the past. My present thesis, briefly, is that a failure to appreciate this and similar facts about time series is mainly responsible for the host of claims advanced by so many writers to have detected "periods" in economic, meteorological and geophysical material.

16. As an illustration of the difficulty of the problem consider the definition of "period" in a time series. Under the older scheme of hidden periodicities we could, of course, define the period to be that of the constituent harmonic if there was only one; but even here there were problems if two or more constituent harmonics existed with incommensurable periods. The series would then be "almost" periodic in the mathematical sense and would possess several periods.

When we come to autoregressive series the difficulties increase. There is now no constituent harmonic term. The basic equation of the linear scheme exemplified by

$$u_{t+2} + au_{t+1} + bu_t = 0 \quad . \quad . \quad . \quad . \quad . \quad . \quad (12)$$

has a solution which consists of a damped harmonic with a certain period; but this *fundamental* period, so called, does not necessarily correspond in length to any of the recurring properties of the series which we normally regard as measuring its "period." * There are two measures which suggest themselves intuitively: the distance between peaks in the series and the distance between upcrosses—*i.e.*, between points where the series, measured about its mean, changes sign from negative to positive.

17. Whichever measure we use, it is no longer possible to speak of *the* periods of a series. The distance between peaks or between upcrosses (and similarly, of course, between troughs and downcrosses) are variable and form frequency distributions. We must replace the concept of a unique period of recurrence by that of a distribution of periods; and we may, of course, use any convenient central measure of such a distribution to summarize its main features. I shall frequently refer to the mean-distance between peaks or upcrosses, the arithmetic mean of the distributions concerned.

It follows that for autoregressive series we cannot speak of "cycles" in the true sense of the term. The word is better avoided altogether unless it can be

* I wish to modify a suggestion in my paper [15] to the effect that for most practical autoregressive series of the simple type of equation (9) the fundamental period and the mean distance between upcrosses are approximately the same. This was true of the experimental series known to me at the time but further experiment has shown it to be false more frequently than I originally thought.

shown that the series do contain elements which are strictly periodic functions of the time. In this sense trade cycles do not, I believe, exist; but there are certainly trade oscillations. The distinction is much more than a mere matter of words. Anyone who believes in the trade cycle must also believe that its recurrences can be closely predicted, whereas one who recognizes trade oscillations may nevertheless consistently hold that the prediction of crises is impossible save within very wide limits.

18. What most economists have in mind, I think, in speaking of cyclical movements, is what I should call the mean-distance between peaks or between troughs. From the point of view of the national life it is the crises and depressions at their worst which exert the greatest influence and arouse the greatest interest. To show how careful one has to be in attempting to base economic conclusions on the existence or mean-period of recurrence of peaks, I prove in the Appendix a result to the general effect that in annual series the mean-distance between peaks is between 3 and 6 years almost irrespective of the nature of the process which generates it. I do not wish to put too much emphasis on this result, which is to be taken as illustrative of the difficulties of economic inference rather than a law of behaviour of economic time series. All the same, it is worth pointing out that Davis ([7], p. 548), summarizing the distribution of 166 " cycles " in 17 countries according to length (I think, between peaks) gives a mean period of 5·2 years.

19. I return to the periodogram and correlogram of the Beveridge trend-free index of wheat prices mentioned at the beginning of the paper. Table 1 gives the first 60 serial correlations of the whole series of 370 values, the correlogram being shown in Fig. 1.

TABLE 1

Serial correlations of the Beveridge trend-free wheat-price series

Order of correlation, k	Value of correlation, r_k	Order of correlation, k	Value of correlation, r_k	Order of correlation, k	Value of correlation, r_k
1	0·562	21	—0·021	41	0·008
2	0·103	22	—0·062	42	0·034
3	—0·075	23	—0·088	43	0·065
4	—0·092	24	—0·084	44	0·099
5	—0·082	25	—0·076	45	0·009
6	—0·136	26	—0·091	46	—0·036
7	—0·211	27	—0·052	47	—0·013
8	—0·261	28	—0·032	48	0·042
9	—0·192	29	—0·012	49	0·062
10	—0·070	30	0·059	50	0·065
11	—0·003	31	0·060	51	0·050
12	—0·015	32	—0·008	52	0·009
13	—0·012	33	—0·039	53	—0·027
14	0·047	34	0·007	54	—0·053
15	0·101	35	0·056	55	—0·073
16	0·158	36	0·010	56	—0·106
17	0·109	37	—0·004	57	—0·084
18	0·002	38	—0·015	58	—0·019
19	—0·075	39	—0·047	59	0·003
20	—0·062	40	—0·047	60	0·010

Owing to the rapid damping of the correlations, the latter has been drawn on an enlarged vertical scale, and it is remarkable how the oscillations persist even when the correlations are damped almost out of existence. To my eye this diagram clearly indicates an autoregressive scheme with a main wave of about

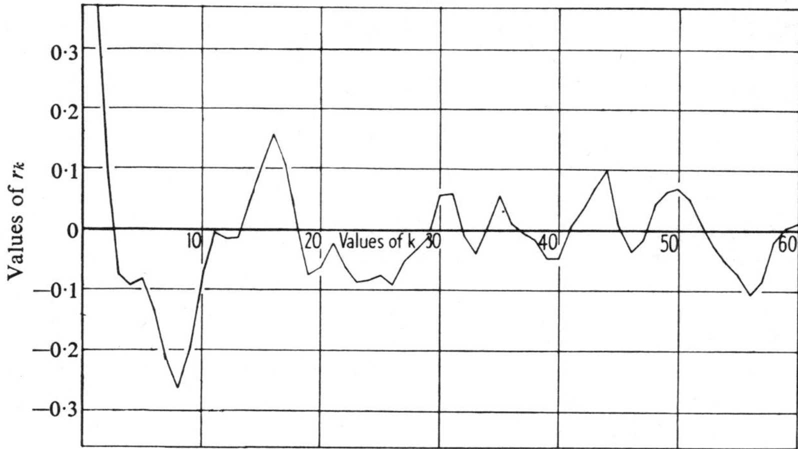

FIG. 1

Correlogram of the Beveridge trend-free wheat-price index

15 years and a ripple of about 5 years. Possibly if we had 300 serial correlations instead of 60 a careful analysis would reveal further effects; but it seems clear

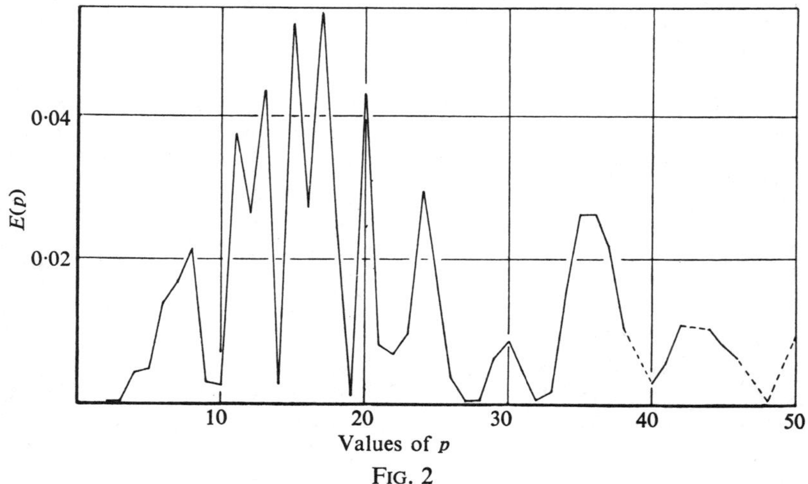

FIG. 2

Periodogram of the Beveridge series for integral trial periods up to 50

that they would only be trifling compared with the two main periods, and I should see no reason whatever on this evidence to suspect more than two.*

* Yule [23] gave the correlogram of the alternative Beveridge series which includes trend and although his form differs from mine it exhibited the same sort of oscillatory effect about a trend line. Wold [21] computed the first 15 serial correlations for the last 100 terms of the trend-free index and concluded that the series suggests a scheme of moving averages. His results are consistent with mine but he did not take them far enough to detect the 15-year oscillation or the persistence of both oscillations in the correlogram.

The paper by Sir William Beveridge gives the values of the periodogram ordinates and the periodogram itself. In Fig. 2 I have drawn, for comparison with other examples given later in the paper, the periodogram with standardized ordinates *for integral values of the trial period.* The resulting figure has 10 peaks, all of which must arise from distinct harmonics on the Schuster theory, if they are significant at all, and Beveridge found nine more by ascertaining the ordinates for a large number of non-integral trial periods and extending the trial period up to 84. However, 10 will be enough for the moment. A somewhat similar diagram, with the unstandardized $\sqrt{I}$ as ordinate instead of the standardized E, has been given by Davis ([7], p. 313). By calculating ordinates in the neighbourhood of peaks, Beveridge found even larger values than some of those in Fig. 2, so that the present form rather under-represents their significance. On a minor point it may be remarked that he also chose as his ordinate $300I/(mp)$ in an endeavour to standardize his figures for the varying numbers of terms in the Buys-Ballot table. This is unnecessary, but as the factor $300/(mp)$ is not very different from unity, I have not bothered to remove the " correction."

20. In concluding the review in this section I should refer to the work of Dinsmore Alter, who has studied the correlogram since 1927 and has recently suggested a method of detecting frequencies which he also calls a periodogram [1]. Alter calculated a series of values typified by the kth value $\Sigma \mid u_{j+k} - u_j \mid$ that is to say, the sum of absolute difference of terms k intervals apart. He concluded that a period in rainfall data which he examined was more definitely established by this method than by previous investigations, a conclusion with which I should agree. But evidently Alter has, in effect, rediscovered a kind of correlogram. Apart from factors which are exactly or approximately constant, the ordinate of the correlogram is proportional to $\Sigma(u_j u_{j+k})$, and hence, except for constants, varies with $\Sigma(u_{j+k} - u_j)^2$. Alter uses a similar function, except that he takes absolute differences instead of squares of differences. In practice his form may be somewhat easier to compute, at least for long series, but for theoretical reasons the Yule correlogram is decidedly preferable; measures dependent on sums of squares being much more easily discussed mathematically than sums of absolute differences.

21. The main problem considered in this paper may then be expressed in this way: apart from unessential variants, there are three methods of determining oscillatory elements in time series in current use, (*a*) direct counting of intervals between peaks or upcrosses in the primary series, (*b*) analysis of the data by the periodogram method and (*c*) analysis by the correlogram method. Although the first requires more investigation, it appears that mean-distances between peaks are insensitive to variations in the nature of the generating process and will reveal very little. The other two give, in some cases at least, quite inconsistent results if the current methods of interpretation are adopted. It is necessary to resolve this apparent inconsistency.

TESTS OF SIGNIFICANCE OF THE PERIODOGRAM AND THE CORRELOGRAM

22. Schuster himself realized the importance of not attaching significance to peaks in the periodogram which could have arisen by chance from a random series. His principal result may be derived very simply from a consideration of the Buys-Ballot table. If the series is random and normal with variance v, the

sums U are independent and normal with variance mv. The functions A and B then have variances

$$\frac{4mv}{(mp)^2}\sum_{j=1}^{p}\left(\cos^2\frac{2\pi j}{p}\right) \quad \text{and} \quad \frac{4mv}{(mp)^2}\sum_{j=1}^{p}\left(\sin^2\frac{2\pi j}{p}\right)$$

or $\frac{2v}{mp}$ in each case. Furthermore, since

$$\sum_{j=1}^{p}\left(\cos\frac{2\pi_j}{p}\sin\frac{2\pi_j}{p}\right) = 0$$

A and B are independent. Their joint distribution is therefore

$$dF \propto \exp\left\{-\frac{mp}{4v}(A^2 + B^2)\right\}dAdB, \quad . \quad . \quad . \quad . \quad (13)$$

which, by a polar transformation, gives for the distribution of the intensity

$$dF = \frac{mp}{4v}\exp\left(-\frac{mpI}{4v}\right)dI. \quad . \quad . \quad . \quad . \quad . \quad . \quad (14)$$

Schuster writes :

$$\kappa = \frac{mpI}{4v} \quad . \quad . \quad . \quad . \quad . \quad . \quad . \quad . \quad (15)$$

and derives the probability that a value of κ fails to exceed a given value κ_0 as

$$P(\kappa \leq \kappa_0) = 1 - e^{-\kappa_0} \quad . \quad . \quad . \quad . \quad . \quad . \quad (16)$$

23. This, however, is not what we require. It gives the probability that a value of κ *chosen at random* will fail to exceed the prescribed value. In practice what we do is to pick out the biggest values in the periodogram for test. If there are s *independent* trial periods, the probability that none of them exceeds the specified value κ_0 is

$$(1 - e^{-\kappa_0})^s$$

and the probability that at least one exceeds that value is

$$P_w = 1 - (1 - e^{-\kappa_0})^s \quad . \quad . \quad . \quad . \quad . \quad . \quad . \quad (17)$$

This result is due to Walker. Fisher [10] added a refinement by ascertaining the corresponding probability when the variance v in (15) has to be estimated from the data and is not given *a priori*. Tables of the Walker and the Fisher probabilities are given by Davis. The problem considered by Fisher is very similar to that of testing the significance of the largest of a set of variances in the analysis of variance, a problem to which contributions have recently been made by Cochran [6] and Finney [9].

24. As applied to tests of significance of peaks in the periodogram, these results are still subject to some very serious limitations:

(*a*) It is assumed that the s ordinates of the periodogram are independent. This is true only of Fourier analysis, when the ordinates correspond to trial periods which are reciprocals of integral multiples of some fundamental value. It is certainly not true if we include all the ordinates of the periodogram at integral trial values, and may be very far from the truth.

(*b*) The distributions on which the tests are based assume that the primary series was random. " Significant " peaks may then only mean that the series is not random. Now, nobody in his right mind would undertake the labour of a periodogram analysis, except for experimental purposes, without satisfying himself by direct and simple tests on the series itself that it was not random. Consequently, all the tests may tell us is something that we knew before we applied them.

25. It therefore seems to me that the existing tests of significance of peaks in the periodogram are not valid. I am quite sure that they are invalid from the strictly logical view-point; but even if we take a more indulgent attitude, and rely on intuitive judgments rather than insist on logical exactitude, the tests may still lead us badly astray. Below I describe the results of my own experiments on the subject, but before doing so may refer to an interesting example given by Davis ([7], p. 281).

Davis took 204 terms of a random series and calculated certain ordinates of the periodogram up to trial periods of 75 units. There are a number of fluctuations, but only four peaks which suggest marked significance to the eye, at 21, 33, 58 and 70. On the Schuster theory, however, something of this kind is to be expected. The Fisher probability for the greatest intensity is greater than 0·5, and experiment conforms to the theory of the random series.

But Davis then takes a moving average of extent 12 (and presumably equal weights) of his random series and recalculates the periodogram. There are now peaks at 21, 32 and 58, *and they are all highly significant.* The smoothing process has generated oscillations in the data (the Slutzky–Yule effect) which produce significant peaks in the periodogram. The expected mean-distance between peaks in the primary series may, in the manner of paragraph 2 of the Appendix, be shown to be 4 units * and that between upcrosses is about 15 units, neither of which is anywhere near the values suggested by the periodogram.

26. The only other periodogram I know of a series whose properties are given beforehand was calculated by Yule [24] in his paper on sun-spots. Yule constructed a series of the type

$$u_{t+2} + (2 - \mu)u_{t+1} + u_t = \varepsilon_{t+2} \quad . \quad . \quad . \quad . \quad . \quad (18)$$

using dice to provide the random disturbance function ε. He analysed the series for intervals of 0·1 or less from trial periods 9 to 11, and found a marked peak about 10, which was according to expectation. It should be noted, however, that the coefficient of u_t in equation (18) is unity, so that the series is undamped. In such circumstances one would expect better agreement between the periodogram of a harmonic term and that of the autoregressive series (cf. para. 41 below).

27. The significance of the correlogram is equally difficult to discuss in theoretical terms. Anderson [2] has recently given expressions for the distribution of a single serial correlation, but as for the periodogram, our real problem is to test the significance of a set of values which are, in general, correlated. It is quite possible for a part of the correlogram to be below the significance level and yet to exhibit oscillations which are themselves significant of autoregressive effects. At the present time our judgments of the reality of oscillations in the correlogram must remain on the intuitive plane.

EXPERIMENTAL RESULTS WITH AUTOREGRESSIVE SERIES

28. A series of 480 terms was constructed according to the formula

$$u_{t+2} = 1{\cdot}1u_{t+1} - 0{\cdot}5u_t + \varepsilon_{t+2} \quad . \quad . \quad . \quad . \quad . \quad (19)$$

* This is true of any series generated from a normal random series by a moving average with equal weights, for the second difference of the serial correlations vanishes. It is approximately true even if the random series is not normal for the generated series approaches normality in virtue of the central limit effect.

TABLE 2.—*Artificial autoregressive series (Series* 1) *constructed according to equation* (19) *of the test*

No. of term, t	Value, u_t	No. of term, t	Value, u_t	No. of term, t	Value, u_t	No. of term, t	Value, u_t	No. of term, t	Value, u_t	No. of term, t	Value, u_t
1	0	81	−3	161	−68	241	−18	321	−17	401	−68
2	15	82	−65	162	−24	242	−70	322	12	402	−73
3	26	83	−102	163	−35	243	−82	323	55	403	−78
4	−28	84	−31	164	−34	244	−68	324	33	404	−55
5	−11	85	42	165	28	245	−9	325	11	405	−24
6	24	86	65	166	39	246	37	326	−31	406	−14
7	41	87	8	167	66	247	9	327	−82	407	−12
8	76	88	−3	168	72	248	31	328	−100	408	−13
9	89	89	36	169	49	249	30	329	−98	409	−36
10	34	90	16	170	50	250	−9	330	−85	410	−80
11	40	91	−38	171	77	251	−40	331	−42	411	−78
12	−15	92	−42	172	70	252	−89	332	−28	412	−66
13	−1	93	−36	173	66	253	−53	333	−44	413	6
14	52	94	−15	174	60	254	5	334	2	414	5
15	30	95	40	175	63	255	75	335	67	415	−38
16	−40	96	22	176	85	256	61	336	96	416	−24
17	−42	97	−40	177	38	257	15	337	47	417	35
18	−32	98	−93	178	−14	258	−36	338	49	418	22
19	−59	99	−121	179	−9	259	−60	339	50	419	−20
20	−45	100	−62	180	34	260	1	340	24	420	−54
21	−15	101	−32	181	18	261	−9	341	29	421	−94
22	6	102	39	182	−14	262	17	342	−11	422	−89
23	7	103	90	183	−38	263	64	343	11	423	−66
24	−35	104	73	184	10	264	101	344	53	424	−73
25	−39	105	19	185	61	265	70	345	59	425	−43
26	−2	106	27	186	28	266	8	346	55	426	−7
27	3	107	−22	187	19	267	−20	347	−3	427	53
28	−38	108	−15	188	−2	268	22	348	−13	428	100
29	−3	109	−24	189	33	269	48	349	−36	429	76
30	26	110	10	190	84	270	40	350	11	430	65
31	−1	111	46	191	111	271	32	351	80	431	46
32	−28	112	27	192	101	272	23	352	96	432	30
33	−36	113	−26	193	50	273	8	353	61	433	−16
34	−66	114	−70	194	−18	274	16	354	38	434	14
35	−8	115	−50	195	−47	275	−17	355	33	435	41
36	−4	116	0	196	−55	276	−20	356	30	436	70
37	−38	117	43	197	−7	277	−59	357	−23	437	27
38	−46	118	47	198	−23	278	−17	358	−78	438	6
39	−68	119	23	199	−63	279	26	359	−99	439	31
40	−15	120	−12	200	−83	280	13	360	−28	440	−3
41	−16	121	−63	201	−87	281	−22	361	−5	441	24
42	−57	122	−25	202	−12	282	−1	362	40	442	43
43	−55	123	13	203	4	283	18	363	61	443	−13
44	−50	124	−12	204	23	284	7	364	32	444	−4
45	−37	125	−69	205	44	285	30	365	21	445	−13
46	−23	126	−56	206	13	286	70	366	22	446	−34
47	5	127	−21	207	−51	287	64	367	58	447	−19
48	−10	128	−22	208	−106	288	80	368	36	448	30
49	−13	129	29	209	−86	289	71	369	32	449	84
50	−55	130	83	210	−8	290	34	370	35	450	122
51	−94	131	31	211	37	291	−13	371	48	451	90
52	−123	132	41	212	39	292	−75	372	3	452	71
53	−116	133	23	213	41	293	−73	373	−4	453	64
54	−105	134	19	214	9	294	−71	374	5	454	29
55	−54	135	−34	215	23	295	−38	375	−40	455	−17
56	−19	136	−57	216	24	296	−47	376	−90	456	−66
57	−36	137	−59	217	−15	297	−59	377	−85	457	−96
58	−46	138	6	218	−36	298	−57	378	−80	458	−117
59	−45	139	66	219	−51	299	−42	379	−51	459	−127
60	21	140	82	220	−78	300	1	380	−29	460	−36
61	62	141	11	221	−29	301	44	381	−44	461	22
62	57	142	−1	222	1	302	8	382	−77	462	−2
63	33	143	−18	223	51	303	8	383	−19	463	11
64	51	144	−44	224	44	304	−9	384	63	464	32
65	−5	145	−9	225	−24	305	−33	385	120	465	55
66	−64	146	6	226	−92	306	−10	386	123	466	61
67	−60	147	52	227	−87	307	52	387	104	467	55
68	−31	148	59	228	−45	308	70	388	68	468	−19
69	24	149	64	229	−5	309	56	389	72	469	−28
70	72	150	2	230	28	310	−14	390	48	470	−6
71	76	151	9	231	31	311	−64	391	28	471	2
72	−1	152	−9	232	59	312	−32	392	50	472	−33
73	5	153	−6	233	74	313	23	393	69	473	−56
74	−12	154	−6	234	30	314	41	394	98	474	−70
75	−24	155	2	235	−7	315	81	395	106	475	−50
76	17	156	−20	236	−26	316	113	396	116	476	−39
77	−3	157	−2	237	−14	317	87	397	78	477	−26
78	33	158	7	238	4	318	39	398	52	478	−23
79	85	159	12	239	25	319	−32	399	1	479	−39
80	58	160	−39	240	35	320	−65	400	−70	480	−39

The disturbance function was derived from the tables of random numbers by Babington Smith and myself, and was a rectangular random variable ranging by units from − 49 to 49. The generated series is given in Table 2 and the first 100 terms graphed in Fig. 3. I shall refer to this series as Series 1. It has a

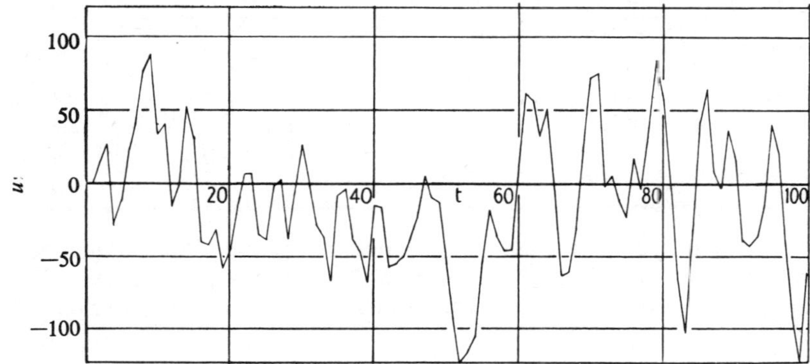

FIG. 3

Graph of first 100 terms of Series I (Table 2)

mean of 0·820,833 and a variance of 2535·109,566. Evidently systematic movements are present although they are obscured to some extent by the random variable. The series is, in fact, highly damped, the damping factor being $\sqrt{0\cdot 5} = 0\cdot 7071$, so that we should expect the disturbance function to exercise considerable influence on the course of the series.

29. The expected mean-distance between peaks in the autoregressive series (9) is given by $2\pi/\cos^{-1}\tau_1$ where

$$\tau_1 = \frac{b^2 - (1+a)^2}{2(1+a+b)} . \qquad (20)$$

(See Appendix, Paragraph 3.) For Series 1, $a = -1\cdot 1$, $b = 0\cdot 5$ and hence $\tau_1 = 0\cdot 3$, $\cos^{-1}\tau_1 = 72\cdot 54°$ and the mean-distance between peaks is $360/72\cdot 54 = 4\cdot 96$. The observed mean-distance in Series 1 was 5·05, an excellent agreement.

The expected mean-distance between upcrosses for the autoregressive series (9) is $2\pi/\cos^{-1}\rho_1$ where ρ_1, the first autocorrelation, is given by

$$\rho_1 = \frac{-a}{1+b} . \qquad (21)$$

(See [15].) In our present case $\rho_1 = 0\cdot 7333$ and the mean-distance between upcrosses reduces to 8·40, the observed value being 8·30. As a matter of interest I give the observed distributions of intervals between peaks and between upcrosses in Table 3 and 4 respectively.

TABLE 3

Distribution of intervals from peak to peak in Series 1

Interval (units)	Frequency	Interval (units)	Frequency	Interval (units)	Frequency
2	10	6	14	10	1
3	17	7	13	11	2
4	14	8	5		
5	13	9	4	Total	93

TABLE 4

Distribution of intervals between upcrosses in Series 1

Interval (units)	Frequency	Interval (units)	Frequency	Interval (units)	Frequency
2	3	9	5	—	—
3	3	10	2	17	2
4	5	11	2	—	—
5	2	12	3	29	1
6	6	13	2		
7	9	14	1	Total	57
8	10	15	1		

30. Series 1 was subjected to a periodogram analysis for all integral trial periods from 2 to 50 inclusive. The standardized ordinates are given in Table 5 and the periodogram in Fig. 4. The results are really rather striking. There

TABLE 5

Standardized periodogram ordinates for Series 1 *and integral trial periods from* 2 *to* 50 *inclusive*

Trial period, p	$E(p)$	Trial period, p	$E(p)$	Trial period, p	$E(p)$
2	0·0001	19	0·0000	35	0·0220
3	0·0002	20	0·0510	36	0·0174
4	0·0000	21	0·0051	37	0·0083
5	0·0040	22	0·0212	38	0·0041
6	0·0007	23	0·0011	39	0·0176
7	0·0145	24	0·0088	40	0·0255
8	0·0136	25	0·0283	41	0·0372
9	0·0062	26	0·0295	42	0·0423
10	0·0022	27	0·0222	43	0·0317
11	0·0052	28	0·0186	44	0·0095
12	0·0163	29	0·0022	45	0·0192
13	0·0097	30	0·0039	46	0·0188
14	0·0199	31	0·0005	47	0·0093
15	0·0061	32	0·0061	48	0·0047
16	0·0188	33	0·0033	49	0·0005
17	0·0107	34	0·0056	50	0·0077
18	0·0120				

are about a dozen peaks, two of which, at 20 and 42, stand out as offering substantial evidence of significant periods. In fact there are periods almost everywhere except in the right place, at 8 or 9. The general appearance of Fig. 4 should be compared with the periodogram of the Beveridge series in Fig. 2.

The test of significance based on the Fisher probability (paragraph 24 above) indicates that the values at 20, 27, 41, 42 and 43 are significant to a 1 per cent. level. At least, I think that is how the test would be interpreted.* Taking $s = 49$, we find for the significance value of κ to the 1 per cent. level, 7·9. In our present case mp is about 480, so that the corresponding significance value of $I/2v$ ($= 2\kappa/mp$) is about 0·032. From this viewpoint, which obviously is

* I ought to make it clear that Professor Fisher has never lent his authority, so far as I know, to the use of his test in periodogram analysis. The problem he was considering was one of Fourier analysis in which the periods *are* independent.

rather moderate, we should attribute significance to three peaks, giving periods at 20, 26 and 42. All three are illusory in the sense that they arise as sampling effects in a series generated by a process which gives no such periods either in the strict sense or as mean-values in distributions.

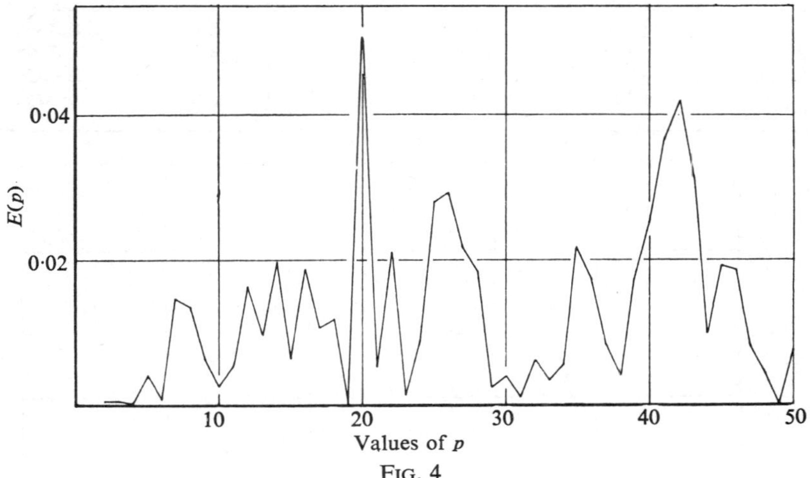

FIG. 4

Periodogram of Series I

31. But this is not the whole story. According to the Schuster theory, there will be, corresponding to a harmonic in the original data, a peak in the periodogram of width $2p^2/(mp)$ or roughly $2p^2/N$. Calculating ordinates at integral trial periods is therefore not enough, and narrower intervals are required at least for smaller values of p. It was for this reason that Sir William Beveridge calculated a large number of ordinates for non-integral values, and remarks, " How close the mesh of analysis must be made . . . is well illustrated by the invisibility of the 5·1-year period if we test at 5·0 years and of the 3·415 period if we test at 3·455." This closer mesh, as already mentioned, gave Sir William seven periods which were not shown up by the analysis for integral trial periods and strengthened the evidence for some of the latter by giving higher intensities for neighbouring non-integral values.

32. The observed distance between upcrosses in Series 1 is 8·3 units, and the fundamental period of the generating equation, given generally by

$$\frac{2\pi}{\cos^{-1}\left(\frac{-a}{2\sqrt{b}}\right)} \quad . \quad . \quad . \quad . \quad . \quad . \quad . \quad . \quad . \quad (22)$$

(see [24], [15]), is 9·25. We might therefore expect some significant peaks to emerge from a closer analysis of the data in the region of 8 or 9 units. I calculated the ordinates for series 1 in the range 8–9, the results being given in Table 6 and the periodogram in Fig. 5.

We now find peaks at 8·33, 8·5 and 8·75, the latter probably being negligible. How to test the significance of the other two I do not know; but even if we take the most favourable view and regard them as significant of a single period ($2p^2/N$ having the value 0·3), the significance is far less than that of the three spurious periods, and indeed less than that of the peaks at 22 and 35 as well.

TABLE 6

Standardized periodogram ordinates for Series 1 and certain non-integral trial periods between 8 and 9 units

Trial period, p	Ordinate, $E(p)$	Trial period, p	Ordinate, $E(p)$	Trial period, p	Ordinate, $E(p)$
8·000	0·0136	8·333	0·0177	8·667	0·0006
8·167	0·0115	8·400	0·0138	8·750	0·0105
8·200	0·0045	8·500	0·0199	8·800	0·0028
8·250	0·0129	8·600	0·0052	9·000	0·0062

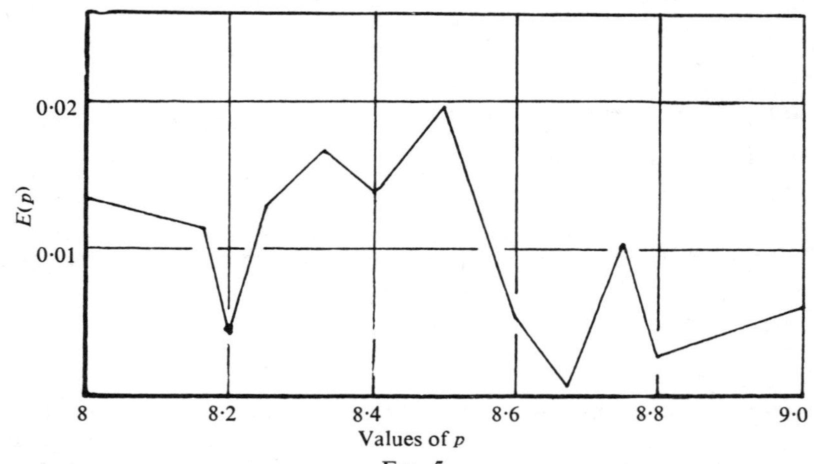

FIG. 5

Periodogram of Series I for non-integral trial values between 8 and 9

33. As a matter of interest I give the ordinates of the Whittaker periodogram in Table 7 and the periodogram itself in Fig. 6 (modified to the extent that I

TABLE 7

Ordinates of the Whittaker diagram for Series 1 and integral trial periods from 2 to 50 inclusive

Trial period, p	η_p^2	Trial period, p	η_p^2	Trial period, p	η_p^2
2	0·0000	19	0·0155	35	0·0632
3	0·0002	20	0·0629	36	0·0582
4	0·0006	21	0·0312	37	0·0843
5	0·0042	22	0·0463	38	0·0492
6	0·0009	23	0·0548	39	0·0563
7	0·0154	24	0·0421	40	0·1156
8	0·0137	25	0·0692	41	0·1207
9	0·0075	26	0·0518	42	0·1076
10	0·0066	27	0·0397	43	0·1146
11	0·0075	28	0·0692	44	0·1249
12	0·0176	29	0·0391	45	0·0639
13	0·0167	30	0·0150	46	0·1165
14	0·0367	31	0·0451	47	0·0441
15	0·0070	32	0·0612	48	0·0810
16	0·0346	33	0·0604	49	0·1103
17	0·0361	34	0·0657	50	0·1044
18	0·0207				

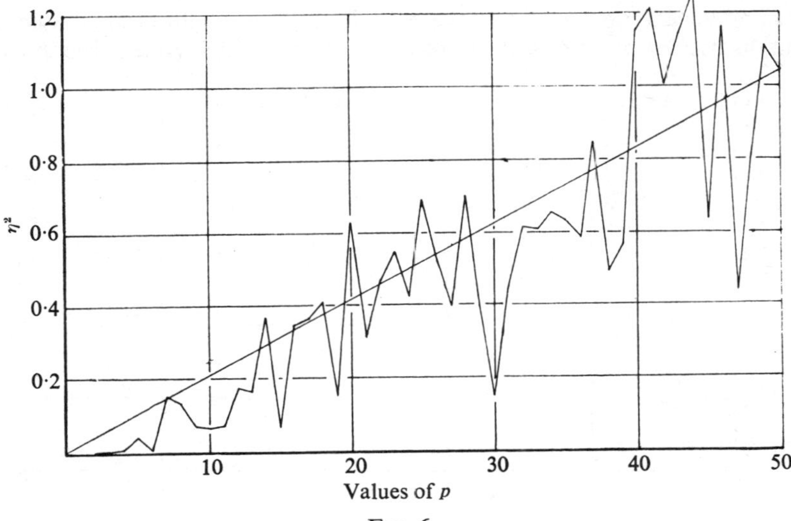

FIG. 6
Modified form of the Whittaker periodogram of Series 1

have used η^2 instead of η, there being no point in adding a further step to the arithmetic by calculating square roots). It will be seen that the ordinate increases systematically and the real test lies in the divergence of particular values from the line $\eta^2 = kp$. There is much the same sort of fluctuation here as in the Schuster periodogram, and it is even more difficult to interpret, for in the latter a peak at trial period p can only correspond to a harmonic with period p (if it indicates a harmonic at all), whereas in the Whittaker form a peak can arise from harmonics with periods which are any integral multiple of p.

34. Now compare the ambiguous and confusing picture presented by the periodogram with the correlogram of Series 1, which is shown in Fig. 7, the values of the serial correlations being given in Table 8. The damped oscillatory

TABLE 8
Serial correlations of Series 1

k	r_k	k	r_k	k	r_k
1	0·762	18	0·032	35	0·008
2	0·377	19	—0·007	36	—0·037
3	0·079	20	—0·025	37	—0·047
4	—0·067	21	—0·005	38	—0·023
5	—0·078	22	0·019	39	—0·017
6	—0·039	23	0·011	40	—0·030
7	—0·007	24	—0·011	41	—0·038
8	0·022	25	—0·041	42	—0·012
9	0·018	26	—0·079	43	0·033
10	—0·036	27	—0·108	44	0·055
11	—0·103	28	—0·126	45	0·048
12	—0·145	29	—0·116	46	0·040
13	—0·128	30	—0·075	47	0·041
14	—0·052	31	—0·018	48	0·035
15	0·029	32	0·039	49	0·022
16	0·069	33	0·069	50	0·007
17	0·065	34	0·048		

effect is now clearly evident, and the only doubt that would occur is that after a point the oscillations do not continue to damp out. This is due to the shortness

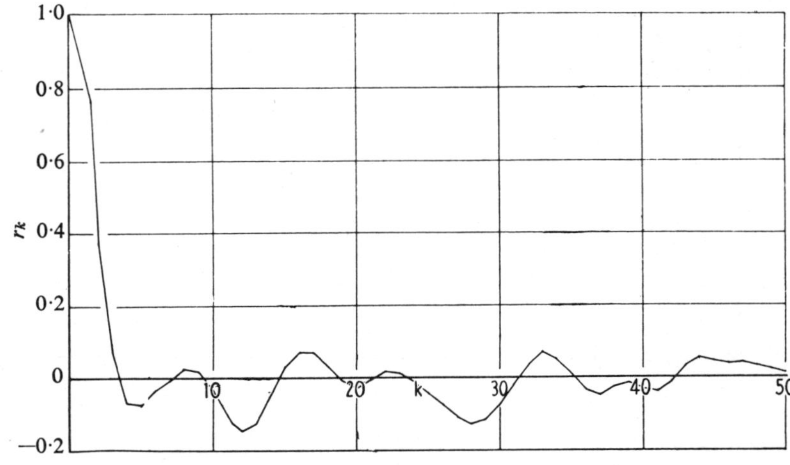

FIG. 7
Correlogram of Series 1

of the series, an effect which I have discussed elsewhere [15] and propose to discuss again on another occasion. The average interval between troughs in the correlogram is 7·2 (or 8·0 if we ignore the doubtful ripple at 41), moderately close to the mean-distance between upcrosses (but considerably longer, one may remark, than the mean-distance between peaks).

It seems undeniable that so far as this particular series is concerned the correlogram gives much better results than the periodogram. Without prior knowledge of the way in which the series was generated, we should be led by the correlogram to suspect a simple autoregressive scheme. Calculation from the observed correlations would lead to the scheme

$$u_{t+2} = 1{\cdot}132u_{t+1} - 0{\cdot}486u_t + \varepsilon_{t+2}$$

as compared with the true generating equation,

$$u_{t+2} = 1{\cdot}100u_{t+1} - 0{\cdot}500u_t + \varepsilon_{t+2}$$

What conclusions would be drawn from the periodogram I can only surmise. On the lines of Sir William Beveridge we should probably claim at least five harmonic terms; and I have no doubt that if further analysis were carried out for non-integral periods the number of "significant" harmonics would be nearer ten.

35. A second experimental series of 240 terms was constructed according to the formula

$$u_{t+2} = 1{\cdot}2u_{t+1} - 0{\cdot}4u_t + \varepsilon_{t+2} \quad . \quad . \quad . \quad . \quad . \quad (23)$$

the disturbance function, as before, being a rectangular random variable ranging from − 49 to + 49. This series will be referred to as Series 2 and is given in Table 9. The first 100 terms are graphed in Fig. 8. The values of the first 30 serial correlations and the correlogram are given in Table 10 and Fig. 9. The

TABLE 9

Artificial autoregressive series (Series 2) constructed according to equation (23) of the text

t	u_t	t	u_t	t	u_t	t	u_t	t	u_t	t	u_t
1	32	41	−28	81	− 70	121	−170	161	17	201	− 90
2	17	42	21	82	− 56	122	−149	162	62	202	− 58
3	42	43	82	83	− 40	123	− 96	163	24	203	− 4
4	78	44	91	84	− 60	124	− 90	164	− 36	204	− 23
5	39	45	56	85	− 91	125	− 42	165	− 60	205	− 7
6	− 13	46	66	86	−108	126	31	166	− 83	206	33
7	− 22	47	10	87	−141	127	82	167	−106	207	58
8	− 51	48	9	88	−128	128	75	168	− 82	208	29
9	− 86	49	23	89	−120	129	39	169	− 23	209	3
10	− 63	50	33	90	− 98	130	1	170	28	210	− 18
11	− 31	51	− 9	91	− 73	131	29	171	12	211	− 57
12	− 6	52	−47	92	− 43	132	8	172	− 15	212	− 16
13	11	53	−15	93	− 28	133	36	173	− 38	213	28
14	60	54	3	94	− 12	134	33	174	− 25	214	80
15	22	55	36	95	− 17	135	18	175	− 26	215	110
16	11	56	48	96	− 16	136	46	176	− 2	216	139
17	14	57	84	97	28	137	4	177	55	217	129
18	9	58	81	98	37	138	9	178	76	218	124
19	− 35	59	73	99	69	139	9	179	38	219	89
20	− 65	60	7	100	116	140	41	180	60	220	71
21	− 89	61	−46	101	132	141	4	181	56	221	57
22	−109	62	−52	102	63	142	− 53	182	29	222	3
23	− 56	63	−89	103	3	143	− 33	183	26	223	− 34
24	3	64	−48	104	− 12	144	− 63	184	− 28	224	− 82
25	19	65	−31	105	− 55	145	− 64	185	− 43	225	− 85
26	35	66	−65	106	− 89	146	− 24	186	− 84	226	−105
27	28	67	−68	107	− 62	147	− 17	187	− 72	227	− 52
28	0	68	−26	108	− 29	148	5	188	− 97	228	25
29	− 23	69	25	109	2	149	56	189	− 38	229	64
30	− 67	70	55	110	25	150	17	190	− 28	230	53
31	− 97	71	65	111	7	151	43	191	6	231	62
32	− 50	72	7	112	− 29	152	51	192	64	232	72
33	− 4	73	2	113	− 54	153	40	193	56	233	21
34	− 28	74	27	114	− 34	154	− 19	194	− 3	234	9
35	− 48	75	62	115	− 53	155	− 35	195	1	235	− 13
36	− 14	76	49	116	− 88	156	− 49	196	− 14	236	18
37	− 15	77	50	117	−123	157	− 62	197	7	237	26
38	− 34	78	28	118	−124	158	− 43	198	− 35	238	54
39	− 14	79	− 4	119	−146	159	− 43	199	− 72	239	9
40	− 52	80	−57	120	−171	160	− 20	200	−103	240	38

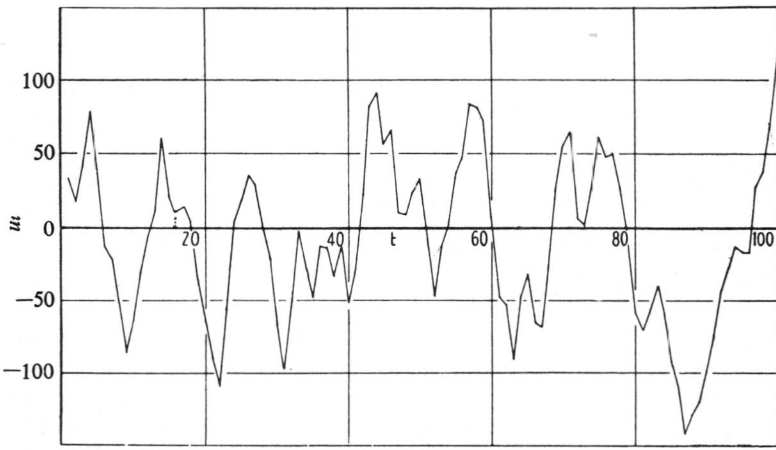

FIG. 8

Graph of the first 100 terms of Series 2

TABLE 10

Serial correlations of Series 2

k	r_k	k	r_k	k	r_k
1	0·850	11	—0·032	21	—0·029
2	0·606	12	—0·015	22	0·049
3	0·350	13	—0·009	23	0·098
4	0·127	14	—0·040	24	0·143
5	—0·044	15	—0·076	25	0·190
6	—0·124	16	—0·128	26	0·186
7	—0·147	17	—0·166	27	0·135
8	—0·138	18	—0·172	28	0·099
9	—0·094	19	—0·151	29	0·097
10	—0·057	20	—0·100	30	0·094

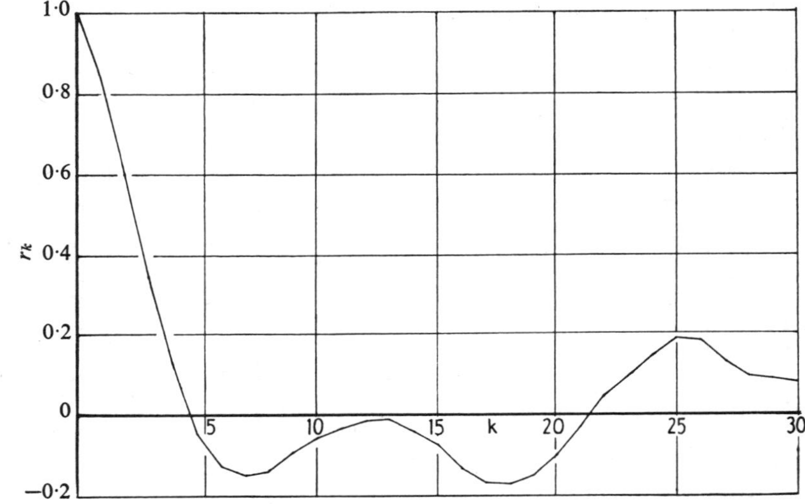

FIG. 9

Correlogram of Series 2

series is even more heavily damped than Series 1, with a damping factor of $\sqrt{0{\cdot}4} = 0{\cdot}6325$. The main constants are as follows:

Mean	—8·179,167
Variance	3414·422,066
Expected mean-distance (upcrosses)	11·61
Observed " "	12·39
Expected mean-distance (peaks)	4·96
Observed " "	5·57
Fundamental period	19·53

The distance between the two troughs in the correlogram is 11 units, in good agreement with the mean-distance between upcrosses.

39. The Schuster periodogram was computed for all integral trial periods from 2 to 40 inclusive, the ordinates being given in Table 11 and the periodogram in Fig. 10. There are fewer serrations in the periodogram, but peaks appear at 13, 17, 20, 28 and 37, the last but one being particularly striking. Possibly the peak at 13 may be regarded as reflecting the recurring effect as measured by the

TABLE 11

Standardized periodogram ordinates for Series 2 and integral trial periods from 2 to 40 inclusive

p	$E(p)$	p	$E(p)$	p	$E(p)$
2	0·0000	15	0·0147	28	0·1448
3	0·0003	16	0·0082	29	0·0953
4	0·0004	17	0·0543	30	0·0425
5	0·0019	18	0·0147	31	0·0321
6	0·0000	19	0·0200	32	0·0151
7	0·0017	20	0·0592	33	0·0293
8	0·0044	21	0·0241	34	0·0465
9	0·0115	22	0·0090	35	0·0561
10	0·0006	23	0·0184	36	0·0635
11	0·0109	24	0·0221	37	0·0787
12	0·0257	25	0·0376	38	0·0747
13	0·0375	26	0·0842	39	0·0598
14	0·0163	27	0·1349	40	0·0529

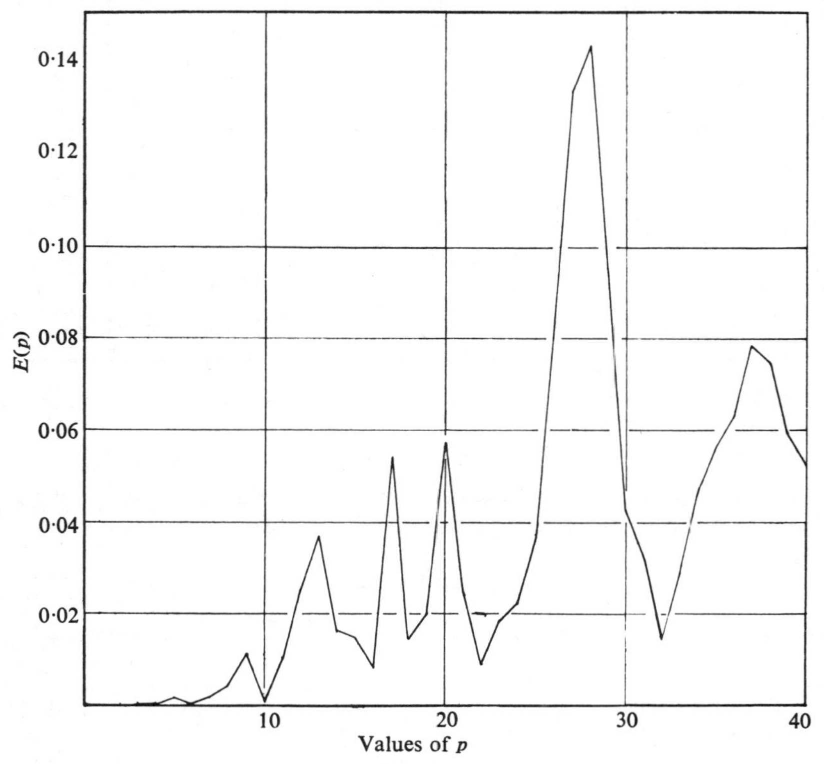

FIG. 10

Periodogram of Series 2. Note that the vertical scale is about one half of that for Figs. 2, 4 and 13

mean-distance between upcrosses; but this is the least significant of the five, and in practice there is little doubt that the other four would be given precedence. Once again we reach the conclusion that the periodogram is grossly misleading.

TABLE 12

Artificial autoregressive series (Series 3) constructed according to equation (24) of the text

t	u_t	t	u_t	t	u_t	t	u_t	t	u_t	t	u_t
1	−102	41	42	81	6	121	−29	161	50	201	77
2	−117	42	−16	82	52	122	54	162	27	202	−20
3	−35	43	−51	83	80	123	123	163	15	203	−85
4	70	44	−9	84	48	124	111	164	22	204	−92
5	133	45	48	85	−3	125	−22	165	57	205	−23
6	117	46	60	86	−49	126	−97	166	14	206	66
7	20	47	22	87	−77	127	−96	167	−53	207	131
8	−89	48	−33	88	−35	128	−11	168	−82	208	111
9	−113	49	−41	89	7	129	17	169	−20	209	20
10	−76	50	−16	90	83	130	39	170	56	210	−81
11	−20	51	−7	91	133	131	−3	171	101	211	−73
12	−9	52	−4	92	114	132	−16	172	63	212	−9
13	−31	53	4	93	26	133	25	173	−35	213	98
14	15	54	−17	94	−35	134	−7	174	−110	214	79
15	36	55	−28	95	−65	135	−65	175	−127	215	41
16	−10	56	−26	96	−23	136	−111	176	−90	216	18
17	7	57	−31	97	42	137	−115	177	17	217	−4
18	16	58	−22	98	36	138	−22	178	99	218	−33
19	−34	59	21	99	−7	139	104	179	69	219	2
20	−41	60	90	100	−47	140	172	180	−32	220	−20
21	30	61	75	101	−69	141	136	181	−86	221	−18
22	−34	62	−38	102	−12	142	35	182	−74	222	9
23	26	63	−93	103	69	143	−119	183	24	223	64
24	43	64	−100	104	55	144	−110	184	128	224	111
25	58	65	−26	105	−15	145	−19	185	165	225	103
26	70	66	15	106	−24	146	63	186	61	226	57
27	12	67	1	107	15	147	93	187	−78	227	24
28	2	68	−7	108	60	148	35	188	−152	228	−18
29	−57	69	0	109	26	149	−59	189	−96	229	−42
30	−90	70	−31	110	−27	150	−61	190	47	230	−25
31	−80	71	−81	111	−5	151	−38	191	96	231	10
32	−3	72	−16	112	34	152	1	192	25	232	−9
33	12	73	46	113	38	153	80	193	−22	233	4
34	−9	74	111	114	33	154	106	194	−23	234	−6
35	−58	75	122	115	−7	155	32	195	−21	235	23
36	−101	76	50	116	8	156	−2	196	−47	236	53
37	−65	77	−13	117	29	157	−39	197	−34	237	42
38	58	78	−77	118	31	158	−61	198	−16	238	−21
39	119	79	−51	119	−8	159	−57	199	47	239	−8
40	98	80	−7	120	−58	160	33	200	106	240	55

40. A third experimental series of 240 terms was constructed according to the formula

$$u_{t+2} = 1{\cdot}1u_{t+1} - 0{\cdot}8u_t + \varepsilon_{t+2} \quad . \quad . \quad . \quad . \quad . \quad (24)$$

the random element being of the same type as in the previous cases. This series, which is referred to as Series 3, is less heavily damped than Series 1 and 2, the

TABLE 13

Serial correlations of Series 3

k	r_k	k	r_k	k	r_k
1	0·623	11	−0·225	21	0·024
2	−0·105	12	−0·135	22	0·141
3	−0·609	13	−0·009	23	0·181
4	−0·596	14	0·125	24	0·094
5	−0·181	15	0·189	25	−0·074
6	0·260	16	0·159	26	−0·199
7	0·402	17	0·055	27	−0·197
8	0·247	18	−0·069	28	−0·078
9	−0·011	19	−0·144	29	0·075
10	−0·201	20	−0·101	30	0·174

damping factor being $\sqrt{0{\cdot}8} = 0{\cdot}8944$. The values of the series are shown in Table 12, the graph of the first 100 terms in Fig. 11, the first 30 serial correlations in Table 13 and the correlogram in Fig. 12.

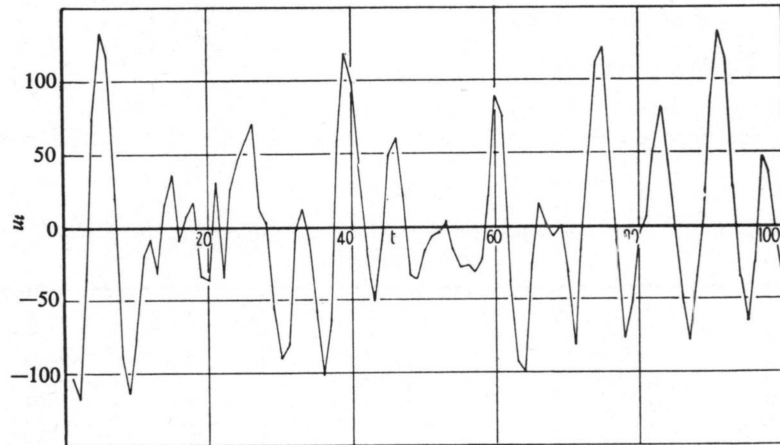

FIG. 11
Graph of first 100 terms of Series 3

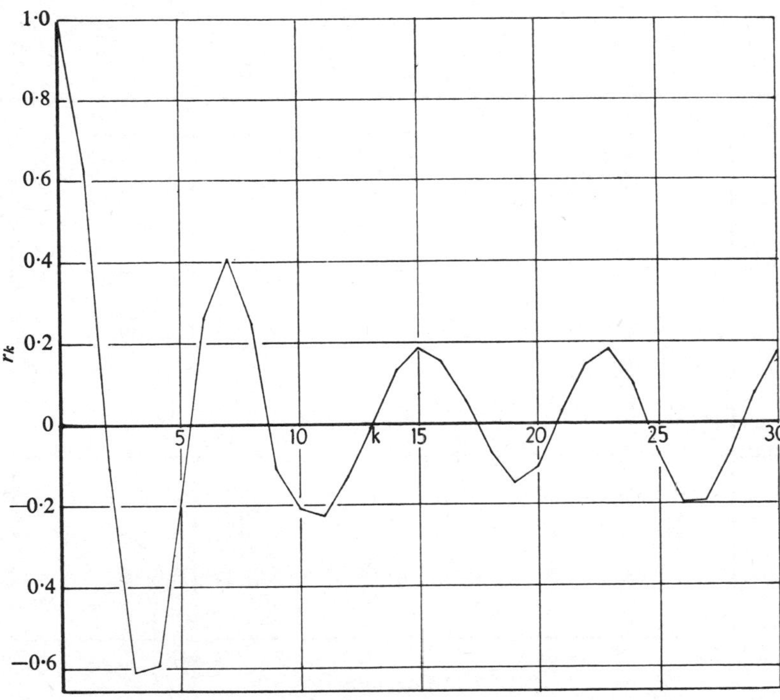

FIG. 12
Correlogram of Series 3

The main constants of the series are:

Mean	3·729,167
Variance	3900·939,149
Expected mean-distance (upcrosses)	6·87
Observed ,, ,, ...	6·21
Expected mean-distance (peaks)	5·69
Observed ,, ,, ...	5·52
Fundamental period	6·92

The mean distance between troughs in the correlogram is 7·67.

41. The Schuster periodogram was computed for all integral trial periods

TABLE 14

Standardized periodogram ordinates for Series 3 *and integral trial values from* 2 *to* 40 *inclusive*

p	$E(p)$	p	$E(p)$	p	$E(p)$
2	0·0000	15	0·0047	28	0·0009
3	0·0004	16	0·0167	29	0·0013
4	0·0003	17	0·0167	30	0·0011
5	0·0047	18	0·0165	31	0·0047
6	0·0457	19	0·0009	32	0·0057
7	0·0533	20	0·0099	33	0·0071
8	0·0277	21	0·0041	34	0·0068
9	0·0057	22	0·0013	35	0·0047
10	0·0135	23	0·0087	36	0·0034
11	0·0210	24	0·0063	37	0·0019
12	0·0009	25	0·0027	38	0·0031
13	0·0020	26	0·0017	39	0·0011
14	0·0068	27	0·0019	40	0·0012

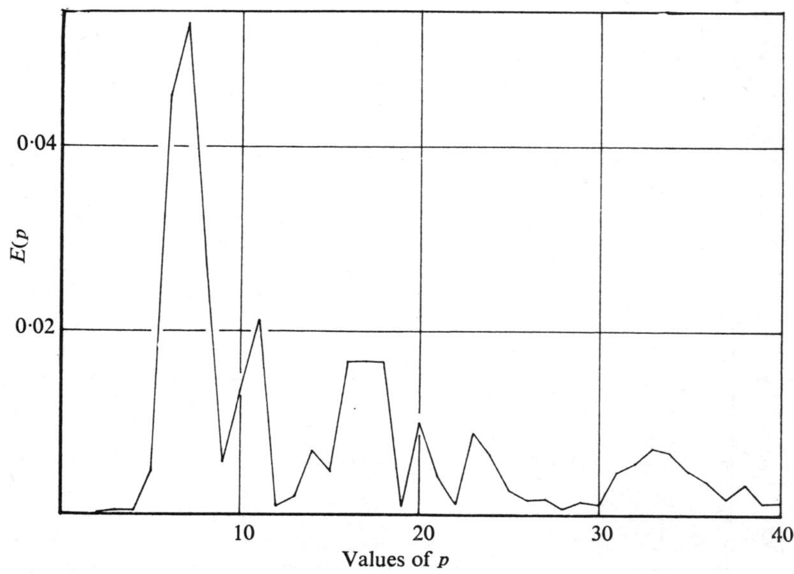

FIG. 13
Periodogram of Series 3

from 2 to 40 inclusive, the ordinates being given in Table 14 and the periodogram in Fig. 13. We now have a major peak at 7 and minor peaks at 11, 17, 20, 23 and 33. This is much nearer the kind of figure we should expect if the periodogram is to be at all reliable.

It is worthy of remark that the periodogram of Series 3, which is the least damped, is the least misleading. General considerations would also lead us to the conclusion, supported by these experiments, that the greater the damping the greater the effect of the disturbance function, and hence the greater liability of spurious peaks in the periodogram due to chance effects.

42. Finally, Series 2 and 3 were added together to make a new series of 240 terms, the *n*th member of which was the sum of the *n*th members of the two series. This new series is no longer linearly autoregressive, but I wanted to see how far a double mean-period would affect the periodogram. Table 15 shows the serial correlations and Table 16 the ordinates of the periodogram, the figure being graphed on the usual lines in Figs. 14 and 15.

TABLE 15

Serial correlations of the sum of Series 2 and 3

k	r_k	k	r_k	k	r_k
1	0·722	11	—0·133	21	0·046
2	0·197	12	—0·086	22	0·150
3	—0·216	13	—0·014	23	0·175
4	—0·325	14	0·050	24	0·110
5	—0·171	15	0·060	25	—0·018
6	0·061	16	0·006	26	—0·131
7	0·180	17	—0·072	27	—0·144
8	0·132	18	—0·133	28	—0·047
9	0·006	19	—0·146	29	0·106
10	—0·106	20	—0·072	30	0·209

TABLE 16

Standardized periodogram ordinates for the sum of Series 2 and 3 and integral trial periods from 2 to 40 inclusive

p	$E(p)$	p	$E(p)$	p	$E(p)$
2	0·0000	15	0·0055	28	0·0777
3	0·0003	16	0·0019	29	0·0494
4	0·0005	17	0·0323	30	0·0170
5	0·0023	18	0·0303	31	0·0079
6	0·0243	19	0·0120	32	0·0051
7	0·0194	20	0·0152	33	0·0125
8	0·0157	21	0·0034	34	0·0187
9	0·0004	22	0·0023	35	0·0198
10	0·0046	23	0·0015	36	0·0249
11	0·0305	24	0·0075	37	0·0285
12	0·0088	25	0·0139	38	0·0260
13	0·0109	26	0·0350	39	0·0219
14	0·0201	27	0·0770	40	0·0173

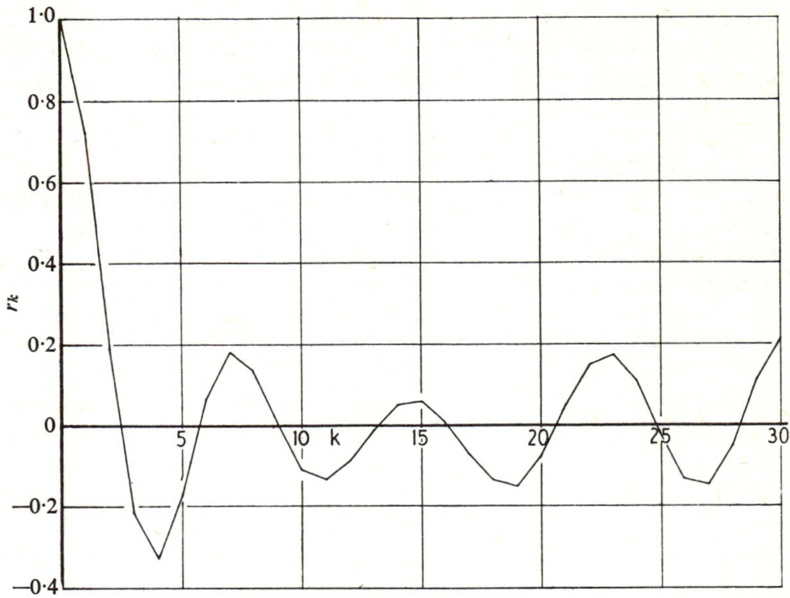

FIG. 14
Correlogram of the sum of Series 2 and 3

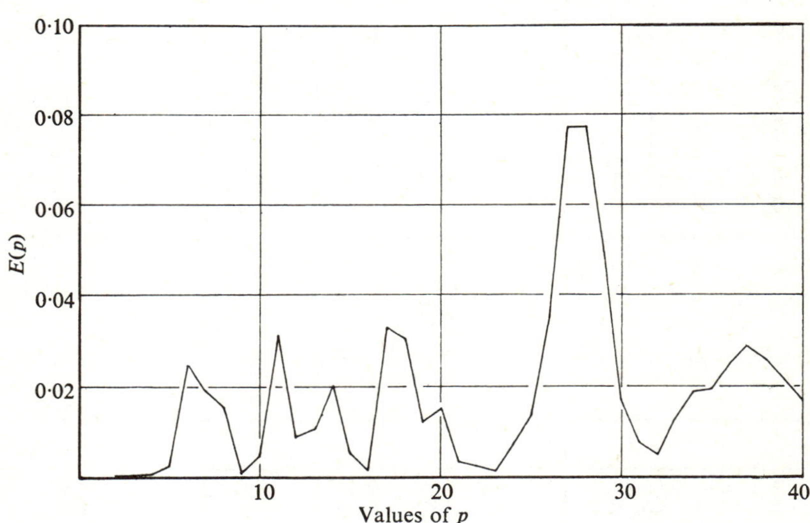

FIG. 15
Periodogram of the sum of Series 2 and 3. Note that the vertical scale is one half of that for Figs. 2, 4 and 13

The main constants of the series are:

Mean	—4·450,000
Variance	7497·905,833
Expected mean-distance (upcrosses)	8·51
Observed " "	7·79
Expected mean-distance (peaks)	5·47
Observed " "	5·13

The mean-distance between troughs in the correlogram is 7·67 units.

43. Since Series 2 and 3 are independent, we have for infinite series, denoting by a single prime the constants of series 2 and by two primes those of the sum,

$$\text{var } u'' = \text{var } u + \text{var } u' \quad . \quad . \quad . \quad . \quad . \quad . \quad . \quad (25)$$

$$\text{cov } (u_j'', u_{j+k}'') = \text{cov } (u_j, u_{j+k}) + \text{cov } (u_j', u_j'{}_{+k}) \quad . \quad . \quad (26)$$

$$\rho_k'' = \frac{\rho_k \text{ var } u + \rho_k' \text{ var } u'}{\text{var } u + \text{var } u'} \quad . \quad . \quad . \quad . \quad . \quad . \quad (27)$$

so that the serial correlations are a weighted average of those for the constituent series. These theoretical relations, however, are not very well realized for series of such length as these, as the following comparisons show—

k	r_k (observed)	r_k (calculated from (27))
1	0·722	0·729
2	0·197	0·227
3	—0·216	—0·161
4	—0·325	—0·259
5	—0·171	—0·117
6	0·061	0·081
7	0·180	0·146
8	0·132	0·067
9	0·006	—0·050
10	—0·106	- -0·134

44. The periodogram of the combined series shows the usual effects, there being appreciable peaks at 6, 11, 17, 28 and 37. The correlogram, on the other hand, does not on the face of it show that the primary series is the sum of two oscillatory series, and on this evidence alone we should probably conclude that it was of the linearly autoregressive type with only one mean-period.

Discussion of Results

45. The above examples show beyond doubt that it is possible for the periodogram of a simple autoregressive series to give very misleading results. Even for a series containing as many as 480 terms there are peaks which do not correspond to any cyclical effects in the generating process, and no peak of prominence where we should expect to find one from consideration of the mean-period between upcrosses or the correlogram. Where, therefore, there is any serious possibility that a series is of the autoregressive type, the periodogram may not only be worthless, but extremely dangerous in suggesting periods of no reality. The effect of this work has been to lead me to the conclusion that for economic series, and probably for meteorological and geophysical series as well, periodogram analysis is simply not worth the trouble of the arithmetic which it involves.

46. On the other hand, the correlogram of the combined Series 2 and 3 fails to discriminate between the sum of two autoregressive series and a single autoregressive series. The correlogram may therefore be insensitive. I should regard it as reliable in indicating the typical damping effect of autoregression and as providing at least an approximate idea of the character of the oscillatory movement. But where damping is heavy the various random elements may obscure the autoregressive effects in short series. The correlogram, so to speak, gives a lower limit to the oscillatory effects. If it oscillates, there is almost certainly some systematic oscillations in the primary series to be explored; but it may not be able to separate these effects, which for short series are liable to be confounded.

47. In dealing with practical series I have constantly found larger differences than were expected between observation and theory. For most purposes a sample of 480 would be regarded as large in the usual statistical meaning of the word. The important question arises whether this is large for the purposes of analysing oscillatory movements in time series, or in other words whether it can be regarded as a " long " series.

On the whole, I doubt whether it can. It is, perhaps, moderately long; but series of 100 terms, or even 200 or 300, I should regard as short. I have given elsewhere [15] an example of a series of 60 terms wherein serial correlation after about the 20th are 100 times as big as the theoretical values for an infinite series. The length of the series is thus a most important factor in the comparison of observed and expected values, and it is necessary, though disappointing, to have to record that conclusions based on short series may have to be extremely tentative.

48. In fact, the material number appears to be not the number of members of the series, but the number of mean periods which it covers. For instance, Series I covers about 57 mean-distances between upcrosses and it is 57, not 480, which is our effective sample number. Even this may be misleadingly high in the sense that there may be correlations between successive intervals between upcrosses, so that we have not got 57 independent observations. My experimental series are a good deal longer than many of the economic series to which periodogram analysis has been applied, and the conclusions above concerning the unreliability of the inferences drawn from the appearance of peaks in the periodogram are all the stronger in such cases.

49. To return to the specific case of the Beveridge series. From the experimental examples it seems clear that a periodogram such as that actually found by Beveridge could arise from a series of the simple autoregressive type and *a fortiori* from a more complicated autoregressive scheme. In short, many of the periods are not significant. All the collateral evidence seems to me to point in the same direction; the variation of intensity in the two halves of the series, the definite indication of damping in the correlogram, the evidence of English wheat prices and acreages since 1867 [14], the fact that the mean-distance between peaks is 4·67 years, and that between upcrosses is 7·08 years, all are perfectly consistent with autoregression and require no more than two mean-periods to explain them. One can never prove mathematically that the series is not the sum of 19 harmonics and a random error of observation; in fact, mathematically it may be. But I have no doubt myself that the underlying scheme of generation is of the autoregressive type and that 17 of Sir William's periods are spurious. Those who are unwilling to go so far will at least admit, I hope, that there is very little statistical evidence for them.

Footprints in the Sands of Time

50. I should be doing Sir William Beveridge an injustice if I gave the impression that he attributed significance to his 19 periods solely on the strength of the evidence of the peaks in the periodogram. On the contrary, he was careful to apply a number of supplementary tests. Unfortunately, as it seems to me, they are of no value.

51. The first of Sir William's supplementary tests is that the functions A and B of equations (2) and (3) should change sign between two neighbouring trial periods which " bracket " a true period. As he himself says, it has " only

subsidiary importance as a makeweight." I would value it even lower, for the condition is obeyed by nearly all the spurious periods in my experimental series.

52. The second test was what Schuster called the test of "continuity." The series is divided into halves, and separate intensities calculated for each, the idea being that if the period suggested has any reality it ought to persist in each half. I have referred above to the anomalies discovered by Schuster himself in applying this procedure to sunspot data. Sir William found much the same effect, some of his intensities for the two half-series being widely different. For instance, at 17·33 the intensity for the first half is 136·19 and that for the second half 11·94, that for the whole being 54·55. This shook Sir William, and he continues, "while it seems most improbable that intensities as high as those near 17·500 do not represent some physical fact of periodicity, uncertainty as to the precise length of the period (including, of course, its phase) and as to its persistence has led to its omission from my final calculation." But he includes it among his 19 proposed periods.

53. Now, it is clear that from the point of view of autoregressive series this kind of situation is not only understandable, but is to be expected on occasion. If by chance a set of intervals between upcrosses, say, happen to fall into step, they will give a high intensity which may disappear in subsequent portions of the series. But it does not follow, conversely, that if the intensities in two consecutive portions of the series are near equality there is any "reality" in the intensity of the whole. The continuity test seems to me to add very little to the periodogram analysis; but so far as the *observed* effects go, they reinforce the case for regarding the series as autoregressive, and militate against the scheme of hidden harmonics.

54. It is, however, when Sir William breaks away from the periodogram and begins to identify his wheat-price periods with rainfall periods that he, in common with many other writers, ventures on the most dangerous ground of all. What he does is to review a number of meteorological periods which had been claimed by other writers and regard any coincidence with the wheat-price periods as confirmation of both. For instance, a period of 2·735 years in the wheat price index is compared with periods claimed by Arctowski and by Hansen and Nansen in tropical temperature, Bigelow in magnetic force in the U.S.A. and one suggested by Shaw in the wheat yields of Eastern England.

55. I hold that this kind of numerical coincidence means practically nothing. So many periods have been claimed by so many writers that it is almost impossible not to find some close agreements for any period within reasonable limits. In his treatise on Meteorology, Sir Napier Shaw gives a list of about 130 periods ranging from 1 to 260 years which have been proposed merely for rainfall, barometric pressure or gradient, temperature and sunspots. If we were to add the series of periods claimed in magnetic variation, economic series and so on, we should probably find an almost continuous range of periods from 1 to 20 and quite a dense set thereafter. How perilously close the identification of periods may be to pure numerology may be illustrated by a few fool's experiments.

56. I took the first group of four numbers from the tables of Babington Smith and myself, 2315, added the height and base of the Great Pyramid * in

* Khufu's at Gizeh. Figures from the *Encyclopedia Britannica*, 13th ed., s.v. "Pyramid." The units are not stated in the table, but they must be inches because the area of the base is about 13 acres.

inches, 5776 and 9068·8, divided by 2 and again by the mystic number of the beast in Revelations, 666. The answer is 12·88. Sir William Beveridge found a very important period at 12·84.

Encouraged by this, I tried another line which, if possible, is even more absurd. The title of Sir William's paper is "Wheat Prices and Rainfall in Western Europe." Counting the letter *a* as 1, *b* as 2 and so on up to $z = 26$ we find, on adding the values for all the letters in the title plus the number of words, 433. This *to the very day* is the number of days in one of the most famous of geophysical periods, Chandler's period in the variation of latitude.

57. Perhaps this is enough to show how unreliable are coincidences with periods which have been claimed in other connections. To show how necessary the warning is, I take a dreadful example from Professor Davis's book ([7], p. 13). Davis gives a graph of the U.S. wholesale commodity price index for each year from 1797 to 1940. He notes that there are inflationary peaks, due to war, in 1814, 1864 and 1920, and proceeds, "This observation, based upon the tenuous example of just three inflations, has led to the assumption of a fifty-year cycle in prices. . . . Some additional evidence is furnished by Sir William Beveridge's periodogram of wheat prices in England. . . . But perhaps the most indirect support of the hypothesis is found in the dates of the three Punic Wars. The dates of the three were 264–241 B.C., 218–201 B.C., and 149–146 B.C. If we presume that the last date in each case was approximately the date of maximum inflation of prices, then the intervals of the cycle would be 40 and 55 years respectively, a fair agreement with the intervals in the cycles of the past century and a half." As if this is not enough, Davis goes one better later in the book (p. 560) by saying, "The 50-year war cycle is undoubtedly connected with the average duration of human life."

Comment is hardly necessary, but I cannot resist one remark. Davis's preface is dated November 1941, a few weeks before the attack on Pearl Harbour. In evident ignorance of the war cycle (or perhaps because of their contempt for the average duration of human life), the Japanese anticipated the next peak by some 29 years.

58. The plain fact is that an investigator into oscillations in time-series nowadays is very much in the position of the detective in the modern crime novel. By the time he arrives on the scene to inspect the corpse, so many feet have trampled all round it that he can easily find a footprint to fit any suspect he likes to choose. The main difference is that under the rules of criminal fiction the detective must not be the culprit. In the theory of time-series he frequently is.

SUGGESTIONS FOR FURTHER RESEARCH

59. The publication of the White Paper on Employment will, I hope, prove to be a landmark in the study of oscillating time-series. For the first time the Government propose to take active steps to anticipate economic crises, in the hope of being able to control them, and it is intended to set up a body of economists and statisticians to carry out a continuous study of economic movements. We may legitimately hope that this study will include research (it will be of little value unless it does), and that therefore the resources of the State will be turned on to the immense amount of investigation which still remains to be done in this field. I am very glad that this should be so, for the labour required even in such matters as arithmetical calculation is almost beyond the capacity of

individual research workers, and in the United Kingdom we have been rather backward in a co-operative attack on the time-series problem. If the findings of the present paper are accepted a great deal of past work on the analysis of oscillatory movements will have to be reconsidered, and I should like to end on a more constructive note by suggesting some lines of possible development which may repay investigation.

60. In the first place, it must be realized that the problem of time series is not peculiar to economics. A great deal of work has been done on parallel problems in many of the physical sciences, and particularly in connection with light and electric waves, sound waves, the discharge of electrical condensers, as well as on meteorological and geophysical series. I understand also that gunnery experience during the war had thrown up a number of problems of the same character, and I know that they arise in the control of the quality of manufactured products. In many cases workers in these fields are to some extent in ignorance of what is being done by others. It seems to me that some machinery is required to co-ordinate the research which is being or will be carried out.

61. Secondly, reference has been made above to certain tests based on mean-distances between peaks or upcrosses in a series. I think this part of the subject could profitably be extended to cover the actual distribution of intervals, which might prove to be characteristic of the type of generating process. Little or nothing has been done on the mathematical theory, except in the case when the series is random. I therefore give in the Appendix the solution of the problem of runs and intervals for the non-random case when variation is normal in the hope that someone may work it out in greater detail.

62. Thirdly, I call attention to the many interesting theoretical problems awaiting the attention of mathematicians in this field. Can we, for example, arrive at an accurate test of significance of the periodogram of an autoregressive series? Is there any tractable scheme of autoregression which will allow us to take into account small disturbances occurring continually at small intervals of time instead, as in Yule's scheme, of finite disturbances occurring only at regular intervals? How far are analyses based on correlogram or periodogram affected by modifications in the unit-interval, *e.g.* by observing the series at monthly instead of annual intervals? What happens if the disturbance function is not a random variable? One can go on asking such questions almost indefinitely, but I have said enough to indicate the kind of domain which awaits exploration by fresh and originative minds.

63. Fourthly, necessity will compel us to attempt forecasts based on short experiences. I have shown above that very long series are required before one can be at all certain about their nature; and although such series may be available to the physicist and the meteorologist, they are rarely so to the economist. What can the economist do to supplement his scanty material? He will, of course, have a number of series of similar character for his own and other countries which will together provide evidence that might be quite inconclusive for any one taken individually. The other resource, I think, is for him to make predictions from time to time on the basis of alternative hypothesis and to see from experience which gives the best prediction. It is here that one would expect the autoregressive scheme to score; for if prediction is possible at all, that is to say if the value at some future time can be forecast from the present and past experience of the series itself, the autoregressive scheme will at least give an approximation.

64. In conclusion I should like to acknowledge the help I have received in the laborious analysis of the experimental series. A good deal of the work was done by Miss Muriel Potter, whose services were put at my disposal by the National Institute of Social and Economic Research. She constructed the artificial series and carried out the correlogram analyses under my supervision. I am very grateful to the Institute for this assistance, and also to Miss Potter for the thorough and accurate way in which she performed these rather tedious computations. The periodogram analysis of Series 3 was done by Scientific Computing Services, Limited, who also checked the last stages of my own analysis of Series 2.

References

[1] Alter, D. (1937). A simple form of periodogram. *Ann. Math. Stats.*, **8**, 121.
[2] Anderson, R. L. (1942). Distribution of the serial correlation coefficient. *Ann. Math. Stats.*, **13**, 1.
[3] Bartels, J. (1935). Zur Morphologie geophysikalischer Zeitfunktionen. *Sitz. Berl. Wiss*, **139**.
[4] Beveridge, Sir William H. (1922). Crop yields and rainfall in Western Europe. *J.R.S.S.*, **85**, 412.
[5] Buys-Ballot, C. H. D. (1847). *Les changements périodiques de température.* Utrecht.
[6] Cochran, W. G. (1941). The distribution of the largest of a set of variances as a fraction of their total. *Ann. Eugen. Lond.*, **11**, 47.
[7] Davis, H. T. (1941). *The analysis of economic time series.* Indiana.
[8] Dodd, E. L. (1939). The length of cycles which result from the graduation of chance elements. *Ann. Math. Stats.*, **10**, 254.
[9] Finney, D. J. (1941). The joint distribution of variance ratios based on a common error mean square. *Ann. Eugen. Lond.*, **11**, 136.
[10] Fisher, R. A. (1929). Tests of significance in harmonic analysis. *Proc. Roy. Soc.*, A, **125**, 54.
[11] Frisch, R. (1933). Propagation problems and impulse problems in dynamic economics. *Economic Essays in honour of Gustav Cassel, London.*
[12] Kendall, M. G. (1941). Proof of relations connected with the tetrachoric series and its generalization. *Biom.*, **32**, 196.
[13] Kendall, M. G. (1943). *The Advanced Theory of Statistics*, Vol. 1. London, Charles Griffin & Co.
[14] Kendall, M. G. (1944). Oscillatory movements in English agriculture. *J.R.S.S.*, **106**, 91.
[15] Kendall, M. G. (1944). On autoregressive time series. *Biom.*, **33**, 105.
[16] Schuster, Sir Arthur (1898). On the investigation of hidden periodicities with application to a supposed 26-day period of meteorological phenomena. *Terr. Mag.*, **3**, 13.
[17] Schuster, Sir Arthur (1906). On the periodicities of sun-spots. *Phil. Trans.*, A, **206**, 69.
[18] Slutzky, E. (1937). The summation of random causes as the source of cyclic processes. *Econometrika*, **5**, 105.
[19] Walker, Sir Gilbert (1931). On periodicity in series of related terms. *Proc. Roy. Soc.*, A, **131**, 518.
[20] Whittaker, E. T., and Robinson, G. (1940). *The Calculus of Observations.* 3rd ed. Blackie & Sons.
[21] Wold, H. (1938). *A study in the analysis of stationary time series.* Uppsala.
[22] Yule, G. Udny (1921). On the time correlation problem. *J.R.S.S.*, **84**, 497.
[23] Yule, G. Udny (1926). Why do we sometimes get nonsense correlations between time series? *J.R.S.S.*, **89**, 1.
[24] Yule, G. Udny (1927). On a method of investigating periodicities in disturbed series with special reference to Wolfer's sunspot numbers. *Phil. Trans.*, A, **226**, 267.

APPENDIX

THE DISTRIBUTION OF RUNS IN CERTAIN NON-RANDOM SERIES

1. If the element ε of a linearly autoregressive series is normal and random, the terms of the series itself are normal; and similarly the series generated by taking a moving average of finite extent of a random normal series is also normal. In what follows it is assumed that the series u_t is normal with auto-correlations ρ_1, ρ_2, . . . etc. Even when, in an autoregressive scheme, the element ε is not normal, one may expect that the averaging process introduced by the scheme will bring the generated series somewhere close to normality (we cannot appeal to the Central Limit Theorem because the Lindeberg conditions are not satisfied). The condition as to normality does not, therefore, appear to be very restrictive.

Mean Distance between Peaks

2. If u_2 is a peak of the series

$$u_2 > u_1 \quad \text{and} \quad u_3 < u_2$$

or

$$\left.\begin{array}{l} u_2 - u_1 > 0 \\ u_3 - u_2 < 0 \end{array}\right\} \qquad (1)$$

Let us take new variables

$$\left.\begin{array}{l} \xi_1 = u_2 - u_1 \\ \xi_2 = u_3 - u_2 \end{array}\right\} \qquad (2)$$

Then ξ_1 and ξ_2 are normally distributed. Their variance is

$$\text{var } \xi_1 = \text{var } \xi_2 = 2(1 - \rho_1) \text{ var } u \qquad (3)$$

and the correlation between them is easily seen to be τ_1 say, where

$$\tau_1 = \frac{-1 + 2\rho_1 - \rho_2}{2(1 - \rho_1)}$$

$$= -\tfrac{1}{2} + \frac{\rho_1 - \rho_2}{2(1 - \rho_1)} \qquad (4)$$

Now the probability that a pair of values of ξ_1, ξ_2 falls in the range $\xi_1 > 0$, $\xi_2 < 0$ is the proportionate volume of the bivariate normal surface with correlation τ_1 lying in the quadrant $\xi_1 > 0$, $\xi_2 < 0$. This, by a result due to Sheppard (cf. Kendall, [13], p. 364), is $\frac{1}{2\pi} \cos^{-1} \tau_1$. (A proof of this result which is capable of generalization is given below.)

Hence, the mean-distance between peaks in a series with the first two auto-correlations ρ_1 and ρ_2 is

$$\text{m.d. (peaks)} = \frac{2\pi}{\cos^{-1} \tau_1} \qquad (5)$$

where τ_1 is given by equation (4).

As a check we note that if $\rho_1 = \rho_2 = 0$, $\tau_1 = \frac{2\pi}{3}$ and the mean-distance is 3, the known result for a random series.

3. For the autoregressive scheme

$$u_{t+2} + au_{t+1} + bu_t = \varepsilon_{t+2}$$

we have

$$\rho_1 = -\frac{a}{1+b}$$

$$\rho_2 = \frac{a^2 - b(1+b)}{1+b}$$

and hence

$$\tau_1 = \frac{b^2 - (1+a)^2}{2(1+a+b)} \quad . \quad . \quad . \quad . \quad . \quad . \quad . \quad . \quad . \quad (6)$$

For instance, in Series I $a = -1{\cdot}1$, $b = 0{\cdot}5$ and hence

$$\tau_1 = \frac{0{\cdot}24}{0{\cdot}8} = 0{\cdot}3$$

$$\text{m.d. (peaks)} = \frac{360}{72{\cdot}54} = 4{\cdot}96.$$

The observed mean-distance is 5·05, an excellent agreement.

4. It is rather remarkable that the mean-distance between peaks is not very sensitive to differences in values of a and b. For some typical values

$a = -1{\cdot}2$, $b = 0{\cdot}4$	m.d. = 4·96
$a = -1{\cdot}5$, $b = 0{\cdot}8$	m.d. = 5·69
$a = -1{\cdot}0$, $b = 0{\cdot}6$	m.d. = 4·96
$a = -0{\cdot}8$, $b = 0{\cdot}8$	m.d. = 5·13

One would therefore expect that in the majority of linear autoregressive series anyone who reckoned his " periods " as the mean-distances between peaks would discover a " cycle " of between 3 and 6 units, mostly somewhere near 5. Reference is made in para. 18 of the paper to the table given by Davis ([7], p. 548), showing the distribution of 166 " cycles " in 17 countries, the mean value of which is 5·2 years.

Distribution of Upruns

5. An uprun is a part of the series with increasing values, and similarly a downrun is a part with decreasing values. I proceed to find the probability that $k + 1$ consecutive terms of a series form an uprun.

Following the device of paragraph 2 of this appendix, let us note that if

$$\xi_j = u_{j+1} - u_j = \Delta u_j \quad . \quad . \quad . \quad . \quad . \quad . \quad . \quad . \quad (7)$$

the sequence $u_1, \ldots u_{k+1}$ is an uprun if all the ξ's are greater than zero. Further, all the ξ's are normal with variance $2(1 - \rho_1)$ var u and their serial correlations τ are given by

$$\tau_j = -\frac{\rho_{j-1} + 2\rho_j - \rho_{j+1}}{2(1-\rho_1)}$$

$$= -\frac{\Delta^2 \rho_{j-1}}{2(1-\rho_1)} \quad . \quad . \quad . \quad . \quad . \quad . \quad . \quad . \quad . \quad (8)$$

Our problem then reduces itself to that of finding the content of a k-dimensional normal hypersurface contained in the positive hyperquadrant $\xi_1 => 0$, $\xi_2 > 0, \ldots \xi_k > 0$.

I have given the solution of an analogous problem in a previous note [12] in the form of an expansion in ascending powers of the correlations. The expansion is, in fact, a generalization of the tetrachoric series.

For instance, if $k = 2$ we have the tetrachoric series ([13], p. 356)

$$\int_h^\infty \int_k^\infty dF = \sum_{j=0}^{\infty} \left\{ H_{j-1}(h)\alpha(h)H_{j-1}(k)\alpha(k) \right\} \frac{\tau_1^j}{j!} \quad . \quad . \quad . \quad (9)$$

where

$$\alpha(x) = \frac{1}{\sqrt{(2\pi)}} \exp\left(-\tfrac{1}{2}x^2\right) \quad . \quad . \quad . \quad . \quad . \quad . \quad (10)$$

$$\alpha(x)H_r(x) = -\left(\frac{d}{dx}\right)^r \alpha(x) \quad . \quad . \quad . \quad . \quad . \quad . \quad (11)$$

and by convention the first term is

$$\frac{1}{2\pi}\int_h^\infty \exp\left(-\tfrac{1}{2}x^2\right)dx \int_k^\infty \exp\left(-\tfrac{1}{2}y^2\right)dy \quad . \quad . \quad . \quad . \quad (12)$$

In our case the limits h and k are zero and since

$$\left.\begin{aligned} H_r(0) &= 0, \ r \text{ odd} \\ H_r(0) &= \frac{(-1)^j(2j)!}{2^j j!}, \ r \text{ even} = 2j \end{aligned}\right\} \quad . \quad . \quad . \quad . \quad (13)$$

the probability required, say P_2, reduces to

$$P_2 = \tfrac{1}{4} + \frac{1}{2\pi}\sum_{j=1}^{\infty}\left\{\frac{\tau^{2j+1}(2j)!}{(2j+1)2^{2j}(j!)^2}\right\} \quad . \quad . \quad . \quad . \quad (14)$$

$$= \tfrac{1}{4} + \frac{1}{2\pi}\sin^{-1}\tau_1$$

$$= \tfrac{1}{2} - \frac{1}{2\pi}\cos^{-1}\tau_1 \quad . \quad . \quad . \quad . \quad . \quad . \quad . \quad . \quad . \quad (15)$$

This checks with the result of Sheppard already used in paragraph 2 of this appendix.

For $k = 3$ we have

$$P_3 = \int_0^\infty \int_0^\infty \int_0^\infty dF =$$

$$\Sigma\left[(-1)^{j+k+l}\frac{\tau_1^{\,j}\tau_2^{\,k}\tau_1^{\,l}}{j!k!l!}\left\{H_{j+l-1}(0)H_{j+k-1}(0)H_{k+l-1}(0)\right\}\alpha^3(0)\right] \quad . \quad (16)$$

Since of $j + l - 1$ and the two similar terms one at least must be odd, the only surviving terms are those for which two of j, k, l, are zero and the other one is odd. We then find

$$P_3 = \tfrac{1}{8} + \frac{1}{4\pi}\left(2\sin^{-1}\tau_1 + \sin^{-1}\tau_2\right)$$

$$= \tfrac{1}{4} - \frac{1}{4\pi}\left(2\cos^{-1}\tau_1 + \cos^{-1}\tau_2 - \pi\right) \quad . \quad . \quad . \quad (17)$$

6. Pausing for a moment, we may observe that (15) and (17) may be obtained by geometrical considerations in a manner employed by Dodd [8]. In fact P_2 is $\tfrac{1}{2}$ less the proportional volume between two hyperplanes of angle $\cos^{-1}\tau_1$; P_3 is $\tfrac{1}{4}$ less the area of a spherical triangle cut off on the unit sphere in three dimensions by planes whose angles are $\cos^{-1}\tau_1$, $\cos^{-1}\tau_2$ and $\cos^{-1}\tau_1$. At one time I had hopes of generalizing this approach by finding the content of a region cut off on the hypersphere in k dimensions by k hyperplanes with given angles of intersection, but the direct geometrical approach baffled me; and from the

complexity of the results for $k > 3$ I have now some doubts whether any simple solution exists.

7. For general k and a multivariate normal form with correlations between the jth and kth variate of χ_{jk} the proportional volume in the positive quadrant is

$$\frac{1}{(2\pi)^{\frac{1}{2}k}}\int_0^\infty dx_1 \ldots \int_0^\infty dx_k \int_{-\infty}^\infty \exp\left(-\tfrac{1}{2}t_1^2 - rt_1x_1\right) \ldots \int_{-\infty}^\infty \exp\left(-\tfrac{1}{2}t_k^2 - it_kx_k\right)$$
$$\sum_{j=0}^\infty (-1)^j \frac{(\chi_{12}t_1t_2 + \chi_{13}t_1t_3 + \ldots + \chi_{jk}t_jt_k + \ldots)^j}{j!} dt_1 \ldots dt_k \quad . \quad . \quad (18)$$

the expressions in the second lot of integrals being expressible in terms of the Tchebycheff–Hermite polynomials. For instance, with $k = 4$ writing $\tau_1 = \chi_{j,j+1}$ etc. we have

$$P_4 = \Sigma\left[\frac{\tau_1^{\ j}\tau_2^{\ k}\tau_3^{\ l}\tau_1^{\ m}\tau_2^{\ n}\tau_1^{\ p}}{j!k!l!m!n!p!} H_{j+k+l-1}(0)H_{j+m+n-1}(0) H_{k+m+p-1}(0)H_{l+n+p-1}(0)\alpha^4(0)\right] \quad . \quad . \quad . \quad (19)$$

$$= \Sigma \frac{1}{4n^2} \frac{\tau_1^{\ j+m+p}\tau_2^{\ k+n}\tau_3^{\ l}(j+k+l-1)!(j+m+n-1)!(k+m+p-1)!(l+n+p-1)!}{j!k!l!m!n!p!2^{j+k+l+m+n+p-2}\left(\frac{j+k+l-1}{2}\right)!\left(\frac{j+m+n-1}{2}\right)!\left(\frac{k+m+p-1}{2}\right)!\left(\frac{l+n+p-1}{2}\right)!} \quad (20)$$

subject only to the condition that the factorials in the denominator must be integral.

8. These expressions are hardly as difficult as they look. For a damped autoregressive series the expressions for the P's converge fairly quickly and are, I think, amenable to calculation. At any rate, I have not been able to simplify them to any considerable extent. Tables of tetrachoric functions can, of course, be used in the evaluation of the separate terms.

9. From the values of P_k we can easily derive the distribution of upruns. In fact if the relative frequency of a complete uprun of exactly k intervals is f_k, the relative frequency of upruns of 1 is

$$\sum_{j=1}^\infty (jf_j) = P_1; \quad . \quad . \quad . \quad . \quad . \quad . \quad . \quad . \quad (21)$$

that of upruns of 2 is

$$\sum_{j=2}^\infty \{(j-1)f_j\} = P_2; \quad . \quad . \quad . \quad . \quad . \quad . \quad . \quad (22)$$

and so on. Thus

$$P_{k+2} - 2P_{k+1} + P_k = \sum_{j=k+2}^\infty \{(j-k-1)f_j\} + \sum_{j=k}^\infty \{(j-k+1)f_j\} - 2\sum_{j=k+1}^\infty \{(j-k)f_j\}$$
$$= f_k$$

or

$$f_k = \Delta^2 P_k \quad . \quad . \quad . \quad . \quad . \quad . \quad . \quad . \quad . \quad . \quad . \quad . \quad . \quad (23)$$

10. For the purpose of comparing theory with experiment in a given series it would probably be simpler to work with upruns and downruns rather than with intervals between peaks, but the distribution of peaks may, if so desired, be derived from that of runs. In fact the probability that an interval of extent k is an interval between peaks is the sum of the separate probabilities (1) that

the first two terms are a complete downrun and the next $k - 1$ a complete uprun, (2) that the first three are a complete downrun and the next $k - 2$ a complete uprun, and so on. The required distribution is thus obtainable by summing appropriate terms of the distribution of runs.

Distribution of Intervals between Crosses

11. The same methods will yield the distributions of intervals from upcross to downcross and from upcross to upcross. The probability that a series of $k + 2$ terms is an interval from upcross to downcross is the probability that

$$u_1 < 0, u_2 > 0, \ldots u_{k+1} > 0, u_{k+2} < 0$$

and may be evaluated from the content of the appropriate quadrant of the multidimensional distribution of the u's themselves; and the probability that an interval is one from upcross to upcross follows as in the previous paragraph.

For the simple autoregressive scheme

$$u_{t+2} + au_{t+1} + bu_t = \varepsilon_{t+2}$$

the mean-distance between upcrosses is easily seen to be

$$\text{m.d. (upcrosses)} = \frac{2\pi}{\cos^{-1} \rho_1} \quad \ldots \ldots \quad (24)$$

where ρ_1, the first autocorrelation, is given by

$$\rho_1 = -\frac{a}{1 + b} \quad \ldots \ldots \ldots \quad (25)$$

Discussion on Mr. Kendall's Paper

Professor Pearson: I have read Mr. Kendall's paper with very great interest. It is the kind of survey which those who would perhaps be too busy at the moment to go into a more detailed mathematical treatment will appreciate highly. It sets us thinking in the same stimulating way as his paper three years ago on the Future of Statistics. I am very much pleased indeed to have been asked to propose the vote of thanks.

I shall not attempt to go into the points at issue between Mr. Kendall and Sir William Beveridge; the latter will no doubt make his own reply in writing. But as one who might very easily have been in the same boat, I must confess to a good deal of sympathy with Sir William. Some fifteen years ago Professor Dinsmore Alter (to whose work Mr. Kendall refers in paragraph 20) spent a year in this country, and persuaded me to share in the authorship of a small tract on periodic analysis. My share in the job was to try to put in order the work on significance tests. But, for better or worse, the tract was never published, and the responsibility for this was mine. Perhaps I was too lazy to finish the work, or perhaps, to put a better complexion on the matter, I realized, like Mr. Kendall, that to get to the bottom of this difficult subject it was necessary to go very much further into the matter than I had originally intended and, unlike Mr. Kendall, I refused to be drawn in.

As I see it, in this field of oscillatory time series we are brought up with particular force against that problem so often present in the application of statistical theory to practice. We can build up a hypothetical model and say what kind of results we shall observe if it holds good—*e.g.*, given a series of superimposed harmonics or a simple autoregressive scheme, we can say what will be the characteristics of the periodogram or correlogram. But we cannot, without far more research than has hitherto been undertaken, be sure that there are not many other systems or forms of build-up which will result, within the limits of the inevitable obscuring errors, in precisely the same periodogram or correlogram features. Or, indeed, worse still, we cannot feel confident that data with no system in them at all may not produce all the typical features of a quite simple hypothetical model.

The first function of statistical theory is, no doubt, to tell us whether the observational data are consistent with a specified mathematical set-up, but the full benefit of the theory only comes into play when its technique has been so developed that it can be used to separate the class of hypotheses which are admissible on the data from those which are not. In the present field there is, as Mr. Kendall says, a great deal still to be done. We have got to collect and analyse material from many sources as well as to clear the ground by constructing more of these hypothetical model series, before we shall really feel confident that statistical time series analysis can tell us so very much more than an intelligent clear-headed person can extract from inspection of the original graphs.

Before asking Mr. Kendall a few questions on specific points mentioned in his paper, I should like to put before the meeting a rather different type of problem to which it seems that correlogram analysis is applicable. Mr. Kendall has referred in paragraph 60 to problems in gunnery. Without any indiscretion, I might formulate one such problem as follows. In its simplest form it

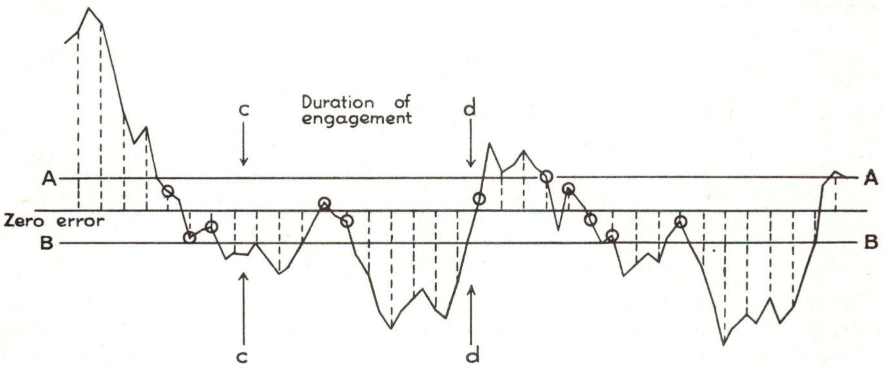

is that which occurs when a man, by turning a handle, follows in his sights a moving target such as an aircraft. There will usually be two men, one following the bearing and the other the elevation. Or the problem may be more complex, the man not directly seeing the target himself, but acting as one of several transmitting agents who each attempts to follow one moving pointer on a dial with another pointer. If, then, we plot the error, say, in elevation against time, we shall get a curve of which that shown in my diagram is typical. Errors at short intervals of time will be highly correlated, since movement of hand or instrument were continuous. Further, there will be swings from side to side of an oscillatory character, rather like those which occur in steering a boat. In proceeding to the statistical analysis of such data we are likely to have the following objectives in view:

1. The series of autocorrelations provides perhaps the best method of *describing* the data, and hence of comparing one set of observations with another. The unit of time used in calculating the correlations may be $\frac{1}{2}$ sec., 1 sec., 5 secs., depending to some extent on the available method of recording consecutive errors.

2. The correlogram, when interpreted, may be used to throw light on the underlying build-up of the error system; for example, in a complex fire-control system the oscillation periods may be traced back to parts of the mechanism as well as to the human factor.

3. Having succeeded in tracing back oscillations to their origin, it may be possible to introduce some method of smoothing. The economic

counterpart will lie in Government action to iron out the peaks and troughs of the trade cycle.

4. Finally, we have another aspect of the problem. Suppose that we are dealing with a rapid-firing gun and that the equidistant dotted ordinates drawn in my diagram represent the errors at successive moments of fire. It may be that, provided the error lies between the limits *AA* and *BB*, the target will be hit so that in the stretch shown there would have been groups of 3, 2, 1, 4 and 1 successive shots on the target. Actually the engagement will be of limited duration, say occurring in the interval of time between the vertical lines *cc* and *dd*. A practical question, then, is this: what is the chance of securing at least one or, perhaps, at least two hits on target per engagement?

It seems probable that, provided the standard error can be regarded as constant during the course of an engagement, the answer to the question raised under (4) will be expressible in terms of the autocorrelations. As in all problems of this kind, a representative supply of good data must be available before a theoretical attempt at the solution can make much progress. Such data are accumulating. There is unlikely to be any neat mathematical answer, and much computing labour must be expected.

It would be interesting to know whether this type of problem occurs in the economic or any other field. Whether the observed correlograms in the gunnery problem will be consistent with a three-term autoregressive scheme like that of Mr. Kendall's equation (9) I do not know, but it seems clear that the underlying mechanical set-up could not lead to so simple a model.

Finally, here are three questions on matters of detail in the paper:

(*a*) Para. 6. Why is Whittaker's η not a correlation ratio? I agree that there may be difficulties in interpreting the periodogram, although I have found that for artificial data based on superimposed harmonic terms plus a random error, the periodogram shows all the features that would be expected of it. But surely η^2 is an appropriate term for the ratio of the variance of array means to the total variance?

(*b*) Para. 18. The moral, I presume, is that the mean distance between peaks should not be used as a measure of periodicity, but rather the mean distance between upcrosses?

(*c*) Suppose in the wheat-price series, and also in artificial series 1 and 2, a Fourier series were fitted to the data, using the two or three main periods picked out by the Schuster periodogram analysis, how far would the resulting curves give reasonable fits to the data? From the point of view of explaining the set-up, or of predicting ahead, the curves may be valueless, but it would be interesting to know whether they would even describe or graduate the series adequately within the time interval covered by the analysis.

Dr. L. Isserlis, in seconding the vote of thanks, said: I agree with Professor Pearson that this paper is one of great interest, and as he has set the example of being autobiographical, I should also like to say that a matter of twenty years ago I was faced with this same problem, and frightened by it. I do not think I can even say that laziness prevented me from going very far; it was rather terror.

My first introduction to periodogram analysis was a study of the papers of the late Dr. Brownlee, who had observed an undoubted appearance of periodicity in measles epidemics, and he applied the analysis to various collections of data in order to obtain the period in days. He obtained periods which, by the Schuster method, were significant, and came to the conclusion that the periods of measles on the southern side of the river Thames were different from the periods on the northern side. He suggested that there was a kind of measles germ in south London which could not cross the bridges. I could not cross that bridge, and doubted whether Schuster's tests of significance had any real value.

To turn to the last of Professor Pearson's questions, I had for many years been calculating the index-number of monthly shipping freights, and when, after a number of years, I had a long series of monthly indices, I fitted long series of index-numbers by a few trigonometrical terms, and found that three or four of Fourier's terms would give a reasonable fit to a series of from 60 to 80 monthly numbers. It did not seem to have any meaning, so I put that on one side.

I am very sorry that illness has prevented Dr. Snow from being here to-night, because I remember his participation in a former discussion on this subject. I examined shipping freights over a course of 60 or 70 years, obtained maxima and minima, upcrosses and downcrosses, and set the major peaks against the peaks in a series of wholesale prices. There was a good deal of agreement. *Post-facto*, one could give an explanation of any particular prominent peak: one could relate it to a war or a financial crisis or something of that sort. But so far as periods were concerned, I said, " Well, I suppose there are periods; only the wave-lengths differ, the amplitudes differ, and you can never be certain what phase you are in." Dr. Snow said, " Does not the author really mean that there are not any periods? " and in my reply I was content to say that I agreed with him.

Mr. Kendall has helped by showing that if the correlogram and the periodogram analyses are combined we can perhaps get upper and lower bounds. His chief contribution is a negative one, but it is very valuable, all the same. To say that something is not established and that we must try again is itself a valuable thing. It was a little optimistic to suggest that the research unit in the new Central Statistical Office should keep pace with this problem, because, after all, the Central Statistical Office would set out to give guidance to statesmen, and the mathematical problem has not yet got to that stage.

Mr. Kendall has written a very important paper, and I hope he will follow it up. I have great pleasure in seconding the vote of thanks.

Mr. J. R. Womersley said that he knew nothing about economics, but he was quite convinced, as the result of many years' industrial research, that systems of the kind envisaged by Mr. Kendall or systems similar to them did exist in the industrial field, and he wished to quote a few examples. One of these was the control of temperature, humidity, and so on, in industry. Most textile products would exhibit periodicities of this kind. After receiving the present paper he went through some old papers of his own to try to discover a reference, which he failed to do, to an American paper some twelve years ago which showed that the Brownian movements of sensitive galvanometers exhibited precisely similar effects. It seemed to him that Mr. Kendall was unfortunately in the position of having to make a very large jump from the theoretical correlogram of the whole population, so to speak, to a very short series, because he only got one point a month or one point a year in the things with which he dealt.

To bridge that gap, he suggested that something might be learned from work in the industrial field. This raised the question as to whether it was the right way to attack these problems. Granting that these particulars and possibilities did exist in the industrial field, where large masses of data awaited attention, why was it that more had not been done? He thought the answer lay in the fact that to complete these serial correlations did need a great deal of elaboration. If progress was to be made with that long series, one must, by some means or other, " take the backache out of the business," and he would accordingly throw out a suggestion that the Society should have a symposium on the methods of calculation of these correlograms.

When he received the paper he felt quite certain—and private enquiries had since confirmed his guess—that work was being done on automatic instruments for the construction of correlograms, and a discussion in that Society on the question of methods would speed up the work very considerably.

Mr. Champernowne asked whether the author would agree that the correlogram and the periodogram could be likened to the general practitioner and the specialist. The periodogram was the " specialist," which dealt with undamped continuing fluctuations which kept punctually to the phase, whereas the correlogram could also deal with fluctuations which were constantly reinforced by irregular influences from outside, and would not necessarily keep their phase. If that was so, the periodogram might still be used in economics if it was desired to isolate those parts of the movement of the series which were due to outside factors fluctuating regularly. The main point of the attack on the orthodox method of using the periodogram was that the significance tests could not rightly be applied to economic time series, because they nearly always exhibited serial correlation between one term and the term following. What he wished to ask the author to look into was whether it was possible to adapt the significance tests for the periodogram by using some estimate of the number of *independent* terms in the series. He thought it could be proved that such a modification of the test was possible in a simple autoregression series, and he indicated on the blackboard equations by which it was possible to obtain a useful significance test for deciding whether there was any regular periodic fluctuation of significantly large amplitude.

In the case of a simple autoregressive series of N terms generated by the law

$$u_{t+1} = ru_t + E_{t+1} \quad \ldots \ldots \quad (1)$$

the effective number of independent terms was reduced by the serial correlation, n, from N to N', given by

$$N' = \frac{1-r}{1+r}N \quad \ldots \ldots \quad (2)$$

This formula was approximate, and held only for fairly large values of N. The usual significance tests could therefore be applied with N' in place of N, if one wished to test the hypothesis that any apparently significant periods shown by the periodogram were in reality due to a simple autoregressive tendency of the type (1) above.

In general, where the autoregressive mechanism was more complicated than the simple type of equation (1), it should still be possible to estimate N', the effective number of independent terms, or the number of degrees of freedom, in terms of the total number of terms N. He suggested the formula:

$$N' = \frac{N}{T} \quad \ldots \ldots \quad (3)$$

where T was the number of terms lag which made the lag-correlation coefficient r_T (between terms T apart) approximately zero.

The following table illustrates the application of this rule to

(1) The Beveridge series.
(2) Mr. Kendall's series of 480 terms.
(3) Mr. Kendall's series of 240 terms.

Series		(1)	(2)	(3)
Symbol	Meaning			
N	Total number of terms.	300	480	240
T	Lag needed to make correlation zero.	2 +	3 +	5 −
$N' = \frac{N}{T}$	Effective number of independent terms.	150 −	160 −	48 +
Max. $E(p)$	Highest peak in periodogram. (*See* Kendall's paper.)	0·055	0·051	0·145
Max. $K = \frac{1}{2}N'E$	Peak standardized for number of independent terms.	4·1	4·1	3·5
$P(k)$	Likelihood of obtaining so great a peak by chance in the first 50 points of the periodogram.	0·57	0·57	0·71

The figures shown in the last line of the table showed that in the case of each series, including the Beveridge series, no peak of the periodogram was higher than would normally result from chance alone, given the effect of serial correlation in reducing the number of effectively independent terms in the series. He would therefore go further than Mr. Kendall, and say that *none* of the periods in the Beveridge series had any significance.

He then demonstrated an alternative method of estimating the number of independent terms by fitting a curve of type

$$y = 50e^{-\alpha x}$$

to show the number y of the first 50 points in the periodogram which gave values of $E(p)$ exceeding x. N' might then be estimated as equal to 2α. Applying this technique to Kendall's 480-term series gave $\alpha = 76 \pm 6$, so that $N' = 152 \pm 12$, which checked the value $N' = 160$ suggested by the earlier method.

Mr. Champernowne was prevented by time from raising various other points; he afterwards submitted them in writing, as follows:—

Mr. Kendall has drawn our attention again to a very important point with which Slutsky, Yule and Frisch have been concerned. Wherever a mechanical system or a system of economic variables, capable of a damped oscillatory motion, is subjected to a series of random shocks, we may expect a motion of fairly constant amplitude which will appear to be periodic. Such a motion may be produced experimentally by bombarding a pendulum with pellets, or mathematically, in the way that Mr. Kendall has constructed artificial autoregressive series. When we draw graphs of these artificial series they bear a very close resemblance to actual series of economic data.

This is only natural, since the economic system contains many parts liable to damped oscillation about an equilibrium, or normal, level, and it is moreover subjected to numbers of irregular shocks. But the existence of this resemblance between economic series and artificially constructed autoregressive series helps very little in explaining the working of the economic system: The trouble is that such a large number of completely different theories can fit facts of this kind. Corresponding to any kind of trade-cycle theory we can build up a system of equations connecting

1. The degree of employment;
2. Other economic variables;
3. Past history and rates of change of variables;
4. Random shocks;

and we shall be unlucky if we cannot find values of the elasticities and regression coefficients in all these equations to fit the facts and statistics available. For unfortunately the supply of available theories is more generous than the supply of reliable statistical series.

I would draw the attention of Mr. Kendall to a type of autoregression equation distinct from those he has investigated, namely

$$u^3_{t+1} = Ax_t - Bx_{t-1} + E_{t+1}$$

where E_{t+1} are random shocks, and A and B are constants. This can lead to alternative periods of " boom " and " slump," in each of which u_t will remain fairly constant about one of two distinct values. Then its value will suddenly change over to the other equilibrium position, remain fairly steadily there for a period, and then, as suddenly, return. This is the kind of autoregressive equation which would be appropriate to a consideration of Mr. Kaldor's model of the trade cycle.

An electrical analogy to a system of this type can be found in the arrangement known as the " flip-flop " circuit. Two radio valves are employed, the anode circuit of each valve being fed into the grid circuit of the other; the electric potential at critical points of the system will experience alternate booms and slumps. Time lags can be allowed for by introducing large condensers and resistances into the circuit.

PROFESSOR HADAMARD, F.R.S., was not sure that he had rightly heard everything that had been said; he hoped that what he had to say had not been already affirmed or contradicted by the previous speakers. After reading Mr. Kendall's paper and realizing its great importance, he looked at Schuster's own diagram in his study of sunspots. He noticed that Schuster's periodograms were much less unreasonable than the periodograms which the author had built on an experimental series. Moreover, there was a discrepancy between the two halves of the period considered by Schuster. The numbers of the figures of Schuster corresponded to a period of 152 years, and Schuster had the idea of studying one part for 75 years and another for 75 years. The last 75 years were in full agreement with the classical period (as the speaker thought it was), and he believed that it was a solid fact in science, but he began to doubt it when he saw what Schuster said of his first period. There was a great peak for the number 11 in his second period, but just the contrary for his first period.

He was struck also by other things. Schuster constructed the periodogram in the case of a full theoretical simple and sinusoidal oscillation. Three graphs are given by him in that part of his work: the graph corresponding to the first half of his period, the one corresponding to the second half, and the theoretical graph. Let us admit that the phenomenon was governed by the sinusoidal law and apply the periodogram. There would be a peak coinciding most strikingly with the observed peak in the second half of his observations, but also two small peaks, less important, but distinctly existing and located exactly at the same place for the same numbers of years as the observed peaks.

There was another feature which ought to be verified on the original numbers utilized by Schuster—namely, that although the growth for the first half of the period was completely discrepant, indeed, contradictory, with the other one: so contradictory that in the place where there was a peak for the number 11 there was actually a depression for that number—nevertheless when Schuster took the period in full, making the periodogram over the whole period of 150 years, the peak remained. He did not understand how that could happen.

Then, could not there be a slow modification in the going on of the phenomenon? It might happen, for instance, that oscillatory phenomena, like sunspots, be always governed by a recurrence equation like Yule's [equation (9) in Kendall's paper], but in which the coefficients a and b, instead of being constant, would be subject to slow progressive modifications, by the action of causes unknown to us.

Even assuming a and b to be constant, it must not be forgotten that the recurrence equation—or " finite differences equation "—(with given ε's, of course) admits of ∞ solutions, depending on two " integrations constants." The effect of the ε-disturbances could be expressed by fluctuations in the values of these constants. *It is not true that these fluctuation changings do not accumulate*: as well known, they will not cumulate proportionally to n, but they will proportionally to $\sqrt{n}$.

For these two reasons, it is not at all surprising that in a series of as many as 480 numbers like the first artificial autoregressive series quoted by the author (Series 1) there were periods almost everywhere except in the right place, because even random disturbances in 480 numbers would accumulate and would have an influence which would be 20 times greater than in the beginning.

This is a general evident reason why the correlogram is preferable to the periodogram. It only compares values of u not too distant from each other —in fact, distant only by K ranks. The comparison of periods located at both ends of the series would be likely to give much more aberrant results (would anybody undertake to verify this theoretically or experimentally?). Not having much confidence as a statistician, but speaking rather as a mathematician, Mr. H. would point out an analogy with a curious fact in pure analysis. It is classical that any continuous function of (x) can be, in a given interval (A, B), approximated as closely as might be wished by a polynomial, and the first idea which had come to people had been to take some N values of x between A and B—say equidistant values—and construct a polynomial of the degree

$(N - 1)$ which would be exactly equal to $f(x)$ for these values of the variable x. The fact was found independently by several workers that such an approximation may *not* converge for $N \to \infty$. There is, according to Weierstrass' theorem, a polynomial which would be as close to x as wanted if taken of a sufficiently high degree; but this polynomial could not be found in the aforesaid way: the best approximating polynomial would be equal to x for N values of x, but those values must not be chosen anyway. He was reminded of the French proverb, " *Qui trop embrasse, mal étreint* " (" Who tries to embrace too much will embrace nothing "). That corresponds to the question of the periodogram and the correlogram.

DR. M. S. BARTLETT (read in his absence by MR. ANSCOMBE): I very much regret being unable to be present at the discussion on Mr. Kendall's valuable paper, and hope that in these circumstances I shall be permitted to comment in writing. In his further researches into the statistical theory of time-series, Mr. Kendall has done a service in drawing attention to the dangers of orthodox periodogram analysis in this field, but an amplification of his discussion at one or two points appears desirable. Thus paragraph 24 (b) I found somewhat ambiguous. Periodogram analysis is similar to other forms of regression analysis in demanding for the validity of the usual tests the randomness of the *residuals*. On the other hand, I quite agree that for most time-series even this condition is not satisfied.

Practically, when unknown autocorrelations are present, Mr. Kendall notes that with either periodogram or correlogram analysis, we are still in a dilemma over tests of significance. But in his theoretical discussion, in which he is dealing with series with known parameters, I am rather surprised that he did not attempt to go a little further. For example, in the important paper to which Mr. Kendall refers, Slutsky has generalized Schuster's results for the case of Fourier analysis of autocorrelated series. He not only gives expressions for the expected intensities, these being no longer equal, as in the case of purely random series, but he also points out that for the intensities with the largest expectations sampling fluctuations are correspondingly the largest. This result is entirely consistent with the wild instability Mr. Kendall finds for independent analysis carried out for different intervals. For the correlogram, too, while I agree with Mr. Kendall that the relations between its ordinates are also important, it might have been useful, and probably not too intractable mathematically, to have evaluated at least the approximate theoretical standard errors for the autocorrelations, and indicated these together with the expected autocorrelations in the diagrams.

Finally, no one is likely to refute Mr. Kendall's remark that economic cycles in the strict harmonic sense do not exist, but it is worth stressing the difference between oscillatory series and just fluctuating series. The simplest autoregressive scheme is a Markoff process connecting two consecutive terms. If we superpose two different Markoff processes, we get a three-term autoregressive scheme at first glance like the Yule-Kendall scheme, but it is intrinsically different in having no fundamental period in its solution (the situation is analogous to that with second-order differential equations). Mr. Kendall has indicated the growing importance of the theory of time-series, a branch of statistical theory in its infancy; in addition to the formidable sampling problems, I think quite a lot remains to be done simply on the " natural history " of time-series, and their classification according to type.

MR. ANSCOMBE said that in para. 16 Mr. Kendall suggested two methods of measuring the period of a time series. One was the distance between successive peaks, and that, the present speaker thought, was a useful measure, and would give some idea of the amount of local unevenness in the results. The other was the distance between consecutive upcrosses. This was rather less significant, or had less physical meaning, and could be less easily interpreted. He suggested, therefore, that possibly other measures might be of more interest or more easy to understand.

In his appendix Mr. Kendall had been able to treat these two measures, but other measures might be more difficult to deal with. For many purposes what was important was whether one had got a high peak or a low trough. He suggested that one might define a high peak as being a certain distance away from the mean and a low trough as a certain distance below the mean, and consider upcrosses from a low trough to a high peak. Alternatively, one might consider upcrosses consisting of a movement from a trough to a peak which was of more than a certain size. The periods of such upcrosses would be considerably longer than those Mr. Kendall found between his upcrosses and rather like the long periods found in the periodogram.

He suggested that this raised the point that in an autoregressive series long-term periods would be observable just as a description of the series. These might have nothing to do with any long-term cause, and might be difficult to observe owing to large sampling errors; but it might nevertheless be useful as a property of an autoregressive series to consider what were the long-term effects and what was the average distance between the low trough and the high peak.

Sir Gilbert Walker (read by Dr. Bradford Hill): The copy of Mr. Kendall's paper has just arrived, and I am very sorry not to be able to express at the meeting my admiration for it and my agreement with his main conclusions. I have spent much time and energy in attempting to diminish faith in very doubtful periodicities.

The following comments were received after the meeting:—

Mr. R. Glenday: Lack of time unfortunately prevented me from adding my tribute to the high quality of Mr. Kendall's paper. I was particularly interested to find on pp. 98–9 statistical confirmation of a view which I have long maintained, on purely empirical grounds, namely, that trade cycles as such do not exist, only trade oscillations. The difference, as Mr. Kendall points out, is an important one. Those who believe in trade cycles must not only, as he says, hold that their recurrences are closely predictable, but also encourage the world to believe that their causes are much simpler, less variable and more readily controllable than they in fact are. But once their real nature is appreciated, namely, that they are essentially oscillations in a process of growth, they take on an entirely different character. Their amplitude and form, for instance, will depend, among other things, on whether the system as a whole in which growth is taking place is a reasonably stable one or whether it is expanding or contracting in size. In the case of an agricultural crop or population of rapidly-growing animals in an approximately stable environment, for example, the oscillations in the main will be roughly about a mean level—like the trade cycles depicted in economic textbooks. On the other hand, if the environment in which the population is growing is in process of undergoing a long-term expansion in size favourable to growth, following some permanent modification of its structure by, say, an exploitation of its economic territory in a new way, longer-term oscillations will also occur. There is in real life no such thing as steady and continuous growth. Always we find periods of acceleration alternating with periods of deceleration. If it is desired to check oscillations even of the first kind, it may be necessary not merely to try to control the behaviour of particular, that is, *individual*, growth factors, but also to control modifications in the *arrangement* of a whole range of other factors as well. As every farmer knows, it is possible for a given agricultural area to be planted in exactly the same way in one year as in another, and to receive exactly the same quantity of sunshine and the same quantity of rainfall, manuring and cultivation, and yet in one year to yield a bumper harvest and in the next to prove a failure. The result will depend on the *order* in which the component factors have been arranged and on their *timing*. Equally the addition or elimination of minute quantities of quantitatively unimportant elements may have a quite startling effect on output, as the action of fertilizers, insecticides and the like bears witness. The same principles, I have suggested, hold good in the case of oscillations in the industrial output of economic systems also (see my *The Future of Economic Society*, Chapter IV, also Appendix I). It is useless, therefore, to hope to understand, much less to prevent, oscillations of this type simply by concentrating on *single* growth factors, as economists are prone to do. Not only are the operative factors not necessarily always the same in all oscillations, but even when they are the same, the proportions and the order in which they operate will vary widely, and this may have a major effect on the final result.

The " trade cycles " of the economist, I suggest, are largely compounded from growth oscillations of the above two types. This is why I regard Mr. Kendall's paper as being of first-rate importance. It is especially timely in view of the emergence of an active school of what is known as the " New Economics," which, without having studied the actual mechanics of trade growth or the structure of economic systems, claims to be able to abolish trade cycles, trade oscillations and the like, wherever and however caused, without seriously interfering with the liberty of the common man.

Mr. C. T. Sutton: I have wondered whether, when Mr. Kendall arrived at his conclusions on Sir William Beveridge's investigations, he considered why what he criticizes should have been accepted. This seems to me a very important part of anything constructive that Mr. Kendall has to add.

In paragraph 59 he expressed his hopes of further research with the active encouragement of the Government. And these hopes must be associated with the recently expressed wish of the Society to encourage the recognition of statisticians in commerce.

If Mr. Kendall reveals a weakness in statistical conclusions, he would help further by revealing how he thinks this weakness is caused, in order to suggest how it may in future be overcome. This seems desirable in order, to encourage the confidence of commercial people in statistics. It may not be wholly a weakness in mathematics, but a weakness that is revealed throughout the history of thought, where conclusions are come to that are either exaggerated in their certainty or exaggerated in their use, and the explanation can be, I suggest, a combination of two things. First, there is an absorption in the particular task undertaken, whereby in the necessary isolation of the subject-matter from other influences in order to study it and come to conclusions, those unavoidable influences are not properly reconsidered when predictions are then made based on the conclusions. It is necessary to stand back from the work and review it broadly. Secondly, and more important, is a seeming lack of appreciation of the more simple rules of reasoning and logic, whereby the conclusions are arrived at by way of induction from an insufficient number of examples. The restriction upon such conclusions should be well known. It is quite proper, and extraordinarily useful, to arrive at intuitive conclusions in this way so long as it is realized how far they should be relied upon before being adopted. The trouble is that they are frequently then used, without qualification, as the major premise from which to follow the usual train of deductive reasoning in order to arrive at other conclusions or predictions. If the unconscious nature of this rather formal expression of an everyday manner of forming ideas and making statements was more clearly realized, such mistakes as Mr. Kendall here suggests might be more easily avoided. This, of course, leads to the suggestion that the teaching of statistics, for which it is the Society's aim to issue diplomas, should incorporate some study of the rules of reasoning and expression.

I should be glad to know how far Mr. Kendall thinks this relevant to his suggestions. In this connection I would point out that he has already suggested part of this in paragraph 63, where he quite properly says: " The other resource, I think, is for him to make predictions from time to time on the basis of alternative hypotheses and to see from experience which gives the best prediction."

I am a little uncertain as to that part of the criticism based on the experiments mentioned. Is it not possible to arrive at even valid mathematical conclusions by unreliable methods without discrediting the conclusions, and should the criticism not rather be that, whatever his methods, the scientific enquirer should seek not for dubious confirmations but for exceptions, when considering his conclusions?

MR. W. R. B. HYND: It was with much interest that I came as a visitor to the Royal Statistical Society to hear Mr. Kendall's lecture; the subsequent discussion touched on points with which a group in the Ministry of Aircraft Production, of which I am a member, has been considerably concerned, and I am, therefore, induced to make the following comments.

Professor Pearson, in proposing the vote of thanks, pointed out the importance of autocorrelation in the calculation of survival chances when a target is attacked by rapid fire from a gun, and propounded to the meeting the problem: " What is the chance of at least one hit on the target in a given interval of time? " This has been answered in work carried out by Dr. L. B. C. Cunningham, myself and others (published some months ago in a secret M.A.P. report numbered A.W.A. 51), the survival chance being obtained in the form of a series, which is satisfactorily convergent if there is reasonably high correlation between all the rounds fired during the time-interval involved: it is thus particularly suited to such cases as the firing of a burst of a few seconds duration from a machine gun. The technique, as developed, is also applicable to many other analogous problems.

That the computation of serial correlations involves much monotonous labour was pointed out by Mr. Womersley, who suggested that some rapid method should be developed using automatic calculating machines. A prototype apparatus for rapidly evaluating $\Sigma x_r x_r + k$ was developed some time ago by Messrs. Shire and Runcorn, at the Ministry of Supply, and an improved model is at present being constructed, with which we believe we will be able to obtain such a sum of products, for a series of several hundred data, in approximately one minute, the result being given in the scale of two—and while we are converting one answer to the scale of ten, the machine will be getting on with the calculation of the next.

In this mechanical process it is necessary to group the data to some degree —in the machine under construction, each value in the series must be an integer between 60 and -60. The effect of this on correlations near unity is important in the type of problem posed by Professor Pearson, and may be compensated by the introduction of corrections analogous to Sheppard's corrections; these have been calculated by Dr. H. E. Daniels for the special case when the underlying random error is normally distributed.

The CHAIRMAN put the vote of thanks to the meeting and it was carried unanimously.

MR. KENDALL said that he would much prefer to reply in writing. He would now only express his thanks to the various speakers who had contributed to the discussion. Some very helpful ideas, many of them new to him, had been put forward.

He subsequently wrote as follows:—

There are so many points on which much could be said that perhaps the contributors will forgive me if I reply in rather telegraphic form.

1. As regards the first of Prof. Pearson's queries on paragraph 6, I agree that Whittaker's form is the ratio of the variance of array means to total variance, and perhaps by an *extension* of previous usage it could be called a correlation ratio. But the latter term was introduced for a bivariate frequency table, not a mere array of values. There is no variate in the Buys Ballot Table in the vertical direction, and the members of the array are not frequencies.

2. As regards his query on paragraph 18, I think the moral is that there are all sorts of measures of periodicity, to be chosen according to taste and particular needs, and that they may give very different results.

3. As regards his third question, I do not know how far a Fourier series would represent the data, but I should expect a substantial residual variance in

the cases of the wheat series and the artificial series. This seems to be the general experience—see, for instance, Greenstein's paper in *Econometrika* (1935), **3**, 170.

4. As Mr. Womersley says, a symposium on correlogram analysis would be very useful. I think the newly-formed Research Section are intending to arrange one in the coming session.

5. I agree with Mr. Champernowne that periodogram analysis might be useful where cycles are known to exist, *e.g.*, in meteorological data where there are diurnal or annual variations. The trouble is that if one is certain about their existence, one usually knows their period and all that the analysis does is to exhibit their amplitude. How far estimates of amplitude are affected by the presence of other types of variation does not appear to be known, but I should expect quite substantial sampling effects.

6. I am not very happy about Mr. Champernowne's suggestion that a correlated series can be reduced, so far as significance tests are concerned, to an effective number of independent terms. It seems to me that the effective number might vary according to the test, and I see no prior reason to suppose that the actual number is a linear function of the effective number. However, it is an interesting approach which I hope he will develop.

7. The form of non-linear autoregressive scheme which Mr. Champernowne suggests was new to me, but there are similar differential equations (of the second, not the third order) in the theory of relaxed electrical oscillations. Just at the moment I am finding linear schemes difficult enough to investigate, but no doubt the more general types will have to be examined later.

8. Before I deal with Professor Hadamard's points, perhaps he will allow me to express my pleasure at seeing such a distinguished French mathematician at the meeting. He rightly stresses the puzzling features of Schuster's original investigation; but I think they are only puzzling to the harmonic analyst—from the point of view of autoregression they are quite understandable.

9. It is quite possible that there are small progressive modifications in the constants of an autoregressive series. They do not exist in my artificial series, of course, but in my paper on Oscillatory Movements in English Agriculture (*J.R.S.S.*, 1944, **106**, 91, I considered the effect in a series of sheep populations extending over about 65 years. In the Beveridge series, I should expect the changes over 370 years to be quite marked.

10. Professor Hadamard's point about the correlogram being more affected by sampling fluctuations for correlations of high order is a sound one, and there is some experimental evidence on the subject. See the correlogram at the end of Tintner's *Variate-Difference Method* and some of the correlograms in my forthcoming brochure on *Researches in Oscillatory Time-Series.*

11. Following Dr. Bartlett's comment on paragraph 24(b), I have looked at it again, but I cannot see where the ambiguity arises. The results of equations (14)–(17) in the text depend on randomness of the primary series. I am not sure that the extraction of cyclical components validates a test on the residuals, at least for short series, because of dependence between the items of the variance analysis; but we do not seem to differ on the main conclusion.

12. The reasons I did not carry the standard error tests further were (*a*) that the paper was already long enough, and (*b*) that I am not satisfied that the generalized formulae giving expectations in an *infinite* autoregressive scheme are much use for practical series of finite length, the serial correlations being so much affected by sampling errors. However, much more remains to be said on the subject and I will not attempt to say it here.

13. I entirely agree with Mr. Glenday's comments on the economic implications of the work, and I have sometimes wondered whether attempts to remove all oscillation from a system might not make it so brittle that it would snap at the first serious strain. There is much to be said for damping springs and for shock absorbers, but not for the elimination of spring altogether.

14. As regards Mr. Sutton's comments, I think the reason why people continually discover cycles in all kinds of time series, is that they are looking

for them. With a little goodwill nothing is easier than to detect a cyclical movement in random material. Logically, the error is simple enough: if hypothesis *H* gives rise to effect *E*, the existence of *E* does not demonstrate the truth of *H* unless it can be shown that there are no other hypotheses giving rise to *E*. What has not been realized in the past is that cyclical effects can be generated by random disturbances as well as by schemes of harmonic oscillation.

15. I am very glad that Mr. Hynd has been able to lift to a small extent the veil of secrecy concerning the M.A.P. work on gunnery and the machine for calculating serial correlations. I hope it will be possible to publish the details of this very interesting work in due course.

As a result of the ballot taken during the meeting, the candidates named below were elected Fellows of the Society:—

Clement Burton.
John Gordon Craig.
Alfred Charles Edwards, F.I.A.
John Foster Holdsworth.
Gerald Victor Owen.
Donald Darnley Reid, M.B., Ch.B.
Harry MacLennan Sheard, C.A.
Charles Maxwell Strachan, C.A.
F. R. Vinter.
Nita Grace Mary Watts, B.Sc.Econ.
Matthew Whyte, B.Sc.

From *The American Statistician*, Vol. **13**, No. 5, pp 23–4 (1959)

HIAWATHA DESIGNS AN EXPERIMENT

by

Maurice G. Kendall

1. Hiawatha, mighty hunter
He could shoot ten arrows upwards
Shoot them with such strength and swiftness
That the last had left the bowstring
Ere the first to earth descended.
This was commonly regarded
As a feat of skill and cunning.

2. One or two sarcastic spirits
Pointed out to him, however,
That it might be much more useful
If he sometimes hit the target.
Why not shoot a little straighter
And employ a smaller sample?

3. Hiawatha, who at college
Majored in applied statistics
Consequently felt entitled
To instruct his fellow men on
Any subject whatsoever,
Waxed exceedingly indignant
Talked about the law of error,
Talked about truncated normals,
Talked of loss of information,
Talked about his lack of bias
Pointed out that in the long run
Independent observations
Even though they missed the target
Had an average point of impact
Very near the spot he aimed at
(With the possible exception
Of a set of measure zero.)

4. This, they said, was rather doubtful.
Anyway, it didn't matter
What resulted in the long run;
Either he must hit the target
Much more often than at present
Or himself would have to pay for
All the arrows that he wasted.

5. Hiawatha, in a temper
Quoted parts of R. A. Fisher
Quoted Yates and quoted Finney
Quoted yards of Oscar Kempthorne
Quoted reams of Cox and Cochran
Quoted Anderson and Bancroft
Practically in extenso
Trying to impress upon them
That what actually mattered
Was to estimate the error.

6. One or two of them admitted
Such a thing might have its uses
Still, they said, he might do better
If he shot a little straighter.

7. Hiawatha, to convince them
Organized a shooting contest
Laid out in the proper manner
Of designs experimental
Recommended in the textbooks
(Mainly used for tasting tea, but
Sometimes used in other cases)
Randomized his shooting order
In factorial arrangements
Used in the theory of Galois
Fields of ideal polynomials
Got a nicely balanced layout
And successfully confounded
Second-order interactions.

8. All the other tribal marksmen
Ignorant, benighted creatures,
Of experimental set-ups
Spent their time of preparation
Putting in a lot of practice
Merely shooting at a target.

9. Thus it happened in the contest
That their scores were most impressive
With one solitary exception
This (I hate to have to say it)
Was the score of Hiawatha,
Who, as usual, shot his arrows
Shot them with great strength and swiftness
Managing to be unbiased
Not, however, with his salvo
Managing to hit the target.

10. There, they said to Hiawatha,
That is what we all expected.

11. Hiawatha, nothing daunted,
Called for pen and called for paper
Did analyses of variance
Finally produced the figures
Showing beyond peradventure
Everybody else was biased
And the variance components
Did not differ from each other
Or from Hiawatha's
(This last point, one should acknowledge
Might have been much more convincing

If he hadn't been compelled to
Estimate has own component
From experimental plots in
Which the values all were missing.
Still, they didn't understand it
So they couldn't raise objections
This is what so often happens
With analyses of variance).

12. All the same, his fellow tribesmen
Ignorant, benighted heathens,
Took away his bow and arrows,
Said that though my Hiawatha
Was a brilliant statistician
He was useless as a bowman,
As for variance components
Several of the more outspoken
Made primeval observations
Hurtful to the finer feelings
Even of a statistician.

13. In a corner of the forest
Dwells alone my Hiawatha
Permanently cogitating
On the normal law of error
Wondering in idle moments
Whether an increased precision
Might perhaps be rather better
Even at the risk of bias
If thereby one, now and then, could
Register upon the target.